Wireless Communications

Wireless Communications

Edited by **Elmer Dowse**

CWILLFORD PRESS
New York

Published by Willford Press,
118-35 Queens Blvd., Suite 400,
Forest Hills, NY 11375, USA
www.willfordpress.com

Wireless Communications
Edited by Elmer Dowse

International Standard Book Number: 978-1-68285-014-5 (Hardback)

The publisher's policy is to use permanent paper from mills that operate a sustainable forestry policy. Furthermore, the publisher ensures that the text paper and cover boards used have met acceptable environmental accreditation standards.

Trademark Notice: Registered trademark of products or corporate names are used only for explanation and identification without intent to infringe.

Printed in the United States of America.

Contents

Preface

Wireless communications have undergone tremendous change in the course of past few decades. It has revolutionized the field of communications in both corporate and individual spheres by bridging the distance and time barriers. The ever growing need of advanced technology is the reason that has fueled the research in the field of wireless communication in recent times. This book elucidates the concepts and innovative models around prospective developments with respect to system models, networking, speech and image coding, frequency regulations, and dynamic channel allocation. It will provide comprehensive knowledge to the readers.

This book unites the global concepts and researches in an organized manner for a comprehensive understanding of the subject. It is a ripe text for all researchers, students, scientists or anyone else who is interested in acquiring a better knowledge of this dynamic field.

I extend my sincere thanks to the contributors for such eloquent research chapters. Finally, I thank my family for being a source of support and help.

Editor

Anti-synchronization of chaotic neural networks with time-varying delays via linear matrix inequality (LMI)

Yousef Farid*, Nooshin Bigdeli and Karim Afshar

Department of Electrical Engineering (EE), Imam Khomeini International University, Qazvin, Iran.

In this paper, anti-synchronization problem of two identical chaotic neural networks with time-varying delays is proposed. By using time-delay feedback control technique, mean value theorem and the Leibniz-Newton formula, and by constructing appropriately Lyapunov-Krasovskii functional, sufficient condition is proposed to guarantee the asymptotically anti-synchronization of two identical chaotic neural networks. This condition, which is expressed in terms of linear matrix inequality, rely on the connection matrix in the drive and response networks as well as the suitable designed feedback gains in the response network. Finally, the anti-synchronization of two chaotic cellular neural network and Hopfield neural network with time-varying delays are considered to illustrate the effectiveness of the proposed control scheme, in which, when compared with the nonlinear feedback control method, the proposed method shows superior performance.

Key words: Lyapunov-Krasovskii functional, chaotic neural networks, anti-synchronization, time-varying delay, linear matrix inequality.

INTRODUCTION

Over the recent decades, existence of chaos has been discovered and reported in different aspects of science and technology, such as electrical circuits, chemical reactions, information processing, lasers, optics and neural networks (Chen and Dong, 1998; Wieczorek and Chow, 2009; Yang and Yuan, 2005; Gutzwiller, 1990). Since Pecora and Carroll (1990) established a chaos synchronization scheme for two identical chaotic systems with different initial conditions, chaos synchronization has attracted a great deal of attention (Sun and Cao, 2007; Sanjaya et al., 2010). Another interesting phenomenon discovered was the anti-synchronization (AS), which is noticeable in periodic oscillators. AS is a phenomenon that the state vectors of the synchronized systems have the same amplitude but opposite signs as those of the driving system. In this case, the sum of two signals is expected to converge to zero. So far, different techniques and methods have been proposed to achieve chaos anti-synchronization, such as, active control method (Ho et

al., 2002), adaptive control (Li et al., 2009), H_∞ control (Ahn, 2009), nonlinear control (Al Sawalha and Noorani, 2009), sliding mode control (Chiang et al., 2008), backstepping control (Hu et al., 2005), adaptive modified function projective method (Adeli et al., 2011), etc.

Recently, the study of dynamical properties of neural networks appears more due to their extensive applications in differential fields, such as signal and image processing, pattern recognition, combinatorial optimization and other areas (Cohen and Grossberg, 1983; Carpenter and Grossberg, 1987; Chua and Yang, 1988). In the electronic implementation of the neural networks, time delay will occur in the interactions between the neurons inevitably, and will affect the dynamic behavior of the neural network models and may lead to instability and/or deteriorate the performance of the underlying neural networks. In some particular cases, it has been shown that these networks can exhibit some complicated dynamics and even chaotic behaviors if the network's parameters are appropriately chosen (Yuan, 2007; Lu, 2002).

An efficient tool for solving many optimization problems is linear matrix inequality approach which has been

*Corresponding author. E-mail: yousef.farid @ikiu.ac.ir.

effectively applied in controller design for nonlinear process (Chen et al., 2010). Linear matrix inequalities (LMIs) have been playing an increasingly important role in the field of optimization and control theory, because a wide variety of problems (linear and convex quadratic inequalities, matrix norm inequalities, convex constraints, etc.) can be written as LMIs (Boyd et al., 1994; Guo et al., 2009; Hencey and Alleyne, 2009).

In addition, LMIs have found many applications in exploring properties of recurrent neural networks, since their stability conditions are often expressed with the aid of LMIs (Liu et al., 2005; Lu and Chen, 2006; Lou and Cui, 2006; Li et al., 2008). The objective of this paper is to prepare a control law based on the LMI approach for anti-synchronization of two identical chaotic neural networks with time varying delays, where the stability of the proposed method is guaranteed using Lyapunov stability theory. It will be shown that the performance of the proposed scheme is improved when compared with a recently published paper.

PROBLEM FORMULATION AND SOME PRELIMINARIES

The chaotic neural network with time-varying delay under consideration is described by:

$$\dot{x}(t) = -Cx(t) + Df(x(t)) + Ef(x(t - \tau(t)))$$
$$x(t) = \varphi(t) \quad t \in [-\kappa, 0] \tag{1}$$

where $x(t) = [x_1(t), ..., x_n(t)]$ is the state vector of the neural network with n neurons, $C = diag\{c_{11}, ..., c_{nn}\}$ is a diagonal matrix with $c_{ii} > 0, i = 1, ..., n$ and the matrices D and E are, respectively, the connection weight matrix and the delayed connection weight matrix. $f(x(t)) = [f_1(x_1(t)), ..., f_n(x_n(t))]$ denotes the neuron activation function, $\varphi(t)$ is the initial condition of state vector and $\tau(t)$ is time-varying delay and satisfying:

$$0 \leq \tau(t) \leq \lambda_1, \qquad \dot{\tau}(t) \leq \lambda_2 \tag{2}$$

where $\lambda_1 > 0$ and λ_2 are known parameters. Suppose that the system (Equation 1) be the drive system. The response system is represented by:

$$\dot{y}(t) = -Cy(t) + Df(y(t)) + Ef(y(t - \tau(t))) + B_u u(t) \tag{3}$$
$$y(t) = 0 \quad t \in [-\kappa, 0]$$

where $y(t) \in R^n$ is the state vector of the response system, $u(t)$ is the control input to be designed and

$B_u \in R^{n \times n}$ is the input matrix. Let the anti-synchronous error be defined as $e = x + y$. The objective of the anti-synchronization is to control the behavior of the response system to follow the inverse behavior of the drive system such that $\lim_{t \to \infty} \|x(t) + y(t)\|_2 \to 0$, where $\|\cdot\|_2$ is the Euclidean norm. Then, the error dynamics, can be expressed by:

$$\dot{e}(t) = -Ce(t) + D\psi_1(e(t)) + E\psi_2(e(t - \tau(t))) + B_u u(t) \tag{4}$$

where $\psi_1(e(t)) = f(y(t)) + f(e(t) - y(t))$ and
$\psi_2(e(t)) = f(y(t - \tau(t))) + f(e(t - \tau(t)) - y(t - \tau(t)))$.

Since the information on the size of $\tau(t)$ is available, the controller of the following form is considered:

$$u(t) = K_1 e(t) + K_2 e(t - \tau(t)) \tag{5}$$

where K_1 and K_2 are suitable feedback gains. Substituting Equation 5 into Equation 4, we have:

$$\dot{e}(t) = -(C - B_u K_1)e(t) + B_u K_2 e(t - \tau(t)) + D\psi_1(e(t)) + E\psi_2(e(t - \tau(t))) \tag{6}$$

Remark 1

From the mean value theorem (Leu, 2010) and the Leibniz-Newton formula, that is, $e(t) - e(t - \tau(t)) = \int_{t-\tau(t)}^{t} \dot{e}(s)ds$, it is easy to see that:

$$\psi_2(e(t)) - \psi_2(e(t - \tau(t))) = \dot{\psi}_2(\sigma)(e(t) - e(t - \tau(t))) = \dot{\psi}_2(\sigma)\int_{t-\tau(t)}^{t} \dot{e}(s)ds \tag{7}$$

where σ is a point on the straight line between $e(t)$ and $e(t - \tau(t))$.

Therefore, the error dynamic (Equation 6) can be represented as follows:

$$\dot{e}(t) = -(C - B_u K_1 - B_u K_2)e(t) - (EZ + B_u K_2)\int_{t-\tau(t)}^{t} \dot{e}(s)ds + D\psi_1(e(t)) + E\psi_2(e(t)) \tag{8}$$

where $Z = \dot{\psi}_2(\sigma)$.

Assumption 1

The neuron activation function $f(\cdot)$ is continuous and satisfy $f(0) = 0$ and the Lipschitz condition, that is,

$\|f_i(a) - f_i(b)\| \le \|U_i(a-b)\|$ for any a,b and U_i are known matrices. Thus, we have:

$$0 \le -\psi_1(e(t))^T \psi_1(e(t)) + e(t)^T U_1^T U_1 e(t) \quad (9a)$$

$$0 \le -\psi_2(e(t-\tau(t)))^T \psi_2(e(t-\tau(t))) + e(t-\tau(t))^T U_2^T U_2 e(t-\tau(t)) \quad (9b)$$

Lemma 1

Let $\alpha(\cdot) \in \Re^n$, $\beta(\cdot) \in \Re^m$ and $N(\cdot) \in \Re^{n \times m}$ be defined in the set Ω, then for any matrices $R \in \Re^{n \times m}$, $S \in \Re^{n \times m}$ and $W \in \Re^{n \times m}$, the following inequality holds (Park, 1999):

$$-2\int_\Omega \alpha(r)^T N\beta(r)dr \le \int_\Omega \begin{bmatrix} \alpha(r) \\ \beta(r) \end{bmatrix}^T \begin{bmatrix} R & W-N \\ * & S \end{bmatrix} \begin{bmatrix} \alpha(r) \\ \beta(r) \end{bmatrix} dr \quad (10)$$

where $\begin{bmatrix} R & W \\ * & S \end{bmatrix} \ge 0$.

MAIN RESULTS

The following inequality lemma is necessary to develop the main theorem in this paper.

Theorem 1

For any given scalars $\lambda_1 > 0$ and λ_2, the error dynamic system (Equation 8) is asymptotically stable, if there exist the matrices $U, T_1, \ldots T_5$ and the positive definite matrices $P, H_1, Q_1, \ldots, Q_3$ such that the following matrix inequalities hold:

$$\Sigma + \lambda_1 T Q_3^{-1} T^T = \begin{bmatrix} \Sigma_{11} & \Sigma_{12} & \Sigma_{13} & \Sigma_{14} & \Sigma_{15} \\ * & \Sigma_{22} & -T_3^T & -T_4^T & -T_5^T \\ * & * & -I & 0 & 0 \\ * & * & * & -I & 0 \\ * & * & * & * & \Sigma_{55} \end{bmatrix} + \lambda_1 T Q_3^{-1} T^T < 0 \quad (11)$$

where

$\Sigma_{11} = P^T A + AP + U + U^T - (P^T(B_u K_2) + (B_u K_2)P) - (T_1 + T_1^T) + \lambda_1 H_1 + Q_1 + U_1^T U_1$,

$\Sigma_{12} = -U + P^T(B_u K_2) - T_1 + T_2^T$,

$\Sigma_{13} = P^T D + T_3^T$,

$\Sigma_{14} = P^T E + T_4^T$,

$\Sigma_{15} = T_5^T$,

$\Sigma_{22} = -(1-\lambda_2)Q_1 + U_2^T U_2 - (T_2 + T_2^T)$,

$\Sigma_{55} = \lambda_1(Q_2 + Q_3)$.

Proof

Construct a Lyapunov-Krasovskii functional of the form:

$$V(t) = V_1(t) + V_2(t) + V_3(t) \quad (12)$$

where

$$V_1(t) = e(t)^T P e(t) \quad (13)$$

$$V_2(t) = \int_{t-\tau(t)}^t e(s)^T Q_1 e(s) ds \quad (14)$$

$$V_3(t) = \int_{t-\lambda_1}^t \int_s^t \dot{e}(\theta)^T(Q_2 + Q_3)\dot{e}(\theta)d\theta ds \quad (15)$$

Taking the time-derivative of $V_1(t)$ along the trajectories of error dynamic (Equation 8) yields:

$$\dot{V}_1(t) = 2e(t)^T P\dot{e}(t) = 2e(t)^T P(Ae(t) + D\psi_1(e(t)) + E\psi_2(e(t))) + \Psi(t) \quad (16)$$

where $A = -(C - B_u K_1 - B_u K_2)$,

$$\Psi(t) = -2e(t)^T P(EZ + B_u K_2)\int_{t-\tau(t)}^t \dot{e}(s)ds$$

Using Lemma 1 and Equation 7, it is clear that:

$$\Psi(t) \le \int_{t-\tau(t)}^t \begin{bmatrix} e(t) \\ \dot{e}(s) \end{bmatrix}^T \begin{bmatrix} H_1 & U - P(EZ + B_u K_2) \\ * & Q_2 \end{bmatrix} \begin{bmatrix} e(t) \\ \dot{e}(s) \end{bmatrix} ds$$

$$\le \int_{t-\tau_1}^t \dot{e}(s)^T Q_2 \dot{e}(s)ds + \tau_1 e(t)^T H_1 e(t) + 2e(t)^T(U - P(B_u K_2))(e(t) - e(t-\tau(t))) \quad (17)$$

$$- 2e(t)^T PE(\psi_2(e(t)) - \psi_2(e(t-\tau(t))))$$

The time derivative of $V_2(t)$ and $V_3(t)$ are respectively, as:

$$\dot{V}_2(t) = e(t)^T Q_1 e(t) - (1-\dot{\tau}(t))e(t-\tau(t))^T Q_1 e(t-\tau(t)) \quad (18)$$

$$\le e(t)^T Q_1 e(t) - (1-\lambda_2)e(t-\tau(t))^T Q_1 e(t-\tau(t))$$

And

$$\dot{V}_3(t) = \lambda_1 \dot{e}(t)^T (Q_2 + Q_3)\dot{e}(t) - \int_{t-\lambda_1}^t \dot{e}(s)^T (Q_2 + Q_3)\dot{e}(s)ds \quad (19)$$

$$\dot{V}(t) = 2e(t)^T P(Ae(t) + D\psi_1(e(t)) + E\psi_2(e(t-\tau(t)))) + \lambda_1 e(t)^T H_1 e(t)$$

$$+ 2e(s)^T (U - P(B_u K_2))(e(t) - e(t-\tau(t))) + e(t)^T Q_1 e(t) - \int_{t-\tau(t)}^t \dot{e}(s)^T (Q_3)\dot{e}(s)ds \quad (20)$$

$$- (1 - \lambda_2)e(t-\tau(t))^T Q_1 e(t-\tau(t)) + \lambda_1 \dot{e}(t)^T (Q_2 + Q_3)\dot{e}(t)$$

According to Leibniz–Newton formula, for any column matrix T, the following equations hold:

$$2\Theta(t)^T T(e(t) - e(t-\tau(t)) - \int_{t-\tau(t)}^t \dot{e}(s)ds) = 0 \quad (21)$$

where $\qquad T = col\{T_1, T_2, ..., T_5\}$ and

$\Theta(t) = col\{e(t), e(t-\tau(t)), \psi_1(e(t)), \psi_2(e(t-\tau(t))), \dot{e}(t)\}$.

By adding the terms on the right sides of Equation 9a to b and left side of Equation 21 to Equation 20 and by the fact that for any $r \ge 0$ and any function $f(t)$,

$$\int_{t-r}^t f(t)ds = rf(t) \quad (22)$$

$\dot{V}(t)$ can be expressed as follows:

$$\dot{V}(t) \le \Theta(t)^T (\Sigma + \lambda_1 T Q_3^{-1} T^T)\Theta(t)$$
$$- \int_{t-\tau(t)} (\Theta(t)^T T + \dot{e}(s)^T Q_3)Q_3^{-1}(\Theta(t)^T T + \dot{e}(s)^T Q_3)^T ds \quad (23)$$

Thus, if the inequality,

$$\Sigma + \lambda_1 T Q_3^{-1} T^T < 0 \quad (24)$$

holds, it follows $\dot{V}(t) < 0$. Therefore, we conclude that under sufficient condition (Equation 24), the error dynamic (Equation 5) is asymptotically stable.

Illustrative examples

The sufficient condition for asymptotically anti-synchronization of a class of delayed neural networks presented in this paper is demonstrated by a couple of examples and numerical simulations.

Therefore, using Equations 17 to 20, the following result can be obtained:

Example 1

A two-dimensional cellular neural network (CNN) with time varying delays is given in (Gilli, 1993) and described by the following equation:

$$\dot{x}_i(t) = -c_i x_i(t) - \sum_{j=1}^2 d_{ij} f_j(x_j(t)) - \sum_{j=1}^n e_{ij} f_j(x_j(t-\tau_j)) + B_u u_i(t), \quad i = 1,2 \quad (25)$$

where $\quad c_i = 1$, $\quad D = (d_{ij})_{2\times2} = \begin{bmatrix} 1.0+\pi/4 & 20 \\ 0.1 & 1.0+\pi/4 \end{bmatrix}$,

$E = (e_{ij})_{2\times2} = \begin{bmatrix} -1.3\sqrt{2}\pi/4 & 0.1 \\ 0.1 & -1.3\sqrt{2}\pi/4 \end{bmatrix}$ and

$f_i(x_i) = 0.5(|x_i+1| - |x_i-1|)$, respectively. The delays $\tau_1(t) = \tau_2(t) = (1-e^{-t})/(1+e^{-t})$ are time-varying and satisfy $0 \le \tau_j(t) \le 1 = \tau_1, 0 \le \dot{\tau}_j(t) \le 0.5 = \tau_2, j = 1,2$. The chaotic behavior of the system with delay varying form 0.845 to 1 has been reported (Gilli, 1993). Figure 1 shows the $x_1 - x_2$ plot of the uncontrolled CNN ($u(t) = [0,0]^T$) with the initial condition $[x_1(0), x_2(0)]^T = [0.1, 0.1]^T$ for delay 0.85.

With considering $B_u = diag\{1,1\}$ and by solving the LMI (Equation 11) via Matlab LMI toolbox, possible solutions for the feedback gains of controller (Equation 5) are as:

$$K_1 = \begin{bmatrix} -3.8471 & -24.4628 \\ -24.4628 & -279.9206 \end{bmatrix}, \quad K_2 = \begin{bmatrix} 0.0004 & 0.0018 \\ 0.0019 & 0.0207 \end{bmatrix} \quad (26)$$

In the following numerical simulation, we take the initial conditions as:

$$[x_1(0), x_2(0)]^T = [0.1, 0.1]^T, \quad [y_1(0), y_2(0)]^T = [0.2, -0.2]^T$$

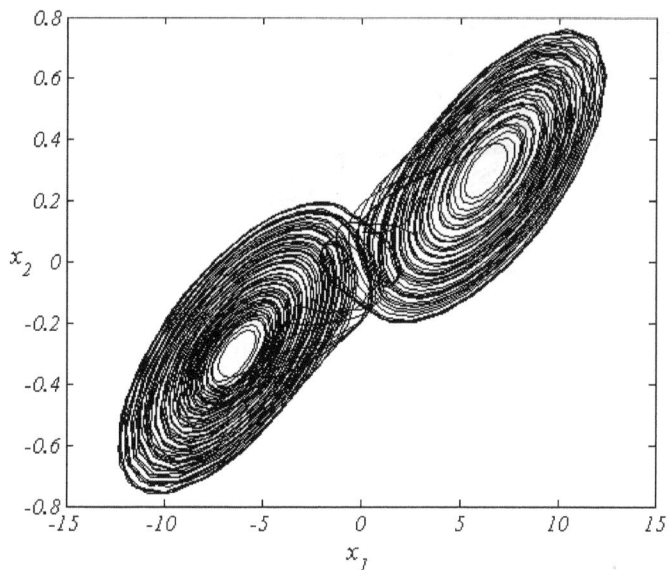

Figure 1. Chaotic behavior of drive system (Equation 25) in phase space.

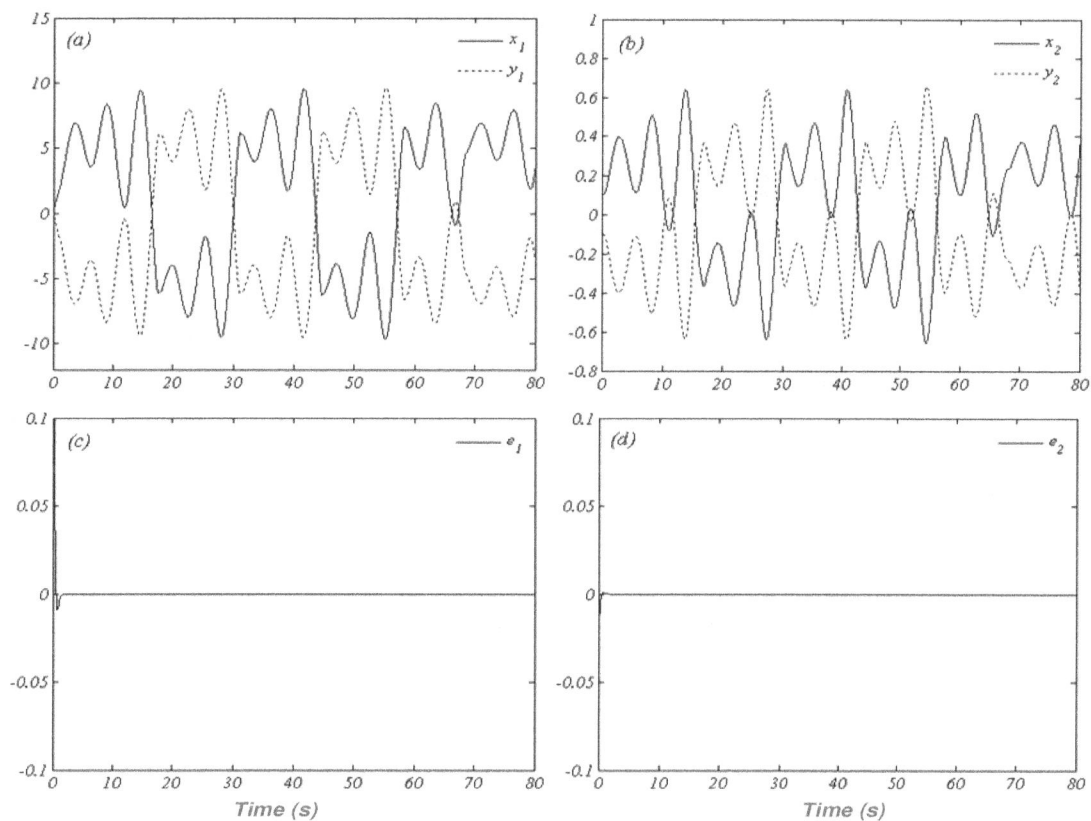

Figure 2. Simulation results of Example 1a and b state trajectories (solid line = master system, dashed line = slave system); (c) and (d) anti-synchronization errors.

Simulation results are as shown in Figure 2. The state responses of the drive and response systems are as shown in Figure 2a and b, respectively. Figure 2c and d, respectively, shows that the anti-synchronization errors

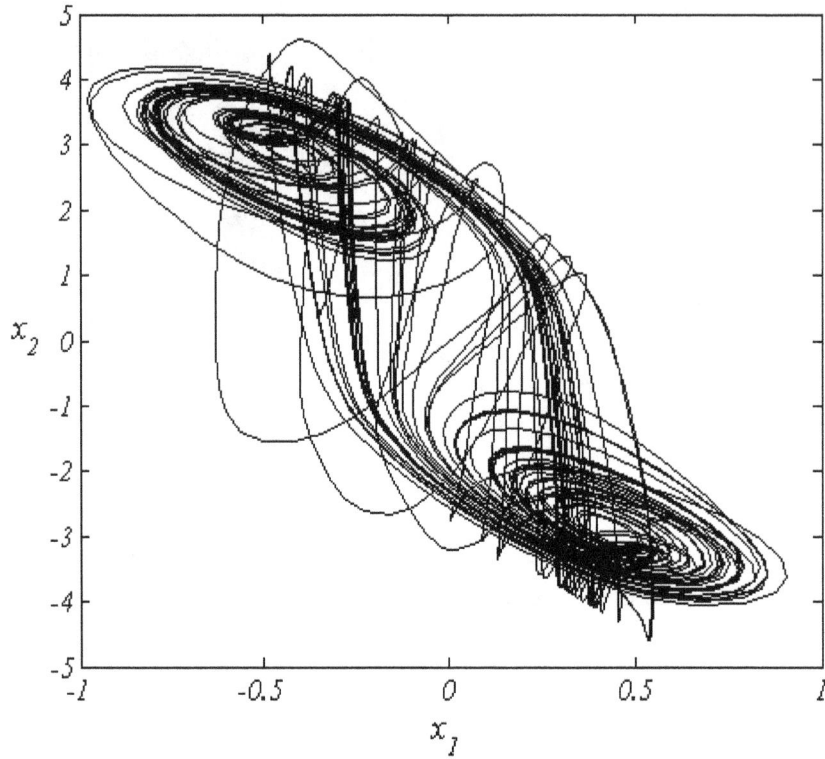

Figure 3. Chaotic behavior of drive system (Equation 27) in phase space.

$e_1(t)$ and $e_2(t)$ between drive and response systems are stabilized to zero, respectively after a short while.

Example 2

Consider the following two-order Hopfield neural network (HNN) with time-varying delay:

$$\dot{x}_i(t) = -c_i x_i(t) - \sum_{j=1}^{2} d_{ij} f_j(x_j(t)) - \sum_{j=1}^{n} e_{ij} f_j(x_j(t-\tau_j)) + B_u u_i(t), \quad i=1,2 \quad (27)$$

where $c_i = 1$, $D = (d_{ij})_{2\times2} = \begin{bmatrix} 2.1 & -0.12 \\ -5.1 & 3.2 \end{bmatrix}$,

$E = (e_{ij})_{2\times2} = \begin{bmatrix} -1.6 & -0.1 \\ -0.2 & -2.4 \end{bmatrix}$ and $f_i(x_i) = \tanh(x_i)$,

respectively. The delays $\tau_1(t) = \tau_2(t) = e^t/(1+e^t)$ are time-varying and satisfy $0 \le \tau_j(t) \le 1 = \tau_1, 0 \le \dot{\tau}_j(t) \le 0.5 = \tau_2, j=1,2$. Figure 3 shows the $x_1 - x_2$ plot of the uncontrolled HNN $(u(t) = [0,0]^T)$ with the initial condition $[x_1(0), x_2(0)]^T = [-0.35, 0.5]^T$ for delay 0.85.

With $B_u = diag\{1,1\}$ and by solving LMI (Equation 11) in Theorem 1, we get:

$$K_1 = \begin{bmatrix} -23.9331 & 10.0806 \\ 10.0806 & -14.9990 \end{bmatrix}, \quad K_2 = 10^{-3} \times \begin{bmatrix} 0.3027 & -0.1208 \\ -0.1224 & 0.1695 \end{bmatrix} \quad (28)$$

The initial conditions drive and response systems are as:

$$[x_1(0), x_2(0)]^T = [0.2, 0.5]^T, \quad [y_1(0), y_2(0)]^T = [-1.3, 2.1]^T$$

The simulation results are as shown in Figure 4. From Figure 4c and d, one can see that the anti-synchronization error between the two drive and response systems state vectors asymptotically converges to zero.

To present a quantitative comparison between the proposed method and nonlinear feedback control method (Cui and Lou, 2009), the two following criteria are used:

1. Synchronization error settling time (SEST). It is the time at which $|e| < 0.005$.

2. Integral of SQUARED synchronization error (ISSE) up to SEST.

As we know, the less the SEST, the sooner the convergence. The less the ISSE the better the synchronization achieved. Table 1 presents the results. Referring to

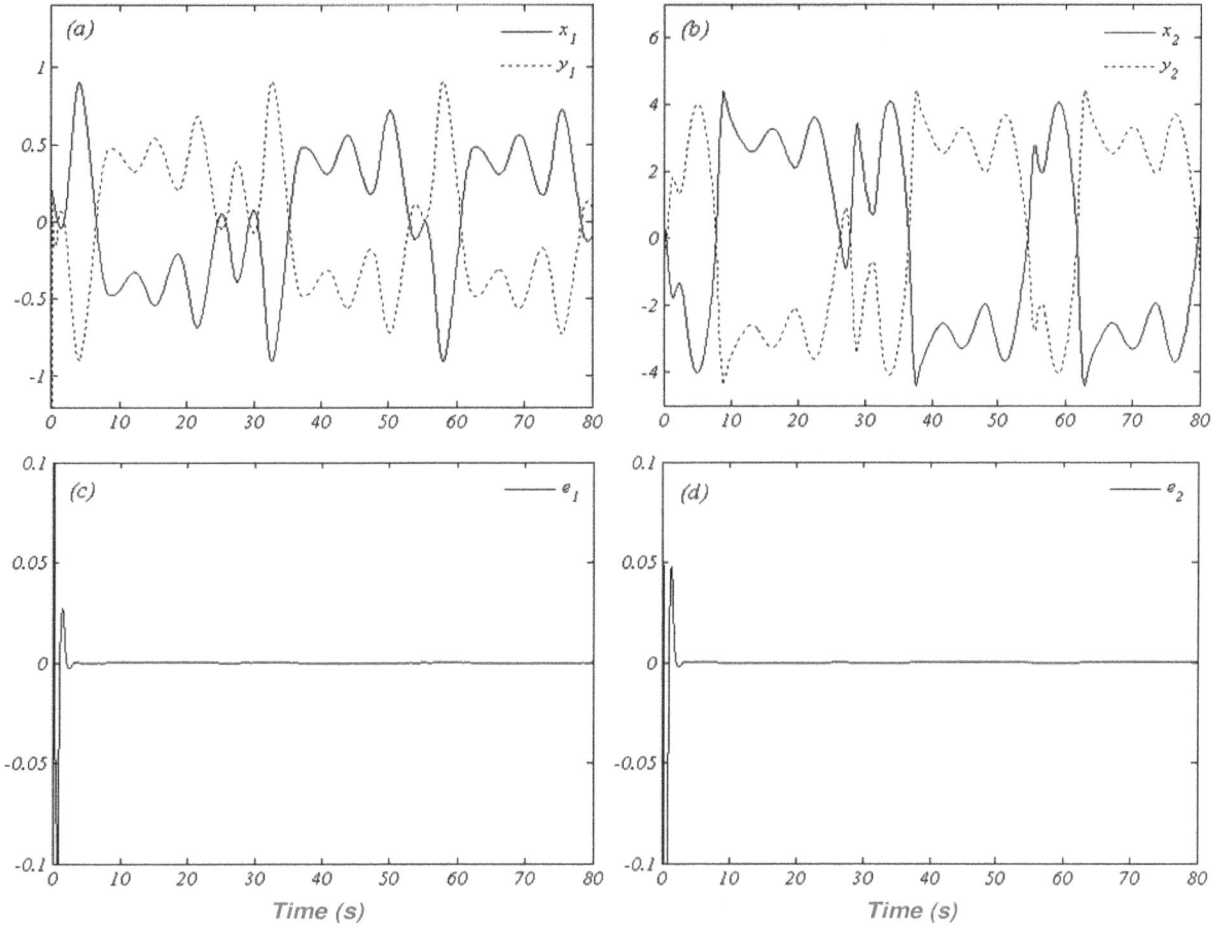

Figure 4. Simulation results of Example 2 a and b state trajectories (solid line = master system, dashed line = slave system). (c) and (d) anti-synchronization errors.

Table 1. Comparison between two different methods of anti-synchronization.

Criteria		ISSE	SEST (Sec.)
Nonlinear feedback control method (Cui and Lou, 2009)	Example 1 (CNN)	ISSE(e_1)= 139.52, ISSE(e_2)= 128.94	SEST(e_1) ≈ 12.78 SEST(e_2) ≈ 10.42
	Example 2 (HNN)	ISSE(e_1)= 1046.08, ISSE(e_2)= 1253.4	SEST(e_1) ≈ 14.56 SEST(e_2) ≈ 15.13
Proposed method	Example 1 (CNN)	ISSE(e_1)= 10.6581, ISSE(e_2)= 0.2044	SEST(e_1) ≈ 1.036 SEST(e_2) ≈ 0.3176
	Example 2 (HNN)	ISSE(e_1)=239.1924, ISSE(e_2)=136.8759	SEST(e_1) ≈ 1.27 SEST(e_2) ≈ 0.7131

the Table 1, we can conclude that the speed of synchronization with the proposed method is better than that with nonlinear feedback control methodology.

Conclusion

By designing time-delay feedback controller, this paper deals with the anti-synchronization problem of a class of chaotic neural networks with time-varying delays. An effective sufficient condition for global asymptotic anti-synchronization between the state vectors of the drive-response chaotic neural networks has been derived. These conditions, which are expressed in terms of linear matrix inequalities, are used to design suitable feedback gains in the response networks. Also, it illustrates that the speed of synchronization of the states is very fast and better than what was obtained by the nonlinear feedback control methodology. Two numerical examples with graphical illustrations are given to illuminate the presented synchronization scheme.

REFERENCES

Adeli M, Saedian A, Zarabadipour H (2011). Anti-synchronization of chaotic system using adaptive modified function projective method with unknown parameters. Int. J. Phys. Sci., 6(32): 7322 - 7327.

Ahn CK (2009). An H_∞ approach to anti-synchronization for chaotic systems. Phys. Lett. A. 373: 1729-1733.

Al-Sawalha MM. Noorani MSM (2009). On anti-synchronization of chaotic systems via nonlinear control. Chaos Solit. Fract., 42: 170-179.

Boyd S, Ghaoui LE, Feron E, Balakrishnan V (1994). Linear Matrix Inequalities in System and Control Theory. SIAM, Philadelphia, PA.

Carpenter MC, Grossberg S (1987). Computing with neural network. Science, 37: 51-115.

Chen Y, Li M, Cheng Z (2010). Global anti-synchronization of master-slave chaotic modified Chua's circuits coupled by linear feedback control. Math. Comput. Model., 52: 567-573.

Chen G, Dong X (1998). From chaos to order: methodologies, perspectives and applications. Nonlinear science. Singapore: World Scientific

Chiang TY, Lin JS, Liao TL, Yan JJ (2008). Anti-synchronization of uncertain unified chaotic systems with dead-zone nonlinearity. Nonlinear Anal., 68: 2629-2637.

Chua LO, Yang L (1988). Cellular neural networks: Applications. IEEE Trans. Circuits Syst. I., 35: 1273-1290.

Cohen MA, Grossberg S (1983). Absolute stability of global pattern formation and parallel memory storage by competitive neural networks. IEEE Trans. Syst. Man. Cynern., 13: 815-826.

Cui B, Lou X (2009). Synchronization of chaotic recurrent neural networks with time-varying delays using nonlinear feedback control. Chaos Solitons Fractals, 39: 288-294.

Gilli M (1993). Strange attractors in delayed cellular neural networks. IEEE Trans. Circuits Syst. I., 40: 849-853.

Guo J, Huang X, Cui Y (2009). Design and analysis of robust fault detection filter using LMI tools. Comput. Math. Appl., 57: 1743-1747.

Gutzwiller MC (1990). Chaos in Classical and Quantum Mechanics, Springer, New York.

Hencey B, Alleyne A (2009). An anti-windup technique for LMI regions. Automatica, 45: 2344-2349.

Ho MC, Hung YC, Chou CH (2002). Phase and anti-phase synchronization of two chaotic systems by using active control. Phys. Lett. A., 296: 43-48.

Hu J, Chen S, Chen L (2005). Adaptive control for anti-synchronization of Chua's chaotic system. Phys. Lett. A., 339: 455-460.

Leu YG (2010). Mean-based fuzzy identifier and control of uncertain nonlinear systems. Fuzzy Sets. Syst., 161: 837-858.

Li R, Xu W, Li S (2009). Anti-synchronization on autonomous and non-autonomous chaotic systems via adaptive feedback control. Chaos Solitons Fractals. 40: 1288-1296.

Li T, Sun C, Zhao X, Lin C (2008). LMI-based asymptotic stability analysis of neural networks with time-varying delays. Int. J. Neural Syst., 18: 257-265.

Liu Q, Cao J, Xia Y (2005). A delayed neural network for solving linear projection equations and its analysis. IEEE Trans. Neural Netw., 16: 834-843.

Lou XY, Cui BT (2006). New LMI conditions for delay-dependent asymptotic stability of delayed Hopfield neural networks. Neurocomputing. 69: 2374-2378.

Lu HT (2002). Chaotic attractors in delayed neural networks. Phys. Lett. A., 298: 109-116.

Lu WL, Chen TP (2006). Dynamical behaviors of delayed neural network systems with discontinuous activation functions. Neural Comput., 18: 683-708.

Park P (1999). A delay-dependent stability criterion for systems with uncertain time-invariant delays. IEEE Trans. Auto. Cont., 44: 876-877.

Pecora LM, Carroll TL (1990). Synchronization in chaotic systems. Phys. Rev., Lett., 64: 821-824.

Sanjaya M, Halimatussadiyah, Maulana DS (2010). Bidirectional chaotic synchronization of non-autonomous circuit and its application for secure communication. Int. J. Phys. Sci., 6(2): 74-79.

Sun Y, Cao J (2007). Adaptive synchronization between two different noise-perturbed chaotic systems with fully unknown parameters. Phys. A., 376: 253-265.

Wieczorek S, Chow WW (2009). Bifurcations and chaos in a semiconductor laser with coherent or noisy optical injection. Opt. Commun., 282: 2367-2379.

Yang XS, Yuan Q (2005). Chaos and transient chaos in simple Hopfield neural networks. Neurocomputing, 69: 232-241.

Yuan Y (2007). Dynamics in a delayed-neural network. Chaos Solitons Fractals, 33: 443-454.

2

Extended reach spectrum-sliced passive optical access network

Bobrovs Vjaceslavs, Spolitis Sandis and Ivanovs Girts

Institute of Telecommunications, Riga Technical University, Azenes Str. 12, Rīga, LV-1048, Latvia, Europe.

The spectrum sliced dense wavelength division multiplexed passive optical network (SS-DWDM PON) is a cost effective and power efficient solution for passive optical access networks to satisfy the growing worldwide demand for transmission capacity. In this work we successfully demonstrate the simulation model of effective spectrum-sliced 16-channel dense wavelength division multiplexed passive optical network (WDM-PON) system with data rate 2.5 Gbit/s per channel with significant optical link reach improvement by implementation of dispersion compensating fiber (DCF) and fiber Bragg grating (FBG) for chromatic dispersion compensation. We also show the realization of a broadband amplified spontaneous emission (ASE) light source with high output power +23 dBm and flat spectrum in wavelength range from 1545.32 to 1558.98 nm (C-band). Proposed system is built on ITU-T recommended DWDM frequency grid, in this way making it more adaptive and convergent to existing optical access systems.

Key words: Amplified spontaneous emission (ASE), arrayed-waveguide grating (AWG), optical fiber communication, spectrum slicing, wavelength division multiplexed passive optical network (WDM-PON).

INTRODUCTION

Spectrum sliced dense wavelength-division-multiplexed passive optical network (SS-DWDM PON) is an attractive and cost effective solution to satisfy the growing worldwide demand for transmission capacity in the next generation fiber optical access networks (El-Sahn et al., 2010; Spolitis and Ivanovs, 2011). Traditional wavelength division multiplexed (WDM) systems have multiple transmitter lasers operating at different wavelengths, which need to be wavelength selected for each individual channel operated at a specific wavelength (Vukovic et al., 2007). It increases complexity of network architecture, cost and wavelength (channel) management (Spolitis et al., 2011; Spolitis and Ivanovs, 2011). The strength of spectrum-sliced WDM-PON technology is use for one common broadband seed light source and its ability to

place electronics and optical elements in one central office (CO), in that way simplifying the architecture of fiber optical network (Choi and Lee, 2011). Such optical systems benefit from the same advantages as WDM, while employing low cost incoherent light sources like amplified spontaneous emission (ASE) source or light-emitting diode (LED) (Kaneko et al., 2006; Bobrovs et al., 2011).

This spectral slicing method is promising and cost-efficient solution for the transmitter in an optical line terminal (OLT) of WDM-PON architecture. Spectrum sliced WDM PON system is more energy efficient than traditional WDM-PON systems because there is employed a common single broadband light source for transmission of large number of wavelength channels, not a one source per user as it is in the traditional WDM-PON (El-Sahn et al., 2010). The optical bandwidth per channel of SS-WDM PON system is large compared to the bit rate. Therefore, dispersion considerably degrades the performance of this system more than it is observed in conventional laser-based systems (Choi and Lee, 2011). The influence of dispersion needs to be studied in order to understand the characteristics of a spectrum sliced WDM PON system employing standard single mode optical fiber.

The novelty of this paper is in fact, that we investigated the influence of chromatic dispersion compensation on overall improvement of multi-channel optical access system's performance as well as compared two different passive chromatic dispersion (CD) compensation techniques for network reach improvement.

Using the optical system's simulation experience from previous works (Spolitis et al., 2011; Bobrovs et al., 2011) we demonstrate a new model of 16-channel SS-DWDM PON system with CD compensation and broadband ASE seed light source with flat spectrum in wavelength range from 1545.32 to 1558.98 nm (C-band). In this system single broadband ASE source using flat-top type arrayed-waveguide gratings (AWG) are shared (sliced) among many user channels in such a way allocating a unique spectral slice to each user. AWG is mostly used as an optical filter because it can simultaneously play two roles as a filter and multiplexer (Bobrovs et al., 2011; Spolitis et al., 2012). In contrast to other studies made in this field before (El-Sahn et al., 2010; Choi and Lee, 2011) our main goal of this research is to evaluate and compare the reach improvement of SS-DWDM PON system by implementing dispersion compensating fiber (DCF) and fiber Bragg grating (FBG) for CD compensation in dispersion compensation module (DCM) as well as investigate the overall performance of this system by building it on the ITU-T DWDM frequency grid, defined in recommendation G.694.1.Based on this recommendation the channel spacing is chosen equal to 100 GHz (0.8 nm in wavelength) (ITU-T Rec. G.694.1, 2012). In addition, our research differs from other researches made in this field with data transmission rate which is chosen 2.5 Gbit/s per channel, and is relatively high for this type of optical access networks. It is very important to build a new type optical system based on widely used frequency grid, recommended by international standards to make it more compatible with other already existing WDM-PON optical systems multiplexer (Spolitis et al., 2012). Our proposed spectrum-sliced dense WDM-PON system with CD compensation (where one seed broadband ASE source is spectrally sliced and used for multiple users) is potentially capable to replace existing classical WDM-PON access system (where one laser source is used for each user) (Leeson et al., 2006; Leeson and Sun, 2008). The main benefit of this system includes the reduction of network architecture complexity as well as cost per one user.

CHROMATIC DISPERSION COMPENSATION TECHNIQUES

Chromatic dispersion significantly degrades the performance of spectrum sliced dense WDM PON system because of stochastic and noise-like nature of broadband ASE light source as well as large optical bandwidth of spectrum sliced channel (Choi and Lee, 2011). Dispersion causes optical signal pulses to broaden and lose their shape as they travel along the optical fiber. When pulses become wider, they have tendency to interfere with adjacent pulses. Eventually this limits the maximum achievable data transmission rate and transmission distance. The broadening of signal pulses causes intersymbol interference (ISI). Due to influence of ISI there can occur problems to restore transmitted information. In this case it is difficult to separate transmitted bit sequence at receiver side and it is resulting bit errors (high BER) or even failure of fiber optical transmission system (Spolitis et al., 2011; Bobrovs et al., 2011).

In order to reduce this negative impact DCM can be used for chromatic dispersion compensation and system's performance improvement in these fiber optical transmission systems (Kani et al., 2009; ITU-T Rec. G.694.1, 2012). Dispersion compensating fiber and fiber Bragg grating is used for CD compensation modules in our investigated SS-DWDM PON system.

Typically DCM is specified by what length, in km, of standard single mode fiber (SMF) will be compensated or by the total compensation value of CD over a specific wavelength range, specified in ps/nm. Dispersion compensation module can be placed before or after standard SMF span (CD pre-compensation or post-compensation). In our research we employ a pre-compensation configuration for effective chromatic dispersion compensation. In previous researches it is found as the most effective dispersion compensation configuration for WDM passive optical networks (Spolitis et al., 2011; Spolitis and Ivanovs, 2011).

DCM containing DCF fiber typically has high fiber attenuation and accordingly the insertion loss resulting from insertion of device in fiber optical transmission link will be high. DCM insertion loss can be compensated by optical amplifier, located in central office (CO). The effective core area A_{eff} of a DCF fiber is much smaller than for standard ITU-T G.652 single mode fiber and thereby DCF experience much higher optical signal distortions caused by nonlinear optical effects (Chomycz, 2009). Typical DCF has a large negative dispersion and produces a negative slope. By lowering optical power the impact of nonlinear optical effects can be reduced. As a second solution for CD compensation are dispersion compensation modules with chirped FBG, (Figure 1). Because of good optical characteristics the chirped FBG is suitable for WDM systems.

Like DCF unit FBG is also a completely passive optical

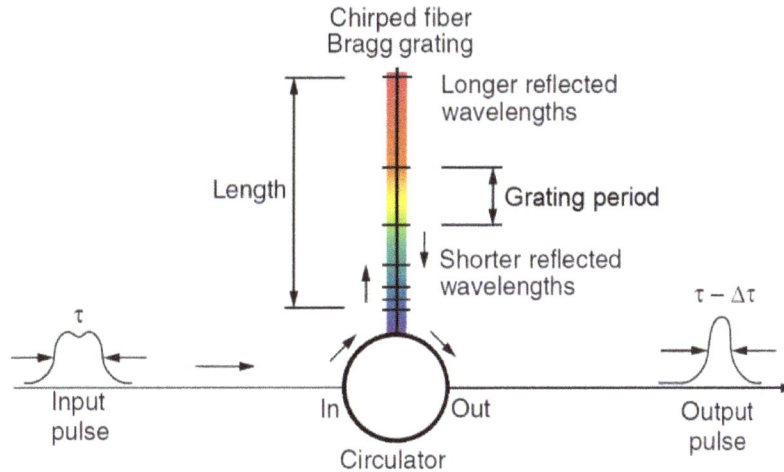

Figure 1. Chirped fiber Bragg grating.

unit. FBG has a grating period which is not constant but changes linearly over the length of the grating, starting with the shorter grating period located at the beginning of the grating. The grating period is a distance between two adjacent maximum values of the refractive index (Chomycz, 2009). The FBG reflects a narrow spectrum of wavelengths that are centered at λ_B and passes all the other wavelengths. Reflected wavelength λ_B can be obtained by the following equation:

$$\lambda_B = 2 \Lambda n_g \qquad (1)$$

where λ_B - reflected wavelength, nm; Λ - grating period, nm; n_g - fiber's effective group refractive index. When dispersion affected input pulse with width τ passes through the chirped FBG, pulse width is decreased by $\Delta \tau$ and its shape is restored on the output of transmission line.

FBG grating has shorter grating periods at beginning but over the length of the grating these periods linearly increases. It means that shorter optical signal wavelengths are reflected sooner and have a less propagation delay through the FBG unit but longer signal wavelengths travel further into the fiber grating before they are reflected back and accordingly have more propagation delay through the FBG. Typically the length of the FBG grating is from 10 to 100 cm (Agrawal, 2010).

A significant advantage of dispersion compensation modules with FBG over the DCF fiber is its relatively small insertion loss resulting from the insertion of a device in fiber optical transmission system. For example, insertion loss of commercially available DCF specified to compensate accumulated CD of up to 120 km standard single mode fiber span is about 10 dB, whereas insertion loss of FBG based DCM capable to compensate the same CD amount is only up to 4 dB (Chomycz, 2009). In contrast to DCF DCM, the FBG based DCM can be used

at higher optical powers without inducing nonlinear optical effects

Spectrum-slicing technique of broadband ASE source using arrayed-waveguide grating

One of basic techniques available in WDM PON systems in order to reduce the cost of components and simplify the passive network architecture is spectrum slicing technique where incoherent broadband light source (BLS) is used. Such a BLS is a good candidate for generating equally spaced multi-wavelength channels (Leeson and Sun, 2008; El-Sahn et al., 2010). The aim of spectrum slicing is to employ a single BLS for transmission on a large number of wavelength channels (Figure 2).

BLS is sliced using arrayed-waveguide grating (AWG). Afterwards optical slices are modulated, multiplexed by second AWG and transmitted over standard single mode optical fiber (SMF) line. Channels are demultiplexed by third AWG located in the end of fiber optical line and transmitted to optical receiver. This receiver can be PIN photodiode or avalanche photodiode (APD). Broadband light sources like LED, SLED or ASE can be used in spectrum sliced systems for data transmission. The transmission power available for each separate optical channel depends on the slice width. It should be considered that a larger slice will increase not only the total channel power but also increase the influence of dispersion and therefore the number of available WDM channels (El-Sahn et al., 2010).

In our research we choose ASE as BLS for spectral slicing because it has the highest optical output power compared to other mentioned broadband light sources. Optical signal from broadband ASE source is stochastic in nature and noise-like, therefore some assumptions and

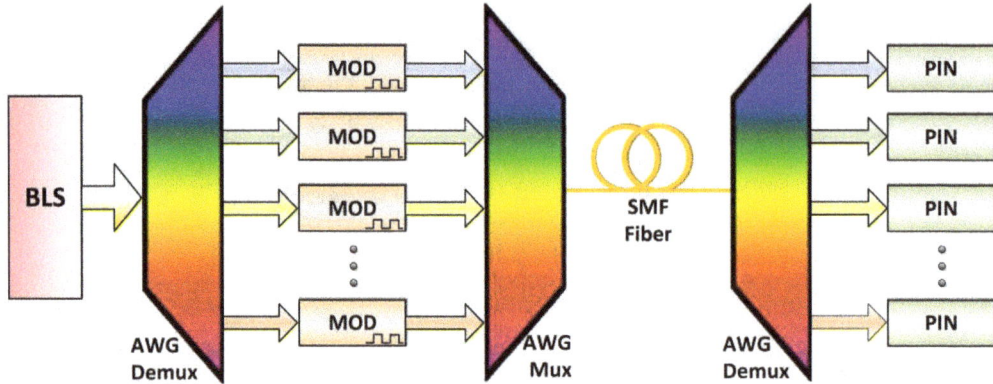

Figure 2. Operational principle of spectrum sliced system using broadband light source sliced by AWG demultiplexer.

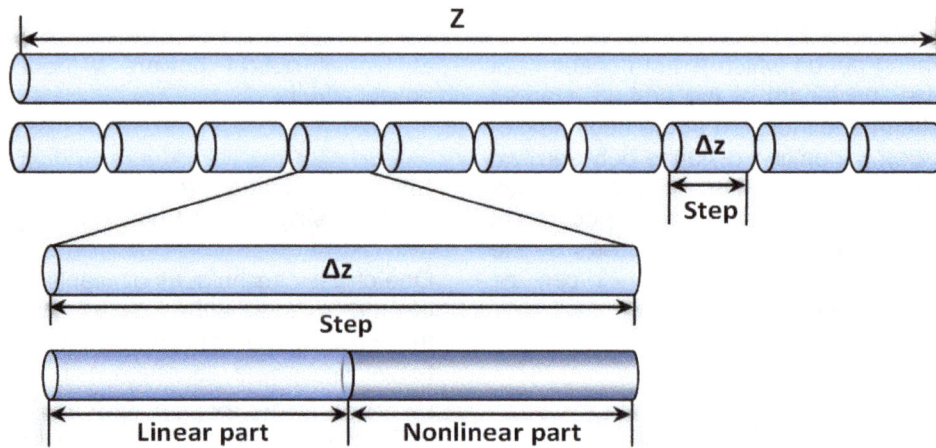

Figure 3. Illustration of time domain split-step (TDSS) algorithm.

formulations that is used in conventional laser based (e.g. distributed feedback (DFB) or Fabry Perot (FP) laser) passive optical systems may not be applicable for SS-DWDM PON systems (Spolitis et al., 2012). In our research we investigate optical system where ASE spectral slicing is realized using AWGs.

Numerical analysis and simulation model

Our accepted research method is a mathematical simulation using newest OptSim 5.2 simulation software, where complex differential equation systems are solved using Split-Step algorithm. In order to study the nonlinear effects in optical fiber the nonlinear Schrödinger equation (NLS) is used. Except certain cases this equation cannot be solved analytically. Therefore, OptSim software is used for simulation of fiber optical transmission systems where it solves complex differential equations using Time Domain Split-Step (TDSS) method. This Split-Step method is being used in most commercial optical system simulation tools. The principle of the method is illustrated in Figure 3 and by the fiber propagation

equation, which can be written as following:

$$\frac{\partial A(t,z)}{\partial z} = \{L+N\}A(t,z)$$

(2)

where $A(t,z)$ – optical field; L – linear operator responsible for dispersion and other linear effects; N – non-linear operator responsible for the nonlinear effects. It is assumed that linear L and nonlinear N effects affect the optical signal independently using the Split-Step method, if the span (step) of simulated optical fiber Δz if enough small.

In Figure 3 is shown that each step Δz consists of two half steps. In the first half step only linear effects (linear part) are taken into account, but in the second half step only nonlinear effects (nonlinear part) are taken into account. All optical fiber length z is divided into steps Δz, and alternately linear and nonlinear effects are considered. For the most accurate results, it is necessary to carefully choose a step Δz. If this step Δz is chosen too small, it will increase the time necessary to perform calculations, but if the step is chosen too large it will decrease accuracy of the calculations in our case the simulation of more than 1024 bits is made to achieve result's estimation accuracy not less than 95%

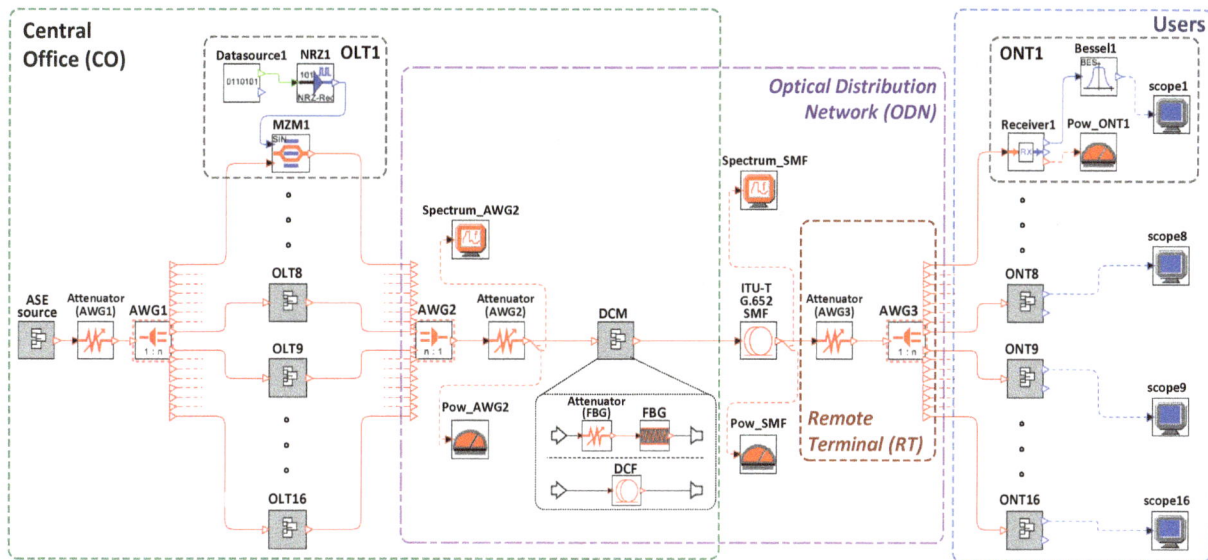

Figure 4. Simulation model of proposed high-speed 16-channel AWG filtered ASE seeded SS-DWDM PON system with DCM module.

(OptSim User Guide, 2012; Spolitis et al., 2011).

The performance of simulated scheme was evaluated by the obtained bit error ratio (BER) value of each WDM channel in the end of the fiber optical link. Basis on ITU recommendation G.984.2 it should be noticed that BER value for fiber optical transmission systems with data rate 2.5 Gbit/s per channel is specified less than 10^{-10} (ITU-T Rec. G.694.1, 2012).

In order to investigate the performance of incoherent broadband ASE light source as seed light for spectrum sliced WDM passive optical networks we created a simulation scheme of high-speed 16 channel SS-DWDM PON system (Figure 4). The design of high performance broadband flat spectrum ASE source on the output of cascaded EDFAs in details is described thus. This section describes the simulation model and parameters of proposed SS-DWDM PON system with flat ASE seed source. As one can see in Figure 4, SS-WDM PON simulation scheme consists of 16 channels. The frequency grid is anchored to 193.1 THz and channel spacing is chosen equal to 100 GHz frequency interval (ITU-T Rec. G.984.2, 2003).

Broadband ASE light source is spectrally sliced using 16-channel flattop AWG filter (AWG1) with channel spacing equal to 100 GHz (0.8 nm in wavelength). ASE source average output power is 23 dBm. Average output power of optical slices after AWG1 are about 5.8 dBm. Using this AWG unit we can obtain spectrally sliced optical channels (slices) with dense channel interval of 100 GHz. Insertion losses of AWG units are simulated using attenuation blocks (attenuators). Simulated a thermal high-performance AWG multiplexers and demultiplexers are absolutely passive optical components (no need for thermal regulation and monitoring electronics) with insertion loss up to 3 dB each (Spolitis and Ivanovs, 2011).

After spectrum slicing by AWG1 optical slices are transmitted to the optical line terminals (OLTs). OLTs are located at central office (CO). Each OLT consists of data source, non-return-to-zero (NRZ) driver, and external Mach-Zehnder modulator (MZM). Each MZM has 5 dB insertion losses, 20 dB extinction ratio, modulation voltage Vπ of 5 Volts and maximum transmissivity offset voltage 2.5 Volts.

Generated bit sequence from data source is sent to NRZ driver where electrical NRZ pulses are formed. It has an electrical output

signal which can assume one of the two electrical levels depending on the transmitted bit. When a "1" is fed into the driver, the output signal is at the low level during the entire bit time. When a "0" is fed into the driver, the output signal is at the high level during the entire bit time. Switching between the two levels may be instantaneous if the field Time Slope is set to zero, or not otherwise, with the desired time slope (OptSim User Guide, 2012). Used NRZ drivers have following parameters: low level -2.5 Volts, high level 2.5 Volts, time slope is 0 and crossing point is chosen 50%. Afterwards formed electrical NRZ pulses are sent to MZM modulator. MZM modulates the optical slice from AWG1 and forms optical signal according to electrical drive signal. These formed optical pulses from all OLTs are coupled by AWG multiplexer (AWG2) and sent into standard optical single mode fiber (SMF) defined in ITU-T recommendation G.652.

Information from OLT is transmitted to an optical network terminal (ONT) or user over the fiber optical transmission link called optical distribution network (ODN). ODN consists of AWG multiplexer, two optical attenuators, dispersion compensation module (DCM), SMF with variable length and AWG demultiplexer. In our research we SMF with large effective core area Aeff = 80 μm², attenuation α = 0.2 dB/km, dispersion D = 16 ps/(nm·km) and dispersion slope Dsl = 0.07 ps/(nm²·km) at the reference wavelength λ = 1550 nm. As one can see in Figure 4 (middle), two simulated CD compensation methods implemented in DCM module are FBG and DCF. In our research we use DCF fiber with Aeff = 20 μm², attenuation α = 0.55 dB/km, dispersion D = -80 ps/(nm·km) and dispersion slope Dsl = 0.19 ps/(nm²·km) at the reference wavelength λ = 1550 nm. Simulated FBG is tunable in terms of both reference frequency and total compensating dispersion value. Additional attenuator simulates 3 dB insertion loss of FBG unit. For the most accurate results we used real parameters of standard DCF fiber and tunable FBG in our simulation setup.

In the end of fiber optical link 16 optical channels are split using AWG demultiplexer (AWG3) located in remote terminal (RT). The receiver section includes ONT units. Each ONT consists of sensitivity receiver with PIN photodiode (sensitivity S = -25 dBm at sensitivity reference error probability BER = $1 \cdot 10^{-10}$), Bessel electrical lowpass filter (3-dB electrical bandwidth B_E = 1.6 GHz),

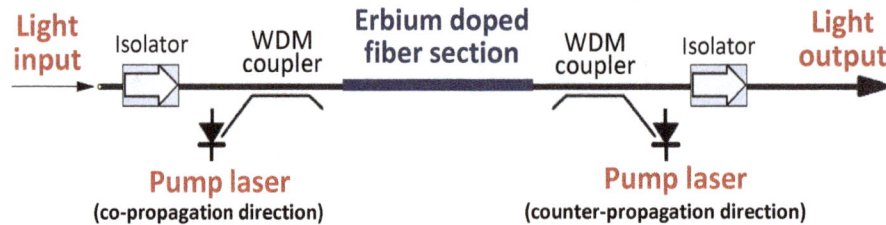

Figure 5. Scheme of erbium-doped fiber amplifier.

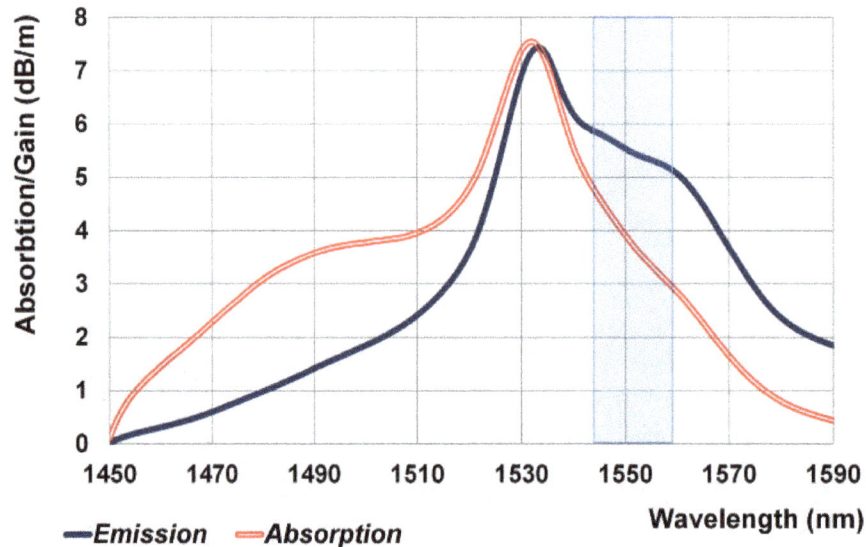

Figure 6. Er^{3+} emission and absorption spectrum in erbium doped fiber.

optical power meter and electrical probe to evaluate the quality of received optical data signal. Optical signal is converted to electrical signal using PIN photodiode and filtered by Bessel electrical filter to reduce noise (Keiser, 2003).

Broadband ASE light source realization by erbium doped fiber amplifiers

Erbium doped fiber amplifier (EDFA) consist of erbium-doped fiber having a silica glass host core doped with active Er^{3+} ions as the gain medium. Basic elements of an EDFA schematically are shown in Figure 5. Erbium-doped fiber is pumped by semiconductor lasers at 980 or 1480 nm in wavelength. By pumping on 980 nm we get low noise figure and low optical gain but pumping on 1480 nm we can get higher optical gain and also higher optical noise. Useful optical signal propagates along short span of a special erbium doped fiber and is being amplified at that time (Keiser, 2003; Agrawal, 2010).

Amplifier is pumped by a semiconductor laser (pump laser), which is coupled by a wavelength selective coupler, also known as WDM coupler. This WDM coupler combines the optical signal from pump laser with the transmitted optical signal which contains useful information need to been transmitted over the fiber optical link. The pump light propagates also in the same direction as the useful signal (co-propagation) or in the opposite direction

(counter-propagation). Optical isolator is used to prevent laser oscillations and excess noise due to unwanted optical reflections (Keiser, 2003).

To make the principle work, erbium atoms needed to be set in excited state. This is done by 980 nm and/or 1480 nm lasers. The pump laser generates a high-powered optical signal beam at a wavelength on which erbium ions absorb and they get to their excited state. Pumping laser power is usually being controlled via feedback control circuit. It is known that EDFA emits high power amplified spontaneous emission (ASE) noise in C- and L- bands if there is no signal to be amplified (Agrawal, 2010). This effect is used to create broadband ASE light source in this research. ASE noise generation and gain occurs along all EDFA fiber length and it depends on erbium ion (Er^{3+}) emission and absorption spectrum shown in Figure 6. The marked area in Figure 6 shows the 1545.32 to 1558.98 nm wavelength range which is used for data transmission in our 16-channel spectrally sliced dense WDM-PON system. Broadband ASE light source realization can be done in different ways: by using one EDFA or connecting several amplifiers in cascade mode (one after another). The latter method allows achieving flatter ASE spectrum and higher output power because of better Er^{3+} ions usage along several amplifiers.

It should be taken into account that by default resulting ASE spectrum on the output of EDFA is not flat and manipulation with EDFA parameters must be done to make it plainer in range necessary for WDM system realization. Such parameters as EDFA

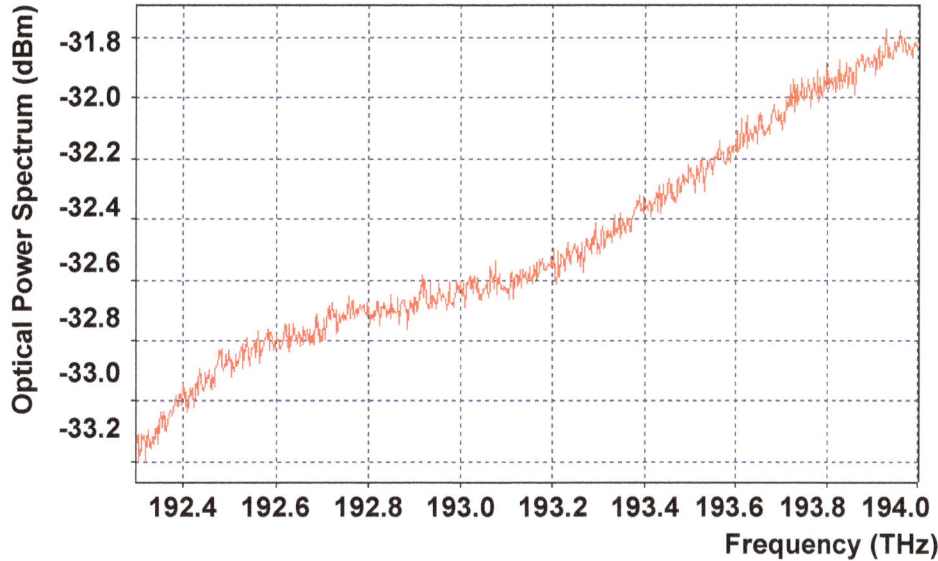

Figure 7. ASE spectrum after first EDFA in proposed SS-DWDM PON system's frequency range.

pumping laser power, its wavelength and length of erbium doped fiber affect Er^{3+} inversion – ratio between number of erbium ions in excited and ground state. This inversion has a direct impact on the shape resulting ASE spectrum. In our research the broadband ASE light source consists of two EDFAs combined in cascaded mode. First EDFA amplifier is realized using 9 m long erbium doped fiber and a 1480 nm co-directional pump laser with output power of 400 mW. As it is seen in Figure 7, such power for the first EDFA is optimal to obtain flat output spectrum of the ASE on the output of cascaded EDFAs system.

First EDFA generates optical ASE signal with irregular spectrum in wide frequency range with total output power -4.16 dBm (0.38 mW). For our research we use the frequency range from 192.3 THz to 193.4 THz (wavelength range from 1545.32 to 1558.98 nm), see marked area in Figure 6. We investigate this particular frequency range because our proposed 16-channel spectrum-sliced dense WDM PON system operates in this range.

The spectrum of broadband ASE source from the first EDFA is being amplified and flattened using second EDFA. In our case the ASE spectrum on the output of EDFA can be flattened by achieving erbium ion inversion when lower frequencies in acquired spectrum are amplified more than higher ones. In this case it means that almost 100% of Er^{3+} ions inversion. For flattening of ASE output spectrum we used bi-directional laser pumping on both - 980 and 1480 nm wavelengths to uniformly distribute the energy along 12 m long erbium doped fiber section in the second EDFA amplifier. As can be seen in Figure 8, by simultaneously increasing the pump power from 100 to 600 mW on all pump lasers (pumping on 1480 nm in first and pumping on 980 and 1480 nm in second EDFA) we obtained six optical output spectra of our ASE broadband light source.

The smoothest ASE broadband spectrum can be achieved if use 400 mW output power for all pump lasers, where first EDFA amplifier is pumped in co-propagating direction on 1480 nm wavelength, and second EDFA is pumped in both co-propagating direction on 1480 nm and counter-propagating direction on 980 nm (Figure 9).

In this manner we construct a broadband ASE source with almost flat spectrum and total output power on the output of cascaded EDFA system about +23 dBm (200 mW). The spectrum of realized

broadband ASE noise-like light source which can be spectrally sliced using AWG unit is shown in (Figure 10). The highlighted area in the figure shows the 1545.322 to 1558.983 nm wavelength range (centered on 1552.524 nm wavelength or 193.1 THz in frequency) which is used for our 16-channel spectrally sliced DWDM-PON system. As one can see, fluctuations of optical power level are minimal and spectrum in this region is almost flat.

RESULTS AND DISCUSSION

There are compared two different CD compensation methods for improvement of maximal reach and performance of 16-channel AWG filtered SS-DWDM PON system with flattened ASE broadband light source. Flat-top type AWG units were chosen for spectral slicing of ASE light source in our optical system because of good filtering performance, excellent WDM channel separation and bandwidth allow passing sufficient high optical power (El-Sahn et al., 2010; Spolitis et al., 2012). To reduce the negative impact of intensity noise as well as cross phase modulation the correct choice of filter's shape and 3-dB pass bandwidth is very important (Ozolins et al., 2010; Ivanovs et al., 2010). As a comparison in Figure 11 are shown two SS-DWDM PON channels (central frequencies 193.1 and 193.2 THz) which are filtered by Gaussian and flat-top type AWGs with 50 GHz 3-dB pass bandwidth.

It is shown that flat-top AWG has narrower spectral filter shape than Gaussian type AWG. Wider filter shape leads to larger slice width, higher resultant slice's power and optical signal dispersion (Leeson and Sun, 2008; Spolitis et al., 2012). In OptSim software we simulated flat-top type AWG filter shape using Raised Cosine optical filter's transfer function but using Super-Gaussian

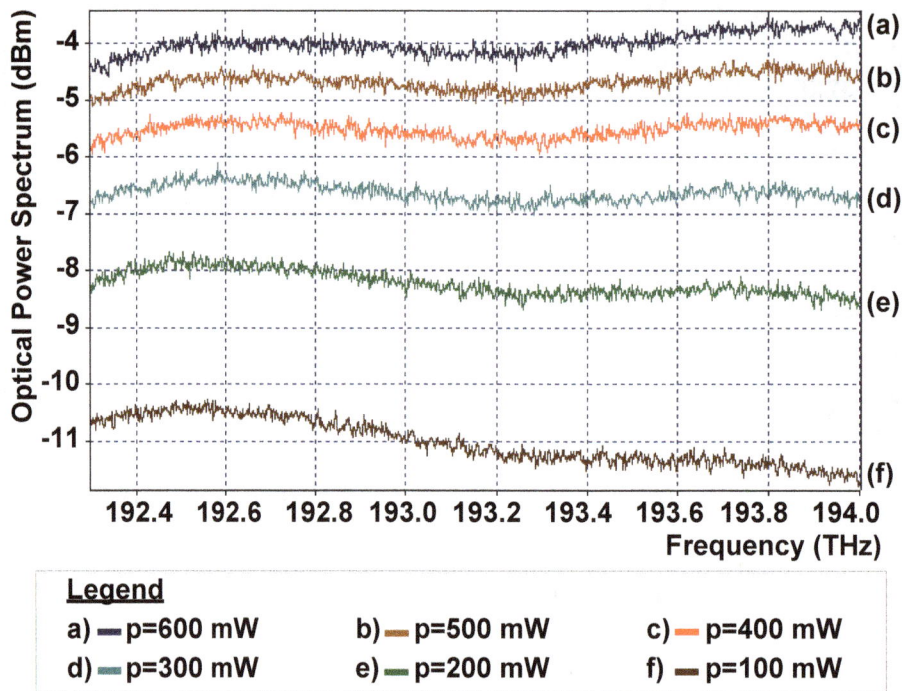

Figure 8. ASE spectrum after second EDFA at different pump laser powers in proposed SS-DWDM PON system's frequency range.

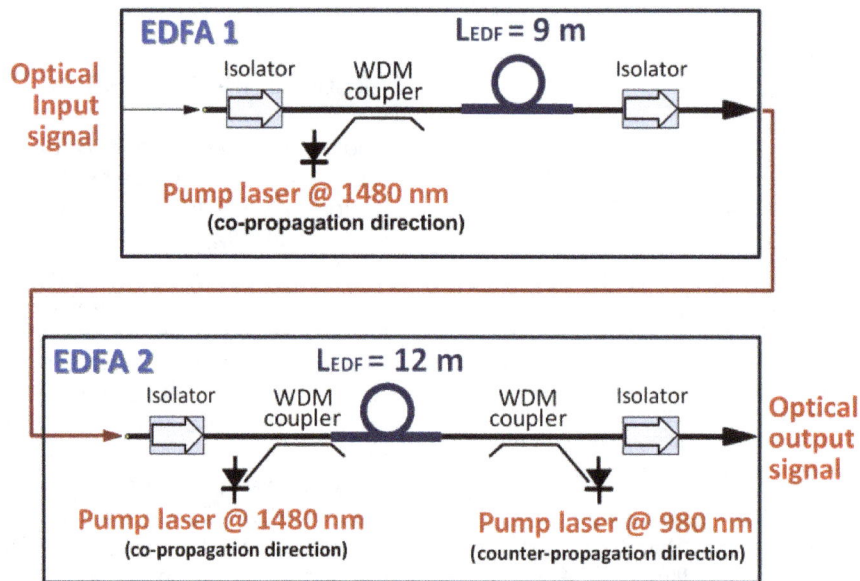

Figure 9. Detailed scheme of cascaded EDFA system with found parameters.

transfer function we approximate Gaussian AWG filter shape (Ozolins et al., 2010). As one can see in Figure 11, in the case of Gaussian filter the crosstalk between channels will be much higher than employing flat-top type filter. On the basis of this we can make a conclusion that AWG with flat-top filter spectral shape has higher optical signal to noise ratio (OSNR) than AWG with Gaussian type filter shape at identical bandwidth (3-dB bandwidth is

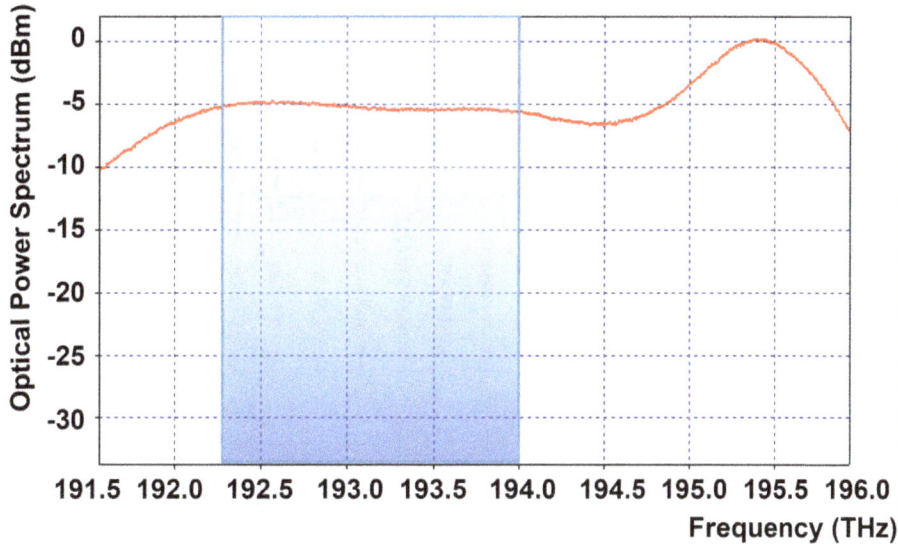

Figure 10. Output spectrum of noise-like ASE broadband light source.

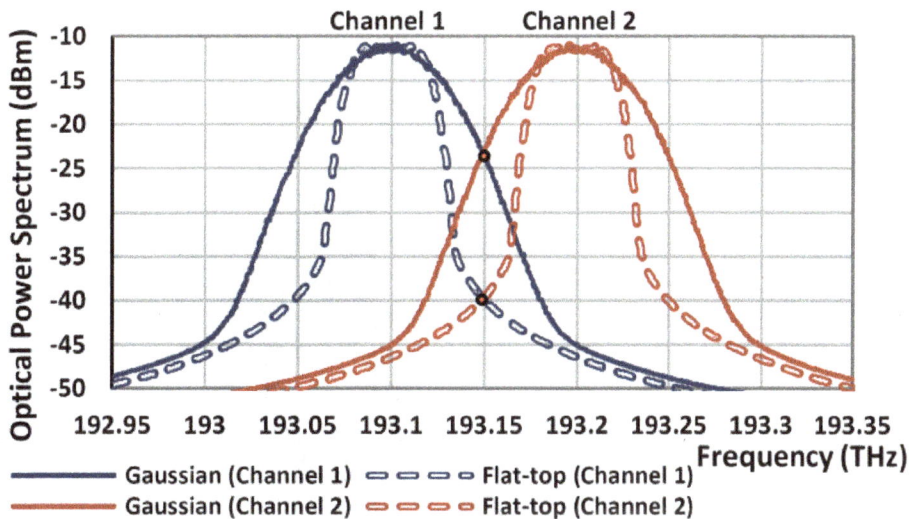

Figure 11. Comparison of Gaussian and flat-top type AWGs with 50 GHz 3-dB bandwidth in case of two adjacent WDM.

50 GHz). Based on the above mentioned facts we chose flat-top AWGs for our investigated SS-DWDM PON simulation scheme.

Figure 12 shows the optical spectrum on the output of ASE source and spectra after each flat-top AWG unit. As one can see, flat-top AWG provides stable performance and good signal filtering on all stages. We found that optimal 3-dB bandwidth value of flat-top type AWG unit for maximal system performance must be about 90 GHz. By slicing the spectrum of proposed ASE source with AWG demultiplexer we obtain 16 separate WDM channels with channel output power variation less than

0.42 dB that is very good result. As one can see in Figure 13, the performance of investigated SS-DWDM PON system was completely sufficient to provide data transmission with BER<$1 \cdot 10^{-10}$ over the fiber optical span of 12 km in length without CD compensation. Theoretically, the value of accumulated CD in this span is about 190 ps/nm. Figure 13 shows the worst channel after 12 km long fiber span we obtained BER = 8.47e-11. The eye opening is good and system performance is sufficient to work under defined BER threshold.

The first realized CD compensation method includes the implementation of DCF in central office. It was found

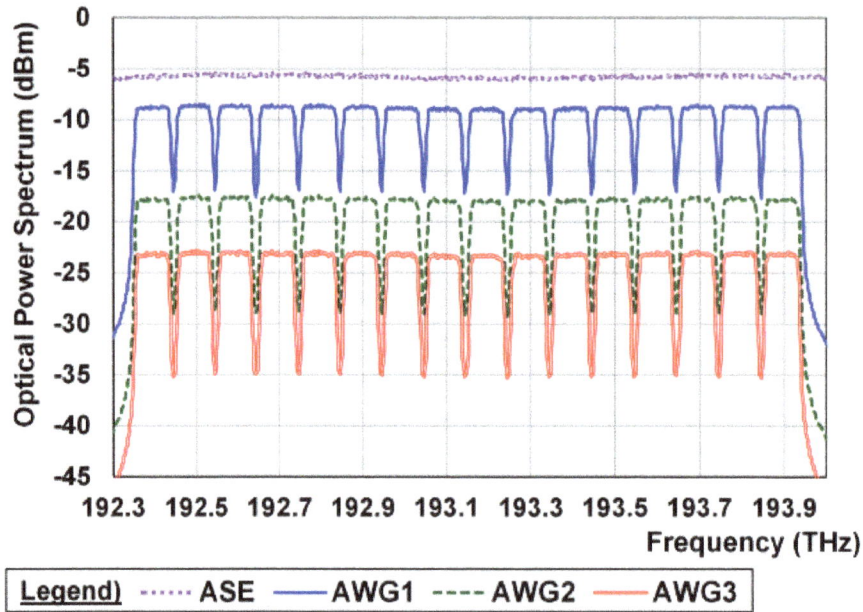

Figure 12. Optical output power spectrum of ASE source and 16-channel SS-DWDM PON system after each stage of flat-top type AWG unit.

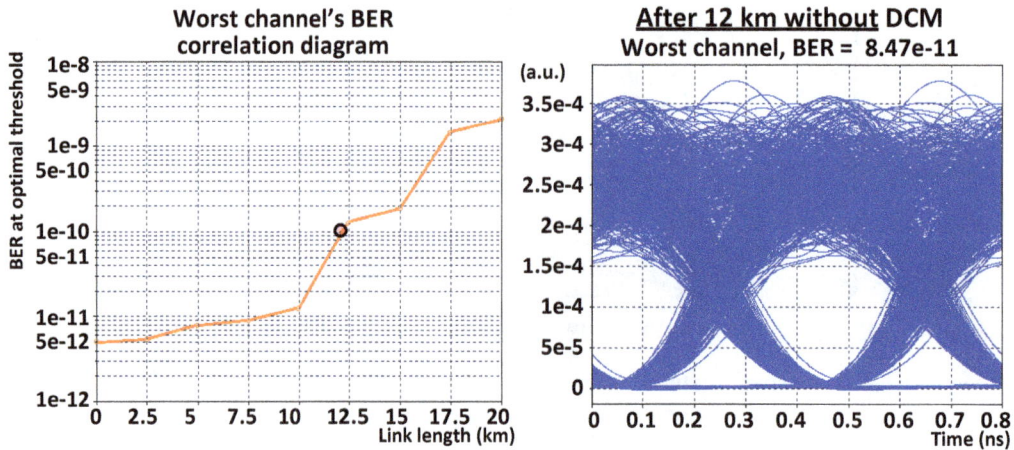

Figure 13. BER correlation diagram and output eye diagram of worst SS-DWDM PON channel after 12 km SMF link length without CD compensation module.

that optimal required DCF fiber length for maximum improvement of our SS-DWDM PON system's performance is 5.1 km. This length of DCF fiber can compensate about 408 ps/nm of accumulated CD. Using DCF fiber with such a length we can achieve maximal 16-channel SS-DWDM PON system's link length up to 24 km with BER = 4.82e-11 of received signal, (Figure 14a). Theoretical value of accumulated CD in the span of 24 km SMF is about 384 ps/nm. By using FBG we achieve better results in the terms of network reach than using DCF for accumulated CD compensation. Implementation of FBG for CD compensation extends the total reach of

SS-DWDM PON system up to 26 km with BER=7.24e-11, (Figure 14b). Theoretical value of accumulated CD in the span of 26 km SMF is about 416 ps/nm. We found that optimal CD compensation amount that must be compensated by FBG is 420 ps/nm and this amount is very close to full CD compensation.

As one can see in Figure 14(a) and (b), that logical 1 level noise is larger than the logical 0 level noise which is due to an ASE beat noise generated by the spectrum-sliced ASE BLS (Keiser, 2003). Using both CD compensation methods eye opening is wide enough for data transmission with BER<$1 \cdot 10^{-10}$ over the fiber optical

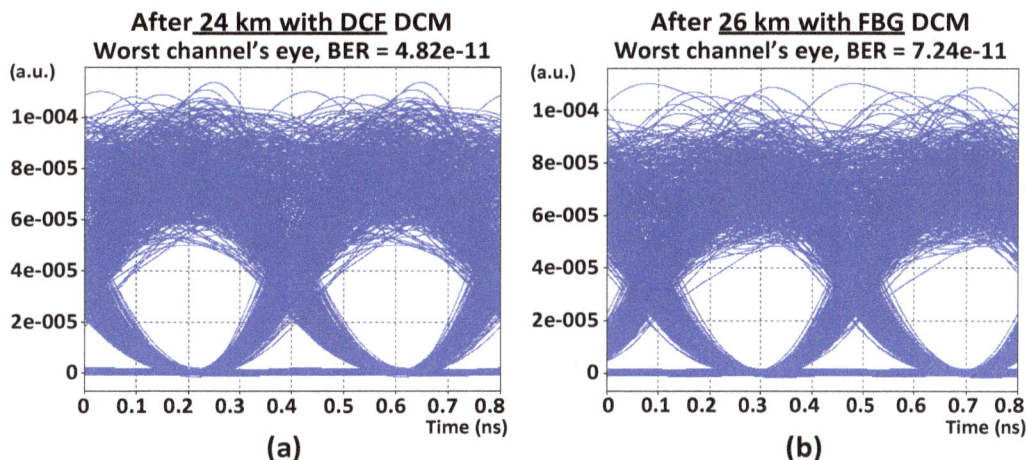

Figure 14. SS-DWDM PON system's eye diagrams and BER values of received signal from worst channels employing (a) DCF fiber and (b) FBG for accumulated CD compensation in DCM unit.

span. FBG provided greater performance improvement because it has relatively small insertion loss (<4dB) and instead of DCF fiber it can be used at higher optical powers without inducing nonlinear optical effects (e.g. cross-phase-modulation-induced nonlinear phase noise (Grobe et al., 2013), which can reduce the system's performance.

Conclusions

In this work we have realized and investigated high-speed SS-DWDM PON system where DCF and FBG are used for accumulated chromatic dispersion (CD) compensation to improve the maximal link reach from OLT to ONT at the same time providing high system performance with BER<$1 \cdot 10^{-10}$. Design of broadband ASE source with +23 dBm output power and flat spectrum in system's operating wavelength range (C-band) using two EDFAs connected in cascade mode is shown and described.

We demonstrated that using DCF for CD compensation SS-DWDM PON reach can be improved by 100% or extra 12 km in length – from 12 to 24 km. But using FBG unit network reach can be improved by 117% or extra 14 km – from 12 to 26 km. On the basis of these results authors recommend to use FBG DCM units for CD compensation in future high-speed 16-channel dense SS-WDM PON systems for maximal system's performance and network reach.

ACKNOWLEDGMENT

This work has been supported by the European Regional Development Fund in Latvia within the project Nr.2010/0270/2DP/2.1.1.1.0/10/APIA/VIAA/002.

REFERENCES

Agrawal GP (2010). Fiber-Optic Communication Systems, 4th edition. John Wiley and Sons, New Jersey, US. P. 626.

Bobrovs V, Udalcovs A, Spolitis S, Ozolins O, Ivanovs G (2011). Mixed chromatic dispersion compensation methods for combined HDWDM systems. Proc. IEEE Int. Conf. on Broadband and Wireless Comput. Commun. Appl. Barcelona. Spain. pp. 313-319.

Choi BH, Lee SS (2011). The effect of AWG-filtering on a bidirectional WDM-PON link with spectrum-sliced signals and wavelength-reused signals. Optics Commun. 284(24):5692-5696.

Chomycz B (2009). Planning Fiber Optic Networks. McGraw-Hill, New York. US. P. 401.

El-Sahn ZA, Mathlouthi W, Fathallah H, LaRochelle S, Rusch LA (2010). Dense SS-WDM over legacy PONs: smooth upgrade of existing FTTH networks. J. Lightwave Technol. 28(10):1485-1495.

Grobe K, Searcy S, Tibuleac S (2013). The Costs of 10-Gb/s and 100-Gb/s Coexistence. The TERENA Networking Conference (TNC). Maastricht, Netherlands. pp. 1-3.

ITU-T Rec. G.984.2 (2003). Gigabit-capable passive optical networks (GPON): Physical media depend (PMD) layer specification. http://www.itu.int/rec/T-REC-G.984.2/en.

ITU-T Recommendation G.694.1 (2012). Spectral grids for WDM applications: DWDM frequency grid. http://www.itu.int/rec/T-REC-G.694.1.

Ivanovs G, Bobrovs V, Ozolins O, Porins J (2010). Realization of HDWDM transmission system. Int. J. Phys. Sci. 5(5):452-458.

Kaneko S, Kani J, Iwatsuki K, Ohki A, Sugo M, Kamei S (2006). Scalability of spectrum-sliced DWDM transmission and its expansion using forward error correction. J. Lightwave Technol. 24(3):1295-1301.

Kani J, Bourgart F, Cui A, Rafel, A, Campbell M, Davey R, Rodrigues S (2009). Next-generation PON-part I: Technology roadmap and general requirements. IEEE Com. Mag. 47(11):43-49.

Keiser G (2003). Optical Communications Essentials. McGraw-Hill Networking Professional, New York. US. P. 372.

Lee K, Lim DS, Jhon MY, Kim HC, Ghelfi P, Nguyen T, Poti L, Lee SB (2012). Broadcasting in colorless WDM-PON using spectrum-sliced wavelength conversion. Optical Fiber Technol. 18(2):112-116.

Leeson MS, Luo B, Robinson AJ (2006). Spectral slicing for data communications. Proc. Symposium IEEE/LEOS Benelux Chapter, Eindhoven, pp. 165-168.

Leeson MS, Sun S (2008). Spectrum slicing for low cost wavelength division multiplexing. ICTON Mediterranean Winter Conference. Marrakech. Morocco. pp. 1-4.

OptSim User Guide (2012). RSoft Design Group Inc., New York. US. P. 459.

Ozolins O, Bobrovs V, Ivanovs G (2010). Efficient wavelength filters for DWDM systems. Latvian J. Phys. Tech. Sci. 47(6):47-58.

Spolitis S, Ivanovs G (2011). Extending the reach of DWDM-PON access network using chromatic dispersion compensation. Proc. IEEE Swedish Communication Technologies Workshop, Stockholm, Sweden. pp. 29-33.

Spolitis S, Bobrovs V, Ivanovs G (2011). Realization of combined chromatic dispersion compensation methods in high speed WDM optical transmission systems. Elect. Elect. Eng. 10(116):33-38.

Spolitis S, Bobrovs V, Ivanovs G (2012). New Generation Energy Efficient WDM-PON System Using Spectrum Slicing Technology. 4th Int. Congress on Ultra Modern Telecommunications and Control Systems. St. Petersburg. Russia. pp. 558-562.

Vukovic A, Savoie M, Hua H, Maamoun K (2007). Performance Characterization of PON Technologies. Int. Conf. Appl. Photonics Technology, Photonics North. Ottawa, Ontario, Canada. P. 7.

3

Experimental study of the elastic multiple scattering by a two-dimensional (2D) periodic array in the case of empty cavities

Ahmed Moumena* and Abderezzak Guessoum

Electronic Institute, LATSI Laboratory, Blida University, Road of Soumaa, PB 270, Blida, Algeria.

In this work, we study the propagation of the plane waves in a solid medium (Aluminium) which contains a periodic two-dimensional (2D) array of cylindrical inclusions using a multiple scattering theory. The aim of this study is to validate the experience of certain theoretical predictions such as the frequential positions of the stop-bands and their widths like their evolution according to the geometrical characteristics of the 2D array. The cavities are excited in normal incidence. Measurements are taken in the case of the empty cavities. The acquisitions of all the transmission and reflection temporal signals are obtained by the use of transducers of contact. All the recorded signals are subjected to a signal processing (FFT, normalization).

Key words: Multiple scattering theory/periodic distributions of inclusions/stop-bands/two-dimensional (2D) array.

INTRODUCTION

The propagation of acoustic waves in the heterogeneous medium equipped with a periodic structure has been the object of an interest growing for a few decades. A great number of periodic structures were studied and various theoretical approaches were employed. The existence of original physical properties such as the presence of stop-bands, that is, frequential bands for which the waves cannot be propagated at long distance in the medium corresponding to a strong attenuation, and pass-bands of less attenuation were all highlighted. The analogy existing with the propagation of electromagnetic waves in the periodic dielectric structures stimulates today's research on the "phononic crystals" composed of elastic inclusions periodically distributed in an elastic matrix. The propagation of elastic waves in such medium is in addition the subject of studies in fields as varied as: Geophysics, medical acoustics and mechanical engineering (let us quote like applications the Non-Destructive-Testing; NDT).

The periodic medium considered in this study is a periodic two-dimensional (2D) array of cylindrical inclusions in aluminium solid.The aim of this work is to validate by the experimental study certain theoretical predictions of Robert et al. (2004) such as the frequential positions of the stop-bands, their widths and their evolution according to the geometrical characteristics of the array. By using the finite element method, Langlet (1993) could characterize the propagation of waves in such an array by analyzing the dispersion and the attenuation of the waves of Lamb being propagated in a periodically bored elastic plate. The theory developed by Robert et al. (2004) is however valid only for infinitely long inclusions with respect to the wavelengths considered. The experimental study of the propagation of Lamb waves in a bored plate was thus not followed. This study will be based on the analysis of the reflection and transmission coefficients of a periodic 2D array finite thickness.

In this study, the fundamental points of the theory are presented. This theory is an extension of the theoretical model developed by Audoly, Dumery, Mulolland and Heckl (Sigalas and Garcia, 2000; Tanaka et al., 2000) for

---

*Corresponding author. E-mail: aamoumena@gmail.com.

the rigid or elastic obstacles arrays immersed in a fluid. This model consists to decompose the periodic 2D array into a series of periodic 1D linear array. If the scattering by each of these arrays is known, the propagation of the waves from one array to another is then deduced from the theorem of Bloch or an iterative method. The approach of these various authors is based on the work of (Mcphedran et al., 2000) on the plane waves scattering by a periodic linear 1D array cylindrical objects. Mcphedran et al. (2000) by using an exact calculation of multiple scattering, obtains an expression of the scattered field in the form of a superposition of plane waves diffracted under various angles. In the context of the scattering in an elastic medium, the formalism of Mcphedran et al. (2000) was generalized, while being based on the matrix theory of transition T, theory intensively exploited by (Mulholland and Heckl, 1994; Heckl and Mulholland, 1995) for similar multi-scatters mediums.

In the experimental study which is proposed, the various arrays are subjected to an incidental plane wave in a perpendicular plane to the axes of cylindrical inclusions. The cavities are excited in normal incidence. There will be thus never conversion between the longitudinal and transversal waves at the time of the reflection or the transmission by the 2D arrays. The study moreover will be led to the low frequencies for which the 2D arrays diffract only one plane wave in the same direction as the incidental plane wave. Consequently, in normal incidence and low frequency, the 2D arrays will have the acoustic behaviour of a stratified fluid medium (and periodic) of which it will be a question of measuring the reflected and transmitted acoustics fields.

To highlight the influence of the geometrical characteristics of the arrays on the reflected or transmitted acoustic field, two arrays are studied. Their implementation consists in practising holes in the thickness of an aluminium block, this thickness being large with respect to the wavelengths considered. Measurements are then taken in the air. The acquisition of the reflected or transmitted signal is obtained then by the use of transducers of contact to the plane interface between Aluminium and the air. The recorded signals are then subjected to a signal processing (Fast Fourier Transform (FFT), filtering, normalization).

THEORY OF THE MULTIPLE SCATTERING FOR PERIODIC 2D ARRAY: CASE OF THE NORMAL INCIDENCE

Reflection and transmission of a periodic and finite array

Figure 1 presents the geometry of the studied array. It is composed of an arbitrary number S of periodic 1D and finite arrays, regularly spaced of a distance D along axis X. The global array is thus finite and thickness $e = (S - 1)$

D according to this axis. We consider in addition that all the 1D linear arrays are identical, d is the period according to the axis y, $2a$ is the diameter of the cylindrical cavities.

The 2D array is subjected by acoustic longitudinal wave in normal incidence. The goal is then to express the reflection and transmission coefficients of the array, noted R and T, in the field of the low frequencies: $f \leq f_1^L$. Under these conditions we recall that there is no conversion mode during the transmission and of the reflection by the periodic 1D linear arrays and that the latter transmit plane modes purely propagates in the same direction as the incidental wave. The periodic 2D array can then be regarded as a "multi-layer" in which only a longitudinal plane wave is propagated perpendicularly with various interfaces.

From now on, the reflection and transmission coefficients of the periodic 1D linear array which are defined by Equations 4 and 6 are supposed to be known. In the continuation, to reduce the notations, we will note them R and T. The goal now is to calculate the two coefficients R and T.

Let us clarify the method for calculation of R. The idea consists to decompose the 2D array as shown (Figure 2) in the following way: the linear array 1 is isolated different from those that form a "sub-array", Numbered 1, of S-1 linear arrays; the array being subjected by a longitudinal plane wave. The field L considered is then the superposition of a wave L reflected by linear array 1, and a series of waves L transmitted by linear array 1 and reflected several times between linear array 1 and sub-array 1. The multiple reflections between the two linear arrays are described by series of Debye. The coefficient of reflection R-can then be written where $k = k_L$ is the number of waves in the elastic medium, and

R_1: the reflection coefficient of sub-array 1;
r_1: the reflection coefficient of the linear array 1;
t_1: the transmission coefficient of the linear array 1;
T_1: the transmission coefficient of sub-array1.

$$R = r_1 + \left(t_1 e^{ikD}\right)\left(R_1 e^{ikD}\right)t_1 + \left(t_1 e^{ikD}\right)\left(R_1 e^{ikD}\right)\left(r_1 e^{ikD}\right)\left(R_1 e^{ikD}\right)t_1 +$$
$$\left(t_1 e^{ikD}\right)\left(R_1 e^{ikD}\right)\left(r_1 e^{ikD}\right)\left(R_1 e^{ikD}\right)\left(r_1 e^{ikD}\right)\left(R_1 e^{ikD}\right)t_1 + \cdots$$
$$= r_1 + t_1 e^{i2kD}R_1 t_1\left\{1 + r_1 R_1 e^{i2kD} + \left(r_1 R_1\right)^2 e^{i4kD} + \left(r_1 R_1\right)^4 e^{i6kD} + \cdots\right\} \tag{1}$$

Geometrical series (Equation 1) is convergent. It can thus be written:

$$1 + \left(r_1 R_1\right)e^{2ikd} + \left(r_1 R_1\right)^2 e^{4ikd} + \left(r_1 R_1\right)^4 e^{6ikd} + \cdots = \frac{1}{1 - \left(r_1 R_1\right)e^{2ikd}} \tag{2}$$

Finally, the reflection coefficient of the total 2D array is written:

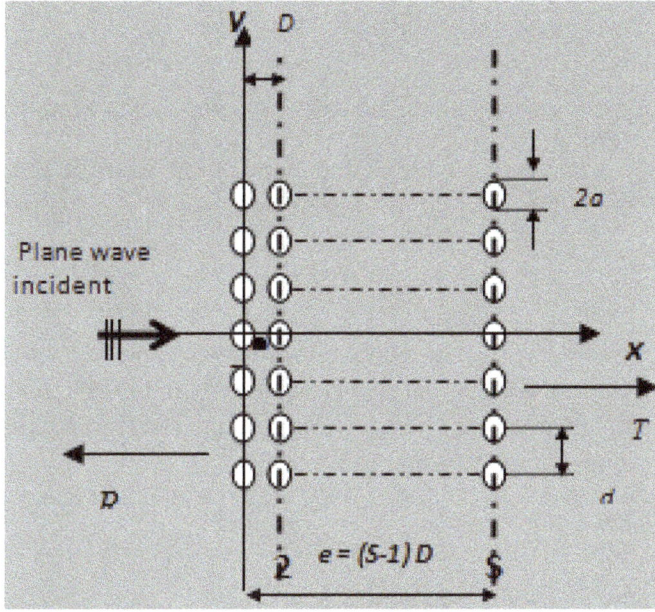

Figure 1. Geometry of the cylindrical cavities array.

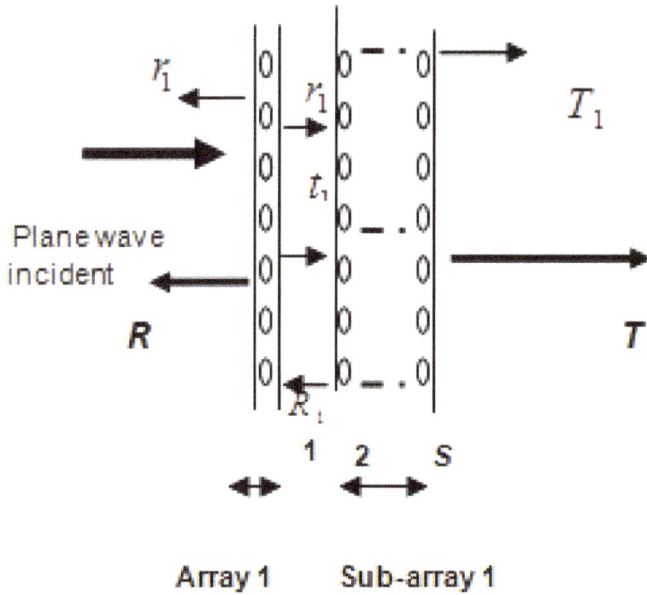

Array 1   Sub-array 1

Figure 2. First decomposition to determine the global reflection coefficient.

$$R = r_1 + t_1 e^{2ikd} R_1 t_1 \frac{1}{1-\left(r_1 R_1\right)e^{2ikd}}$$

(3)

That is to say:

$$R = r_1 + \frac{t_1 R_1 t_1 e^{2ikD}}{1-r_1 R_1 e^{2ikD}}$$

(4)

We now search the global transmission coefficient T: in the same way that previously we can calculate T as follows:

$$T = T_1\left(t_1 e^{ikD}\right)+\left(t_1 e^{ikD}\right)\left(r_1 e^{ikD}\right)\left(R_1 e^{ikD}\right)T_1 + \\ \left(t_1 e^{ikD}\right)\left(r_1 e^{ikD}\right)\left(R_1 e^{ikD}\right)\left(r_1 e^{ikD}\right)\left(R_1 e^{ikD}\right)T_1 +\cdots \\ =T_1 t_1 e^{ik\,D}\left\{1+\left(r_1 R_1\right)e^{2ikD}+\left(r_1 R_1\right)^2 e^{4ikD}+\cdots\right\}$$

(5)

The geometrical series (Equation 5) is convergent. It can thus be written:

$$\left\{1+\left(r_1 R_1\right)e^{2ikD}+\left(r_1 R_1\right)^2 e^{4ikD}+\cdots\right\}=\frac{1}{1-r_1 R_1 e^{2ikD}}$$

Finally, the transmission coefficient can be written in the form:

$$T = \frac{T_1 t_1 e^{ik\,D}}{1-r_1 R_1 e^{2ik\,D}}$$

(6)

In Equations 4 and 6, the reflection/transmission coefficients R_1 and T_1 are unknown (Figure 3). It is thus necessary to repeat this decomposition for sub-array 1. In its turn, this last can be decomposed as follows: the linear array 2 is isolated from S-2 other linear arrays which then form a sub-array 2. We then note R_2 and T_2 the reflection and the transmission coefficients of this last. This second decomposition then provides:

$$R_1 = r_2 + \frac{t_2 R_2 t_2 e^{2ikD}}{1-r_2 R_2 e^{2ikD}}$$

(7)

$$T_1 = \frac{T_2 t_2 e^{ikD}}{1-r_2 R_2 e^{2ikD}}.$$

(8)

In the same way, the reflection /transmission coefficients R_2 and T_2 are unknown. It is thus necessary to decompose in its turn sub-array 2 like the precedents. This method is thus a recurrence. With the decomposition S, Equations 7 and 8 become

$$R_{S-1} = r_S + \frac{t_S R_S t_S e^{2ikD}}{1-r_S R_S e^{2ikD}}$$

(9)

$$T_{S-1} = \frac{T_S t_S e^{ikD}}{1-r_S R_S e^{2ikD}}$$

(10)

In this last decomposition, the last sub-array S does not

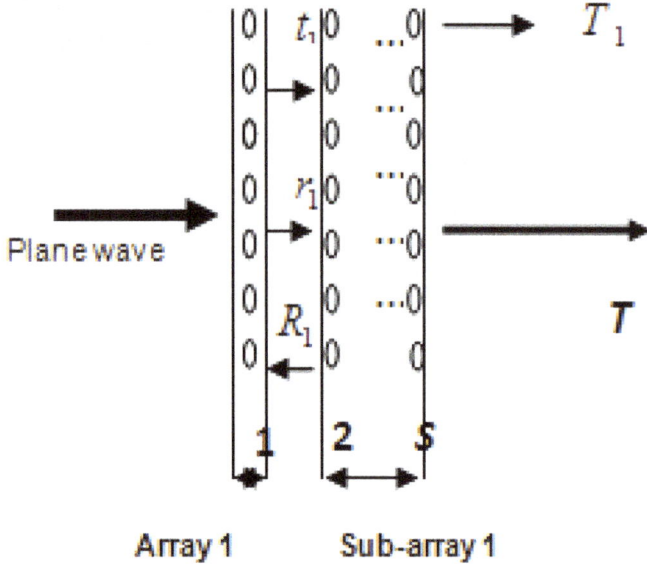

Figure 3. First decomposition to determine the global transmission coefficient.

contain any more that only one linear array whose coefficients r_s and t_s are supposed to be known:

$$R_S = r_S = r, \quad T_S = t_S = t \tag{11}$$

Consequently, the two relations in Equation 11 are relations which "close" the recurrence. To calculate R and T, we start these two last relations, and then we "descend" until the recurrence of Equations 4 and 6. This method of calculation is, numerically, very rapid when we know once and for all the reflection/transmission coefficients of all the linear arrays. In addition, the method can be generalized easily with the case of an oblique incidence (by taking in account mode conversions between the longitudinal and transversal waves) and if all the linear arrays are different from/to each other.

Characteristic equation

If the 2D array comprises a sufficiently large number of linear arrays, it can be regarded as being infinite so that the propagation of waves can be described by the theorem of Bloch. This theorem is applicable to the fields of potentials, displacements or constraints, binds the fields defined in two separate points of the array of a distance D applied to the fields of potentials and displacements, the theorem of Bloch is written:

$$u_i(x+D, y) = e^{i\gamma D} u_i(x, y) \tag{12}$$

$$\phi(x+D, y) = e^{i\gamma D} \phi(x, y) \tag{13}$$

$$\sigma_{ij}(x+D, y) = e^{i\gamma D} \sigma_{ij}(x, y) \tag{14}$$

Where indices i and j indicate the components according to x or y.

The complex size γ is the number of wave of Bloch defined by:

$$\gamma = \gamma' + i\gamma'' \tag{15}$$

The real part γ' is the component according to x of the vector of a wave being propagated in the periodic array, and the imaginary part γ'' is its attenuation. That is to say a linear array S taken randomly.

Between the arrays S - 1 and S, the scalar potential can be written;

$$\phi(x, y) = Ae^{ikx} + Be^{-ikx}$$

$$= \widetilde{A} + \widetilde{B} \tag{16}$$

Between the arrays S and S + 1, at a distance D. we have

$$\phi(x+D, y) = e^{ikD}Ce^{ikx} + e^{-ikD}De^{-ikx}$$

$$= e^{ikD}\widetilde{C} + e^{-ikD}\widetilde{D} \tag{17}$$

Projection according to x and y of the relation of Bloch on displacements lead to;

$$e^{i\gamma D}\widetilde{A} - e^{i\gamma D}\widetilde{B} - e^{i\xi}\widetilde{C} + e^{-i\xi}\widetilde{D} = 0 \tag{18}$$

With $\xi^D = kD$, the relation of Bloch on the constraints σ_{xx}, σ_{xy} provides directly;

$$e^{i\gamma D}\widetilde{A} + e^{i\gamma D}\widetilde{B} - e^{i\xi}\widetilde{C} - e^{-i\xi}\widetilde{D} = 0 \tag{19}$$

Remark

There are two equations to determine four unknown factors, namely: $\widetilde{A}, \widetilde{B}, \widetilde{C}, \widetilde{D}$. The two missing equations are obtained in the following way: if we refer to Figure 4, we can observe that the field B results from the reflection of field A and the transmission of the field D:

$$\widetilde{B} = r\widetilde{A} + t\widetilde{D} \tag{20}$$

In a similar way it is found that;

$$\widetilde{C} = r\widetilde{D} + t\widetilde{A} \tag{21}$$

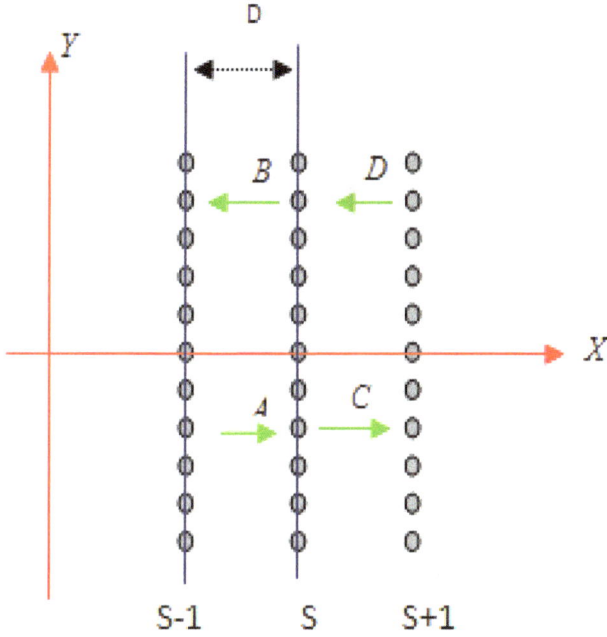

Figure 4. 2d array composed of an infinite number of identical arrays and periodically spaced; the amplitudes of the waves are presented.

There are now four equations for four unknown factors, the problem is thus well posed. There is finally the system of equation according to;

$$e^{i\gamma D}\left[\widetilde{A} + \widetilde{B}\right] = \widetilde{C}e^{ikD} + \widetilde{D}e^{-ikD} \quad \text{(I)}$$

$$e^{i\gamma D}\left[\widetilde{A} - \widetilde{B}\right] = \widetilde{C}e^{ikD} - \widetilde{D}e^{-ikD} \quad \text{(II)} \tag{22}$$

$$\widetilde{B} = r\widetilde{A} + t\widetilde{D}$$

$$\widetilde{C} = r\widetilde{D} + t\widetilde{A}$$

That we can still write in the matrix form;

$$\begin{vmatrix} e^{i\gamma D} & e^{i\gamma D} & -e^{ikD} & -e^{-ikD} \\ e^{i\gamma D} & -e^{i\gamma D} & -e^{ikD} & e^{-ikD} \\ r & -1 & 0 & t \\ t & 0 & -1 & r \end{vmatrix} \begin{vmatrix} \widetilde{A} \\ \widetilde{B} \\ \widetilde{C} \\ \widetilde{D} \end{vmatrix} = \begin{vmatrix} 0 \\ 0 \\ 0 \\ 0 \end{vmatrix} \tag{23}$$

The determinant associated with the system is thus;

$$D = \begin{vmatrix} e^{i\gamma D} & e^{i\gamma D} & -e^{ikD} & -e^{-ikD} \\ e^{i\gamma D} & -e^{i\gamma D} & -e^{ikD} & e^{-ikD} \\ r & -1 & 0 & t \\ t & 0 & -1 & r \end{vmatrix} = 0 \tag{24}$$

The system of Equations 22 has a solution different to zero provided that the determinant associated with the system is null, we obtain the characteristic equation which gives γ.

We can develop calculations further. If we eliminate $\widetilde{B}, \widetilde{C}$ in (I) and (II) of Equation 22, it becomes:

$$\left[te^{ikD} - (1+r)e^{i\gamma D}\right]\widetilde{A} + \left[re^{ikD} + e^{-ikD} - te^{i\gamma D}\right]\widetilde{D} = 0 \tag{25}$$

$$\left[te^{ikD} - (1-r)e^{i\gamma D}\right]\widetilde{A} + \left[re^{ikD} - e^{-ikD} + te^{i\gamma D}\right]\widetilde{D} = 0 \tag{26}$$

The determinant associated with this new system is thus;

$$\begin{vmatrix} te^{ikD} - (1+r)e^{i\gamma D} & re^{ikD} + e^{-ikD} - te^{i\gamma D} \\ te^{ikD} - (1-r)e^{i\gamma D} & re^{ikD} - e^{-ikD} + te^{i\gamma D} \end{vmatrix} = 0 \tag{27}$$

And by simplification, we have:

$$\begin{vmatrix} -re^{i\gamma D} & e^{-ikD} - te^{i\gamma D} \\ te^{ikD} - e^{i\gamma D} & re^{ikD} \end{vmatrix} = 0 \tag{28}$$

The characteristic equation $D = 0$ can be reduced to:

$$ch(\gamma D) = \frac{1}{2t}\left\{\left((t)^2 - (r)^2 + 1\right)\cos\xi^D + i\left((t)^2 - (r)^2 - 1\right)\sin\xi^D\right\} \tag{29}$$

Because of the absence of mode conversions, this equation is identical to that established by Hecklin as in the case of submerged tubes array in water.

EXPERIMENTAL DEVICE

Description of the material

Figure 5 shows the experimental device. The measurements are performed in the case of empty cavities and the two types of arrays 2D are excited at normal incidence. The array consists of cylindrical cavities dug in an aluminum block to form an elastic matrix. Firstly, the excitation and reception of temporal signals reflected and transmitted is obtained by the use of contact transducer with a center frequency of 0.25 MHz. The bandwidth of this transducer is the order of twice the center frequency. The transmitter transducer is excited by a pulsed electric signal delivered by a pulse generator. The latter includes amplification and filtering for processing the signal received by the receiver transducer. This pulse generator comprises a swingable door on two positions. In the first position, it is used as a single transmitter and receiver transducer at a time. In the second position, it uses two transducers of the same center frequency as a transmitter and the other as a receiver. The received signal is displayed on an oscilloscope type Lecroy 9430. Average of 200 scans is then performed in order to filter the signal. The averaged signal is sampled and thereafter stored in a computer. All signals are recorded and then subjected to signal processing FFT.

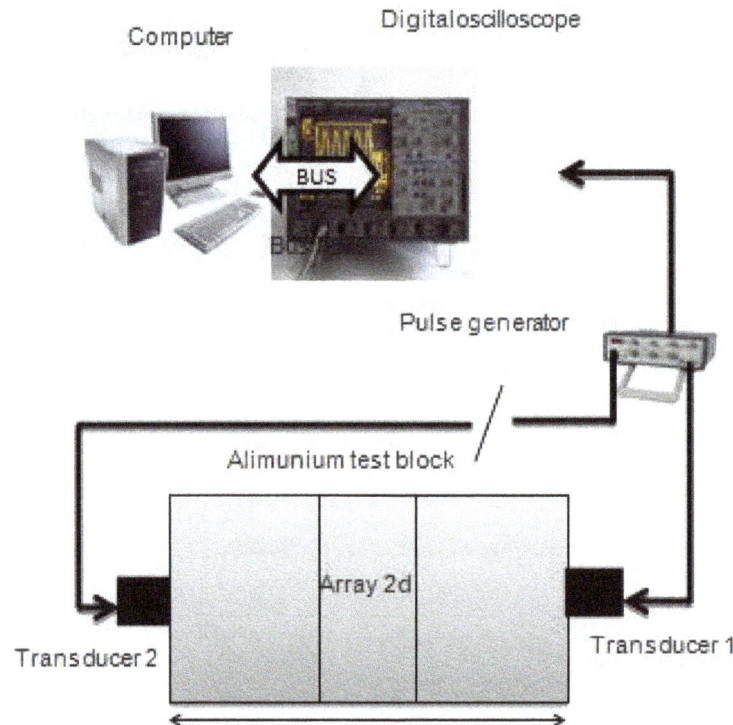

Figure 5. Experimental device.

Table 1. Extraction of all the normalization transmission and reflection temporal signals and with diffusers.

Stop band	Numerical (MHz)	Experimental (MHz)
S.B	0.25 - 0.52	0.26 - 0.47

THE COMPARISON OF RESULTS BETWEEN NUMERICAL AND EXPERIMENTAL STUDY: CASE OF THE EMPTY CAVITIES AND NORMAL INCIDENCE

Results obtained on the first 2D array with $S = 6$ linear arrays and finite

i) The ray of the cylindrical cavities: $\alpha = 2$ mm,
ii) The period of the cavities according to the axis y: d = 5 mm,
iii) The period of the 1D arrays according to the axis x: $D = 6$ mm,
iv) The thickness of the 2D array: $e = (S - 1) D = 30$ mm,
v) The cut-off frequency is given by: $f_c = 0.620$ MHz,
vi) Curve in red color represent reflection coefficient in energy,
vii) Curve in blue color represent transmission coefficient in energy.

In the experimental study, we extracted all the normalization transmission and reflection temporal signals and with diffusers (Table 1). Then, they underwent several processing like the elimination of the undesirable parts (echoes), retiming, the FFT in order to obtain the normalized function of form, so that we can compare it with the result obtained numerically, by using several transducers.

According to the curves (Figures 6 and 7), we see well that the stop-band obtained theoretically and the stop-band obtained in experiments are superimposed, which means that this theoretical result was checked in experiments.

Results obtained on the second 2D array with $S = 5$ 1D arrays and limited

i) The ray of the cylindrical cavities: $\alpha = 2$ mm,
ii) The period of the cavities according to the axis: d = 6 mm,
iii) The period of the 1D arrays according to the axis: D = 20 mm,
iv) The thickness of the 2D array: $e = (S - 1) D = 80$ mm,
v) The cut-off frequency is given by: $f_c = 0.520$ MHz,
vi. Curve in blue color represent reflection coefficient in energy,
vii. Curve in red color represent transmission coefficient in energy.

According to the curves (Figures 8 and 9), we see well

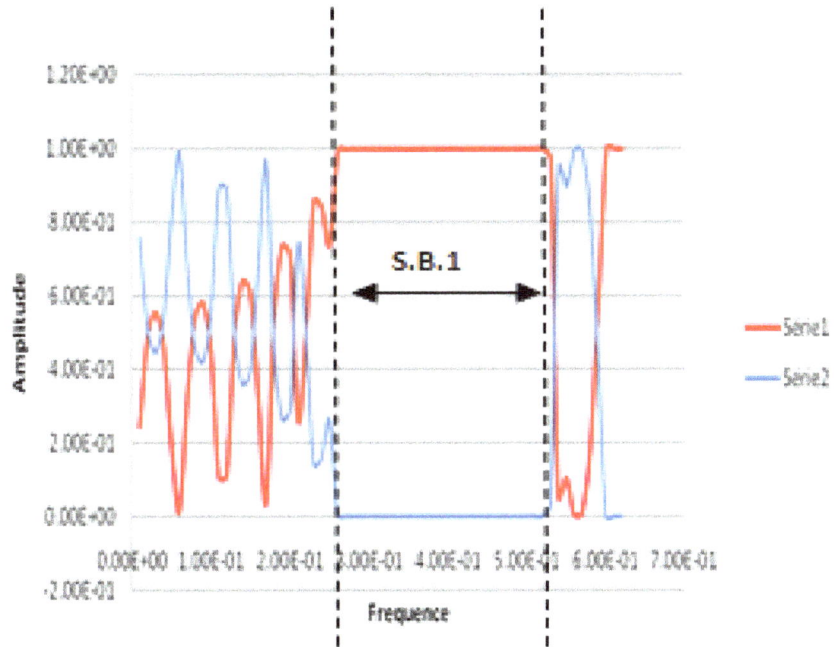

Figure 6. Frequential evolution of the reflection and transmission coefficients in energy $\left|R^{LL}\right|^2$ and $\left|T^{LL}\right|^2$ empty cavities of the array in normal incidence with $d/a=2.5$, $D/a=3$, the cut-off frequency f_c = 0.620 MHz.

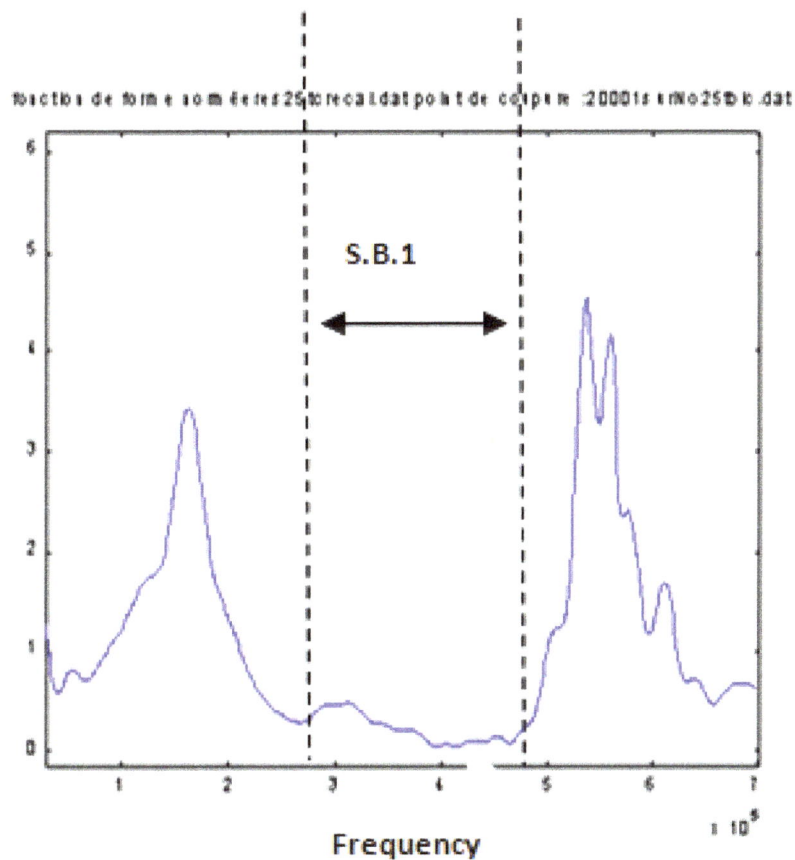

Figure 7. Function form normalized.

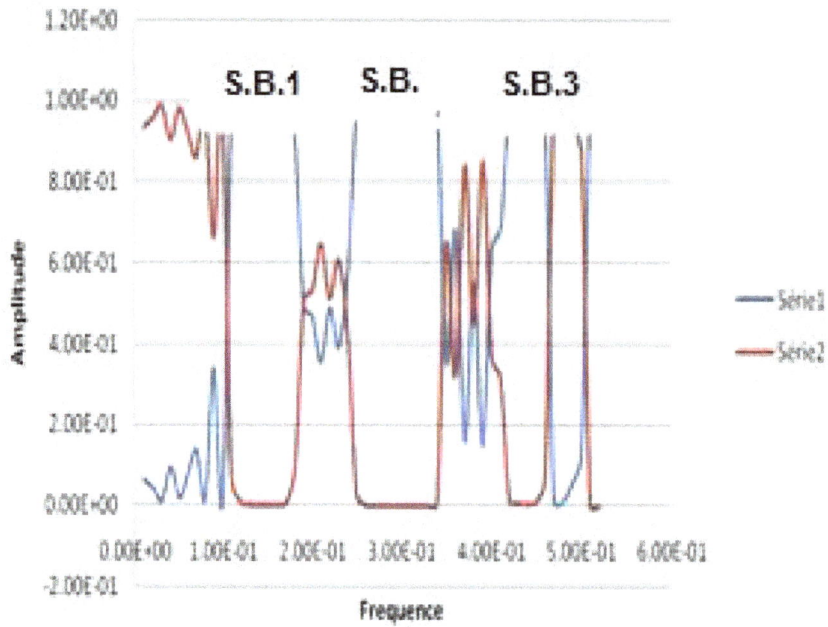

Figure 8. Frequential evolution of the reflection and transmission coefficients in energy $\left|R^{LL}\right|^2$ and $\left|T^{LL}\right|^2$ the empty cavities of an array in normal incidence with d/a=3, D/a = 10, the cut-off frequency f_c = 0.520 MHz.

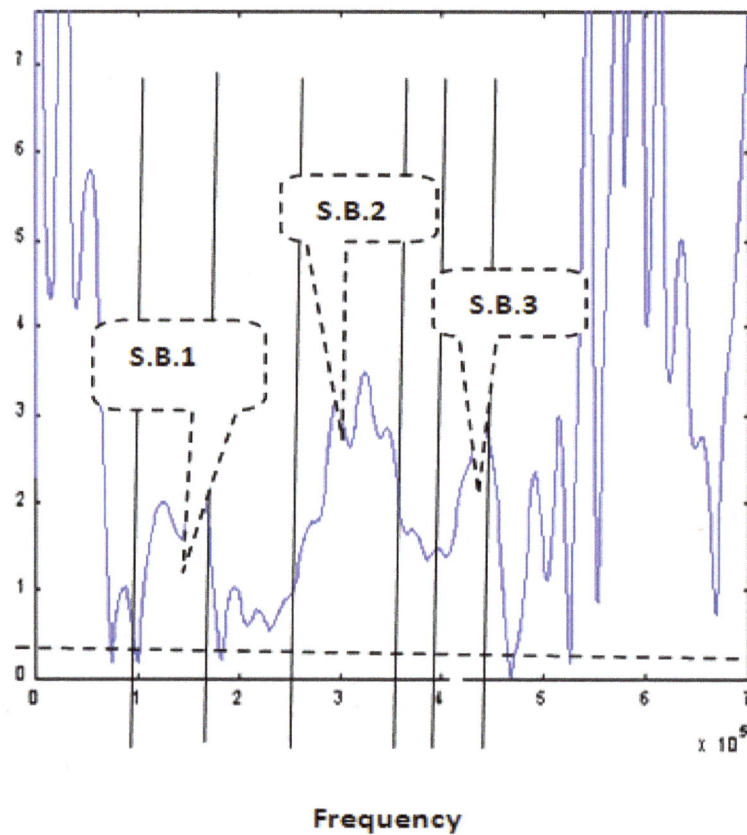

Figure 9. The function form normalized.

Table 2. The three stop-bands obtained theoretically and the three stop-bands obtained in experiments.

Stop-band	Numerical (MHz)	Experimental (MHz)
S.B.1	0.12 - 0.18	0.11-0.17
S.B.2	0.25 - 0.35	0.25-0.38
S.B.3	0.42 - 0.45	0.41-0.46

that the three stop-bands obtained theoretically and the three stop-bands obtained in experiments are superimposed (Table 2).

CONCLUSIONS AND PERSPECTIVES

A method of calculation of the scattering by a periodic 2D array in an elastic medium was presented. This method rest on the decomposition of the 2D array in a limited number of periodic 1D arrays. The scattering by each 1D array is determined by an exact calculation of multiple scattering which was adapted here to the scattering in an elastic medium. The scattering plan after plan is calculated using the theorem of Bloch.

The numerical and experimental study verified the existence of two interesting properties which are: Stop-bands and pass-bands. The appearance of one or more stop-bands is due to the choice of the geometry of the studied 2D array in the case of empty cavities and normal incidence.

The theoretical study developed here can be the later development object and can be applied to an array with variable, a periodic or random porosity, according to the direction of propagation of plane waves.

REFERENCES

Robert S, Franklin H, Conoir JM (2004). Scattering by two-dimensional periodic gratings composed of cylindrical cavities embedded in an elastic matrix, 75th Anniversary Meeting of the Acoustical Society of America, New York, New York. pp. 24-28.

Langlet P (1993). Analyse de la propagation d'ondes acoustiques dans les matériaux périodiques à l'aide de la méthode des éléments finis, Thèse de doctorat, Université de Valenciennes C5-921-C5-924.

Heckl MA, Mulholland LS (1995). Some recent developments in the theory of acoustic transmission in tubes bundles. J. Sound Vib. 179(1):37-62.

Mcphedran RC, Nicirovici NA, Botton L, Grubits KA (2000). Lattice sums for gratings and arrays, J. Math. Phys. 41(11):7808-7816.

Mulholland LS, Heckl MA (1994). Multi-directional sound wave propagation through a tube bundle. J. Sound Vib. 176(3):377-398.

Sigalas MM, Garcia N (2000). Theoretical study of three dimensional elastic band gaps with the finite-difference time-domain method. J. Appl. Phys. 87(63):22-3125.

Tanaka Y, Tomoyasu Y, Tamura S (2000). Band structure and acoustic waves in phononiclattices: Mismatch. Phys. Rev. B 62(11):57387-7392.

4

Decentralized management of a multi-source electrical system: A multi-agent approach

Abdoul K. MBODJI, Mamadou L. NDIAYE and Papa A. NDIAYE

Centre International de Formation et de Recherche en Energie Solaire (C.I.F.R.E.S), ESP BP 5085
Dakar-Fann, Sénégal.

The objective of this paper was design and implementation of a self-adaptive management system of a set of production sources in a changing and unpredictable energy demand environment. The strategy proposed made it possible to achieve optimal management of the energy resource production of the electrical system facing the changing demand. After showing the need to follow «intelligently» the behavior of different entities of the electrical system by a distributed, collaborative and self-adaptive model, the emphasis was placed on the modeling of an energy management multi-agent system. The proposed model allowed the overall production to be optimized in relation with the demand profile and in function of a cost or greenhouse gas reduction criterion. The flexibility of this model could in priority allow both the integration of multi-objectives optimization and that of information.

Key words: Energy, modeling, complex system, multi-agent system, optimization, multi-source system, greenhouse gas.

INTRODUCTION

Electrical energy management goes with the protection, monitoring and control of the entire electrical network. For the operator, the question also concerns the optimizing of the energy consumption cost of different production sources without any prejudice for the activity. This requires effective and real-time control of the overall electrical system parameters. Modern solutions to this control need are products and services using information and communication technologies based on the paradigm of–smart systems, such as data loggers and supervision and control software. Research has been done on the multi-source decentralized power grid management optimization. Logenthiran et al. (2012), present a Multi-Agent System (MAS) for the real-time operation of a microgrid. The multi-agent model proposed in this paper, provides a common communication interface for the entire components of the microgrid. Implementation the MAS allows not only to maximize energy production from local distributed generators, but also to minimize the

microgrid operating cost to be minimized. The recent studies by Monica et al. (2012); Mao et al. (2011); and Pipattanasomp et al. (2009) present an optimal design and implementation method for the intelligent management of electrical distribution networks. The research mentioned focuses on microgrids especially in the electrical distribution part. It would be interesting to enlarge the fieldwork and integrate both transport and production parts into energy management in order to take advantage of more room for maneuver and flexibility in the management system.

Other MAS applications allowing a diagnosis of disturbances on the grid to be made were presented in the work of Nagata and Sasaki (2002) and Wang (2001). An application that makes it possible to monitor the power system is presented in the work of Cristaldi et al. (2003), a secondary voltage control system in that of Phillips et al. (2006) and a visualization power system in that of Dimeas and Hatziargyriou (2005a).

Some other studies, to mention only those of Dimeas and Hatziargyriou (2005b); Dimeas and Hatziargyriou (2004) and Butler-Purry et al. (2004), focus on the control of the micro network operation from a MAS. However, most of that work was applied to power grid using PV generators, batteries and controllable loads. Besides, the emphasis was more on technical than on economic and environmental aspects such as the reduction of operating costs and the amount of greenhouse gases (GHG) emitted by the electrical network.

The present paper proposes a model of generic system management of electrical systems that can be applied to a micro grid as well as to a macro grid. This model can be implemented in a real system thanks to advanced communication techniques, software agents can be embedded in different sources of power generation and loads. These agents cooperate and make decisions together to optimize system management both in technical and economic terms, taking into account technological constraints and resources availability. So, the contribution of this paper is to implement an optimal management platform of decentralized electrical systems minimizing the production cost or the amount of GHG emissions released by the power plants.

MATERIALS AND METHODS

Presentation of power grid (PG)

PG as shown in Figure 1 is a distributed system on several sites S_k. Each site consists of several power plant $C_{j\,(Sk)}$ and each plant site consists of several production generators $G_{i(Cj(Sk))}$. Each generator produces power P_i and the total power supplied by the power grid is given by Equation (1):

$$P_f^T = \sum_{k=1}^{S} \left(\sum_{j=1}^{C} \left(\sum_{i=1}^{G} P_i \right) \right) \tag{1}$$

Where, G: is the number of generators in power plant $C_{j(Sk)}$; C: is the number of power plants in a site S_k; S: is the number of network sites.

The total power demand is given by the following Equation (2):

$$P_D^T = \sum_{k=1}^{H} \left(\sum_{j=1}^{M} \left(\sum_{i=1}^{B} P_{\delta_i} \right) \right) \tag{2}$$

With: B: is the number of clients managed by a low-voltage departure d_i; M: is the number of low-voltage departures; H: is the number of high-voltage departures of the grid. The total power supplied P_S^T by the network at t time is then given by Equation (3):

$$P_S^T = P_D^T + Losses \tag{3}$$

Where, the losses are due to the technical or not technical losses. The production cost of energy sources as shown in Equation (4) takes into account the costs of fuel, oil and maintenance.

$$C_P = C_f + C_o + C_m \tag{4}$$

C_f, cost of fuel consumption; C_o, cost of oil consumption; C_m, cost of maintenance.

Estimation of the quantities of greenhouse gas emissions

The estimation of emissions from fossil fuels combustion in fossil energy sources is presented in three levels of approaches in the 2006 Guidelines (Amit et al., 2006).

Level 1 approach

It requires the knowledge of data such as the quantity of fuel burned per unit of energy and a default emission factor. The associated equation is:

$$E_{G,F} = Q_F \times F.E_{G,F} \tag{5}$$

The total emission of greenhouse gases due to combustion (E_G) is obtained by adding the GHG emissions attributable to the combustion of each fuel (Richalet, 1987). This results in the following Equation (6):

$$E_G = \sum_F E_{G,F} \tag{6}$$

Level 2 approach

In Level 2, the default emission factors from Level 1 are simply replaced by specific emission factors of the corresponding country (Amit et al., 2006).

Level 3 approach

The Level 3 approach considers an emission factor per fuel and per technology (Amit et al., 2006). The mathematical model associated with this approach is given by Equation (7):

$$E_{G,F,T} = Q_{F,T} \times F.E_{G,F,T} \tag{7}$$

The total emission of GHG generated by different technologies is given by Equation (8):

$$E_{G,F} = \sum_T E_{G,F,T} \tag{8}$$

Typology and structure of agents

The approach is to translate the problem of vector processing exchanges of energy flow in an agent space where the system entities cooperate with each other. A situated approach, cooperative and decentralized, is proposed for power system management. This is an approach into which an agent «Source Agent» (Ag_S) is associated with each energy source and an agent

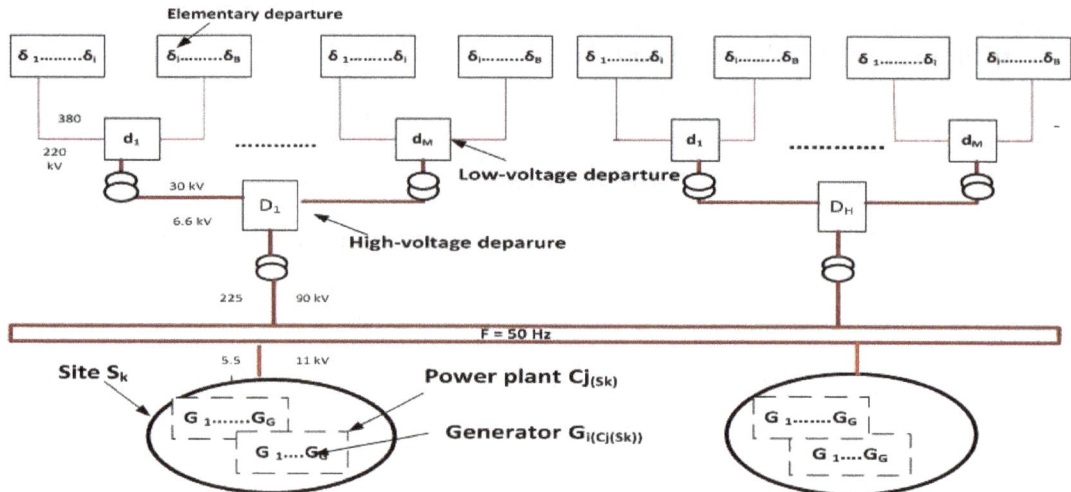

Figure 1. Power grid.

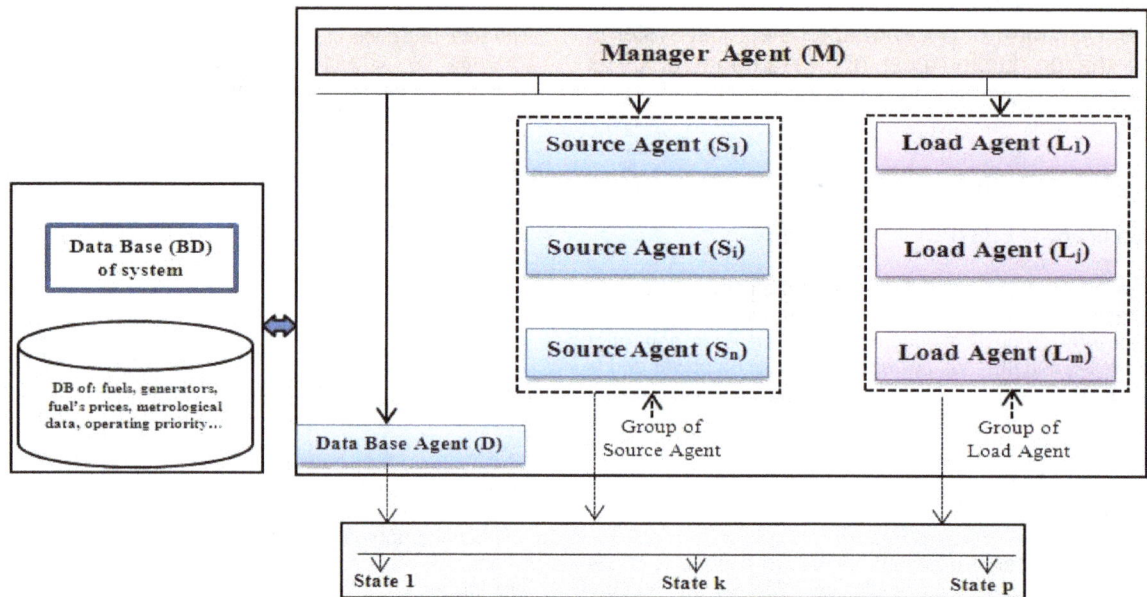

Figure 2. General architecture of the platform developed by a multi agent development kit (MADKIT).

«Load Agent» (Ag_L) is associated with each low-voltage departure.

The proposed architecture (Figure 2) is hybrid and is divided in two layers: the first layer consists of two types of reactive agents (Source Agent and Load Agent). Each Ag_{S_i} (i ∈ [1 - N]) and each Ag_{L_k} (k ∈ [1 - K]), have their own characteristics (Table 1). The second layer consists of a cognitive agent called «Manager Agent» (Ag_M) and a reactive agent called «Data Base Agent» (Ag_D), which manages the database of information handled by Source Agent and Load Agent.

Agent priority is a decisive parameter in the working of the management system. It allows the agents, depending on their priority, to participate or not in meeting the demand. Priority (p_{Si}) of Source Agent is a parameter which depends on the optimization criterion, the availability of the source, the source production cost and / or the amount of GHG released (Equation 9). This is a real value between zero (0) and one (1). A production source has all the higher priority as the value of its priority is closer to one (1).

The priority of Load Agent varies between zero (0), one (1), two (2) and three (3). A Load Agent has all the higher priority as the priority value of its priority is greater. Departures supplying sensitive areas (major national institutions, hospitals, etc.) have a higher priority equal to three (3). Departures supplying secondary areas (industrial, etc.) have a priority equal to two (2). Departures supplying non-priority areas (residential, etc.) have a priority equal to one (1).

Table 1. Attribute of source agent and load agent.

Source agent (Ag_{S_i})	Load agent (Ag_{L_k})
I_{S_i} : Identification number	I_{D_k} : Identification number
Comb : consumed fuel	P_{L_k} : Power demand (MW)
P_{S_i} : Power supply (MW)	p_{Lk}: operation priority
A_{Si} : availability	
C_{Si} : cost per kWh (FCFA/kWh)	
Q_{Si} : quantity of CO_2 released to produce 1 kWh (g/kWh)	
p_{Si} : operation priority	

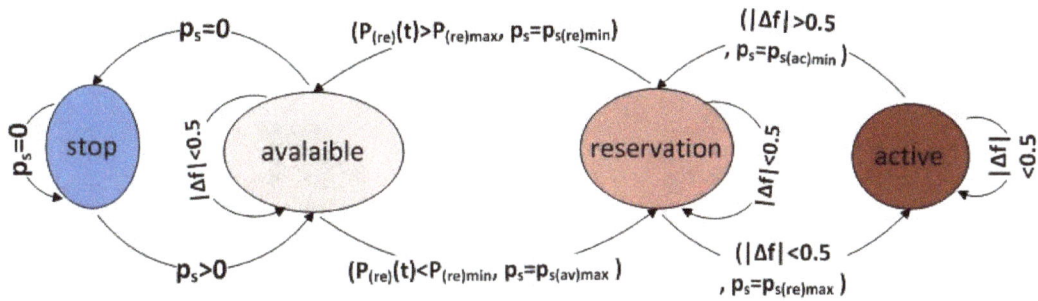

Figure 3. Behavioral model of source agent.

Source Agent can be in four (4) different states (Figure 3): active (ac), reserve (re) or standby, available (av) and stop (st). In each state, the instantaneous power of the source agent is delimited by a minimum and a maximum allowable power ($P_{(state)min} \leq P_{(state)}(t) \leq P_{(state)max}$). The numbers of source agents in state ac, re, av and st are respectively denoted N_1, N_2, N_3 and N_4. Equation 10 gives the instantaneous power reserve which is the sum of the powers of sources agents in the reserve state at t time. The reserve instantaneous power $P_{(re)}(t)$ should always be remaining between a minimum value $P_{(re)min}$ and a maximum value $P_{(re)max}$. Production sources that are in the reserve state can regulate the frequency around 50 Hz.

$$f : \{1,-1\} \times \{0,1\} \times [0\ 1] \rightarrow [0\ 1]$$

$$v_1, v_2, v_3 \rightarrow p_s = f(v_1, v_2, v_3) = v_2 \left(1 - |v_1 \times v_3|\right) \tag{9}$$

v_1 is optimization criterion, $v_1 = 1$ or -1, $v_1 \in \{-1\ 1\}$; v_2 is the source availability, $v_2 = 1$ if the source is available otherwise $v_2 = 0$, $v_2 \in \{0\ 1\}$; Let $\{C_{S_i}\}$ be the set of kWh costs associated with energy sources of the system and $Max\{C_{S_i}\}$ the maximum of this set. Let $\{Q_{S_i}\}$ be the set of

all the amounts GHGs released associated with energy sources of the system and $Max\{Q_{S_i}\}$ the maximum of this set. v_3 is equal to the following value:

$$v_3 = \frac{C_{S_i}}{Max\{C_{S_i}\}} \text{ or } v_3 = \frac{Q_{S_i}}{Max\{Q_{S_i}\}} \quad i = \{1, 2, ..., N\} \text{ by}$$

construction $v_3 = [0\ 1]$.

$$P_{(re)min} \leq P_{(re)}(t) = \sum_{i=1}^{N_2} P_{(re)i}(t) \leq P_{(re)max} \tag{10}$$

Model system management

The main objective of the management system is to minimize the cost per kWh or reduce the amount GHG generated by the system with the constraint of maintaining the equilibrium of the system which requires the frequency in a fixed interval. Satisfaction function of system (SF) is a Boolean variable, it is equal to one (1) if the frequency of the electrical system is in [49.5 50.5] otherwise it is equal to zero (0). The Manager Agent is a cognitive agent and supervises staff departures and production sources, and their associated states (Figure 4). It plays a major role in the timing and coherence of the activities of different agents. It is involved in the cooperation between the different agents of the system. It supervises and coordinates the operation of the system agents.

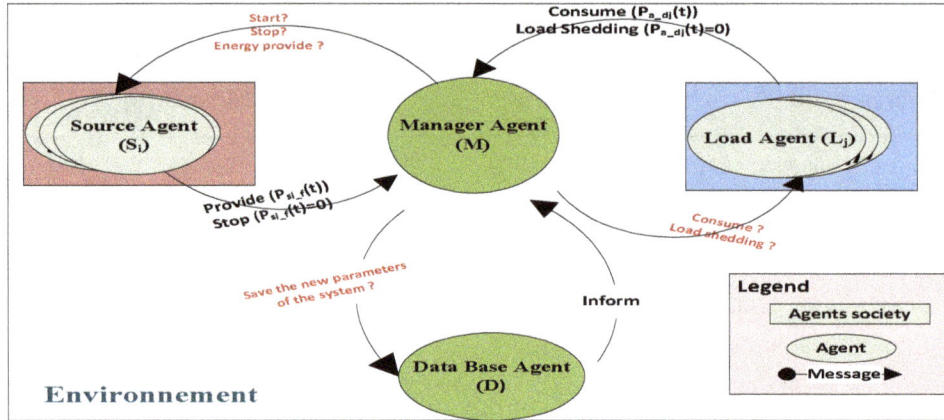

Figure 4. Model of energy management system.

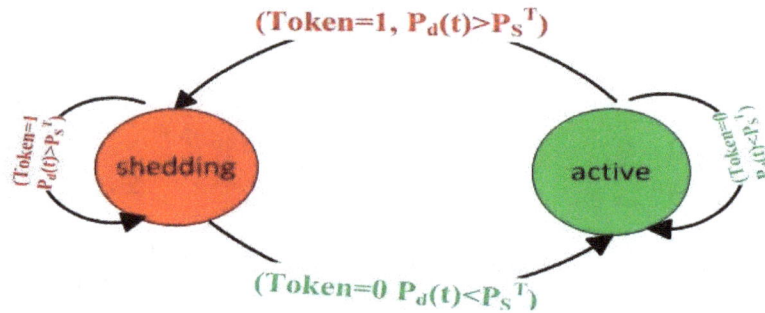

Figure 5. Behavioral model of load agent.

Source agent

According to the demand and the characteristics of the source, a Source Agent can be in four (4) different states: active, reserve, available and stop (Figure 3). A Source Agent

(i) is in the «stop» state when it cannot run because of a maintenance case following a breakdown, incident, etc., Its priority is then zero ($p_s = 0$). It goes into the available state as soon as it is functional again.

(ii) changes from the stop state to the available state when it can run, its priority is then evaluated and is positive. It may at any time switch to the reserve state according to the request and its priority.

(iii) is in the reserve state when it is ready to provide energy. Its production is equal to zero (0). This allows regulation of the frequency by continuous adaptation of the production level to that of the consumption. It goes into the active state according to the demand and its priority.

(iv) is in the active state when it supplies energy. In this state its priority is the greatest of all those of the agents that are in other states.
With:

(v) Δf (Hz) is the variation of the system frequency, it is equal to the absolute value of the difference between the frequency (f) (Hz) of the system at t time and the reference frequency ($f_0 = 50$ Hz) ($\Delta f = | f - f_0 |$ Hz). It is the direct image of the imbalance between the production and the consumption.

(vi) $p_{s(re)min}$ is the lowest source priority among the sources that are in the reserve state.

(vii) $p_{s(re)max}$ is the priority of the most favorable source among the sources that are at the reserve state.

(viii) $p_{s(av)max}$ is the priority of the most favorable source among the sources that are in the available state.

(ix) $p_{s(ac)min}$ is the lowest source priority among the sources that are in the active state.

Load agent

Load Agent (Ag_L) of the system is in an active state or in a shedding state as shown in Figure 5. The management of load shedding is done following attribution of a token «shedding». Load agents having the token « shedding» can pass from the active state to the load shedding state.
Where: P_d (t) is equal to the instantaneous power called by all departures; P_s^T is the total power available in the electrical network.

Power system

An electrical grid similar to that of Senegal was deliberately chosen (Table 2). This table provides a description of the power plants and generators, types of fuel, power installed. More than 88% of the production are of thermal origin. This makes it possible to get closer to reality.

Table 2. Production park of power network.

Power plants	Generators	Fuels	Installed power (MW)	Power plants	Generators	Fuels	Installed power (MW)
C1	Source1 1	Heavy fuel	15.95		Source5 2	Diesel oil	18
	Source1 2	Heavy fuel	15.95	C5	Source5 3	Kerosene	36
	Source1 3	Heavy fuel	15.95		Source5 5	Diesel oil	30
	Source1 4	Heavy fuel	15.95		Source6 1	Diesel oil	15
	Source2 1	Heavy fuel	18	C6	Source6 2	Diesel oil	15
	Source2 2	Heavy fuel	18		Source6 3	Diesel oil	15
C2	Source2 3	Heavy fuel	18		Source6 4	Diesel oil	15
	Source2 4	Heavy fuel	15	C7	Source7 1	Hydraulic	60
	Source2 5	Heavy fuel	15		Source8 1	Heavy fuel	20
C3	Source3 1	Heavy fuel	67.5	C8	Source8 2	Heavy fuel	13
C4	Source4 1	Naphtha	50		Source8 3	Heavy fuel	20

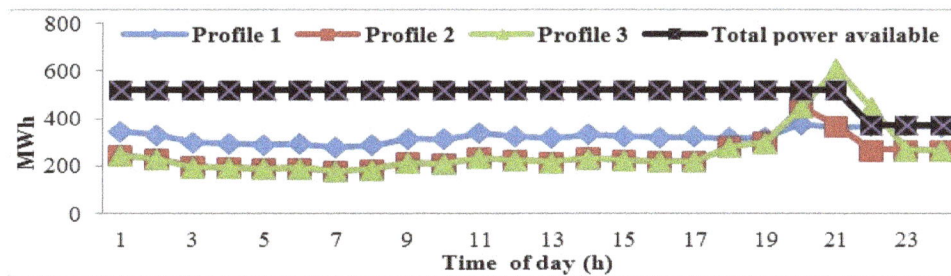

Figure 6. Profiles of the power demand of the network.

RESULTS AND DISCUSSION

To test the simulation model presented previously, three scenarios were set up to show the behavior of the system as clearly as possible.

Load profiles

To test the simulation model presented previously, three different load profiles were used (Figure 6). The first load profile is characterized by a nearly constant demand around 350 MW. The second profile is characterized by a peak power high demand of more than 37% within an interval of one hour. The third profile is characterized by a power demand exceeding the available power equal to 522.3 MW. Several scenarios were simulated in order to show the behavior of the system as clearly as possible.

Scenarios

The simulations were carried out following three criteria:

An optimization criterion based on minimizing the production cost, a criterion based on the minimization of the amount of GHG emissions released and a criterion based on random rules. A test scenario consisted in simulating the behavior of the system when the power system presented below, a given load profile (1, 2, 3) and a given optimization criterion (minimizing the production cost or minimization of the amount of GHG) were applied. Several simulations of each load profile were also performed with a generator assessment criterion based on random rules. The results of various simulations were then compared. Those scenarios made it possible to highlight the robustness and effectiveness of the management system.

Simulation results

The results obtained with the first two criteria (cost and GHG) are compared to the third criterion (random). Figures 7, 8, 9, 10, 11, 12 and 13 show the results of different simulations with the platform.

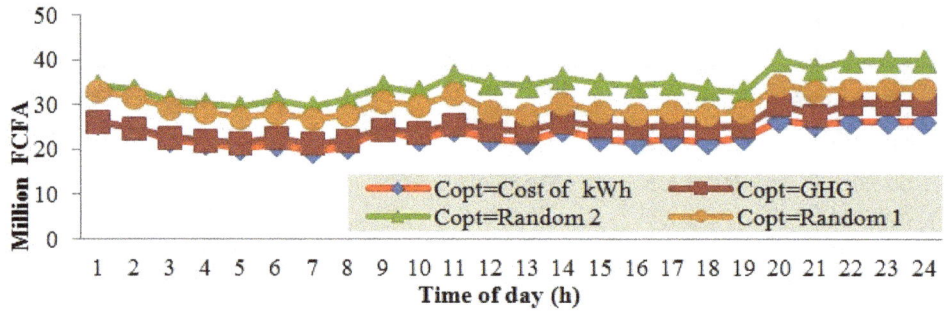

Figure 7. Production sources operating costs (Profile 1).

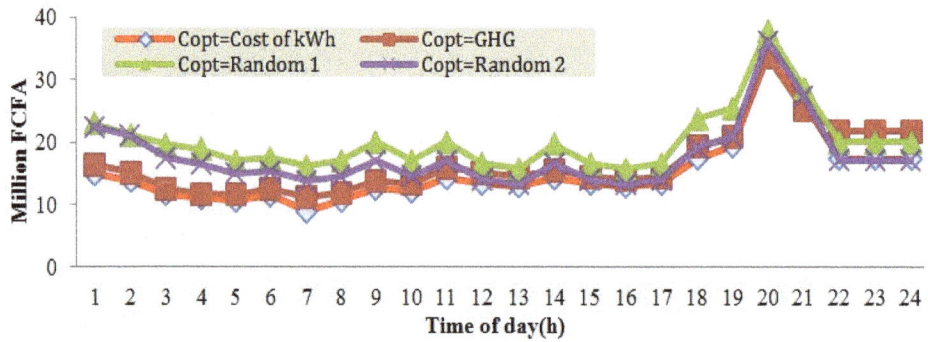

Figure 8. Production sources operating costs (Profile 2).

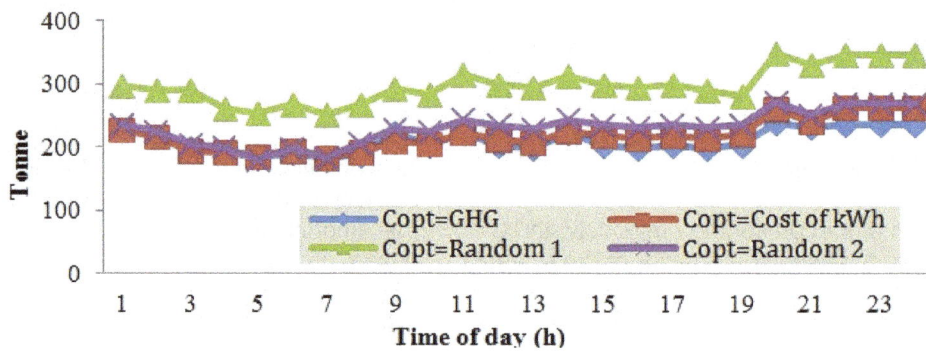

Figure 9. CO_2 quantity (Profile 1).

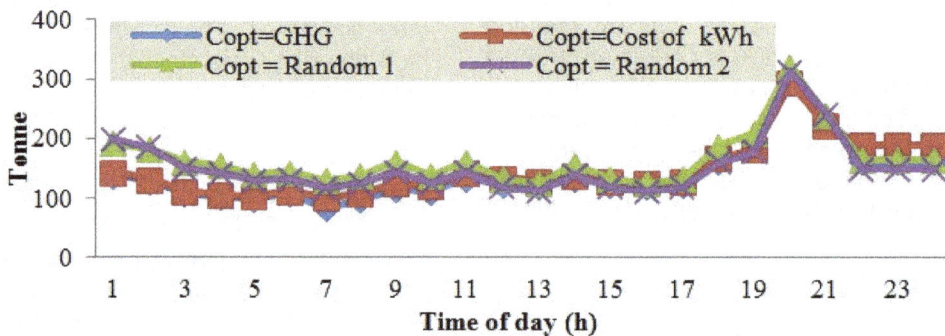

Figure 10. CO_2 quantity (Profile 2).

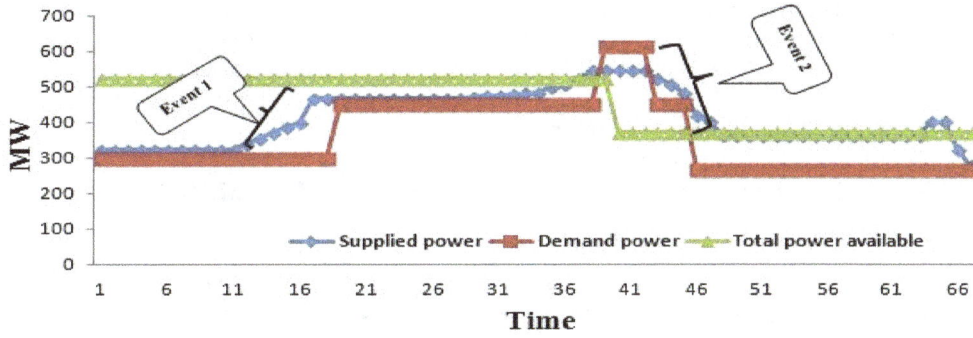

Figure 11. Behavior of management system on the critical moments of the day (between: 19 to 23 h).

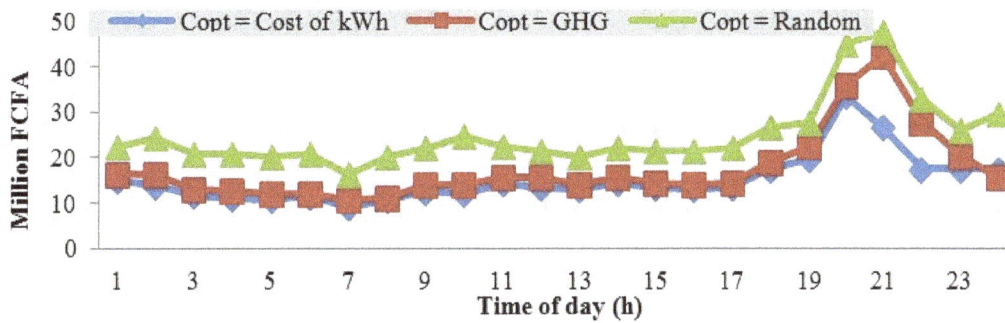

Figure 12. Production sources operating cost (Profile 3).

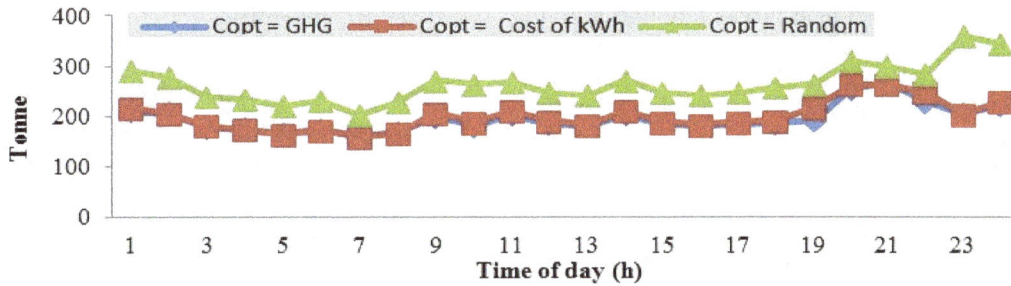

Figure 13. CO_2 quantity (Profile 3).

Results obtained with the first and second scenarios

Optimization criterion: The kWh production cost: Figures 7 and 8 show the operating costs of generation sources obtained respectively with Profile 1 and 2. The results show that the optimal cost is always lower than the cost obtained with the criterion based on random rules.

Optimization criterion: The amount of greenhouse gas emissions: Figures 9 and 10 show the amount of CO_2 released by the sources of production respectively with Profile 1 and 2. The difference between the results

obtained by random criterion and optimized criterion remains significant for the amount of GHG released.

Results obtained with the third scenario

The third load profile shows the behavior of the system facing load shedding. Figure 11 shows many fluctuations between 19 to 23 h with a power peak of the demand exceeding the total power available (P_s^T) of the electrical network, a power loss in the production facilities and a departure opening (load shedding). The management system always tries to balance the inbalance between the energy production and consumption quickly. For the disturbances brought to the electrical network, the reaction time always remains in the interval [0 50] seconds.

Event 1: satisfaction of the 20 h demand (duration of the event 35.39 s); Event 2: peak (exceeding the total power PT), shedding load, satisfaction of demand, (duration of the event 48.9 s).

Optimization criterion: The kWh production cost

Figure 12 shows the operating cost of generating sources obtained with the simulation. The input data of the system are those of Profile 3. Costs obtained with the optimization criterion kWh cost are still lower than all other costs.

Optimization criterion: The amount of greenhouse gas emissions

Figure 13 shows the amount of CO_2 emitted by the sources of production respectively with Profile 3.

Conclusion

The work presented in this paper proposed a design and implementation of a simulation platform for decentralized management based on a multi-agent system. Using a located multi-agent paradigm built model seemed then to be an innovative and promising option for the development of decision support tools. The methodology adopted in this paper set up a control strategy over the agents in order to organize, schedule and interpret the amount of information exchanged between the different entities of the system. The results later achieved with the established platform showed that the optimized production costs of the arrangements of the energy sources by the platform were always better than any other arrangement. The platform also made it possible to assess and minimize the amounts of GHG released by the electrical system. Some incompatibility on the simultaneous satisfaction of the two optimization criteria was noted. The explanation is that generators, whose production cost is cheapest, are the most polluting of the power source.

One of the prospects of this work is to find an arrangement of production sources for optimal operation, taking into account two optimization criteria: the network operation cost and the quantity of greenhouse gas emissions. The other prospect focuses on the diversification of the production sources by integrating the renewable energies. Indeed, new energy sources such as wind, solar generators, are getting into our electrical systems and the user will increasingly be confronted with energy prices varying according to the supplier, the date

and the time. It is in this varied and dynamic context of production and energy consumption that a «smart» control system takes all its importance from both consumption and pricing points of view (production).

ACKNOWLEDGEMENTS

The authors wish to thank the Direction de la Qualité, de la Sécurité et de l'Environnement (DQSE) of Sénégal and particularly Mr DIOP Moussa.

NOMENCLATURE

Ag_D Database Agent

Ag_L Departure Agent

Ag_M Manager Agent

Ag_S Source Agent

C_{S_i} Cost of kWh of the energy source

Q_{S_i} Quantity of fuel burned in the energy source

E_G Total emission of greenhouse gases (kg)

$E_{G,F}$ Emission of a given GHG per fuel type (kg)

$E_{G,F,T}$ Emission of a given GHG per fuel type and per technology (kg)

$E.F_{G,F}$ Emission factor of a given GHG per type of fuel used (kg/GJ)

$E.F_{G,F,T}$ Emission factor of a given GHG per fuel and per technology (kg/GJ)

P_D^T Total power Demand (W)

P_S^T Total power Supplied (W)

$P_{(state)min}$ Minimum power allowable in a state (W)

$P_{(state)max}$ Maximum power allowable in a state(W)

p Operating priority of a system agent (GJ)

Q_F Quantity of fuel burned in energy unit

$Q_{F,T}$ Quantity of burned fuel per energy unit per technology type (GJ)

GHG Greenhouse Gases

REFERENCES

Amit G, Kainou K, Tinus P (2006). Chapitre 2: Combustion stationnaire. Lignes directrices du GIEC, vol. 2 Energie, P. 53.

Butler-Purry KL, Sarma NDR, Hicks IV (2004). Service restoration in naval shipboard power systems. IEE Generation, Transmission, Distribution, 151:1.

Cristaldi L, Monti A, Ottoboni R, Ponci F (2003). Multi-agent based power systems monitoring platform: a prototype. IEEE Power Tech. Conf. 2(5).

Dimeas AL, Hatziargyriou N (2005a). Operation of a multiagent system for microgrid control. IEEE trans. power syst. 20(3):1447-1455.

Dimeas AL, Hatziargyriou ND (2004). A multi-agent system for microgrids. IEEE Power Eng. Soc. General Meeting, 1: 55-58.

Dimeas A, Hatziargyriou N (2005b). A MAS architecture for microgrid control . In Proc. the 13th International Conference on Intelligent Systems Application to Power Systems, November, P. 5.

Logenthiran T, Srinivasan D, Khambadkone AM, Aung HN (2012). Multiagent system for real-time operation of a microgrid in real-time digital simulator. IEEE Transactions on Smart Grid, Article number 6180026, 3(2):925-933.

Mao Meiquin, Dong W, Liuchen C (2011). Design of a novel simulation platform for the EMS-MG Based on MAS.IEEE Energy Conversion Congress and Exposition: Energy Conversion Innovation for a Clean Energy, pp. 2670-2675.

Monica A, Hortensia A, Carlos AO (2012). Integration of renewable energy sources in smart grids by means of evolutionary optimization algorithms. Expert Syst. Applications, 39(5):5513–5522.

Nagata T, Sasaki H (2002). A multi-agent approach to power system. IEEE Transactions on Power Systems, May 2002, 17:457-462.

Phillips L, Link H, Smith R, Weiland L (2006). Agent-based control of distributed infrastructure resources. Sandia National Laboratories, SAND2005-7937. Available: www.sandia.gov/scada/documents/sand_2005_7937.pdf.

Pipattanasomp M, Feroze H, Rahman S (2009). Multi-Agent Systems in a Distributed Smart Grid: Design and Implementation. IEEE PES Power Systems Conference and Exposition (PSCE'09), Seattle, Washington, USA.

Richalet J (1987). Modélisation et identification des processus . Techniques de l'ingénieur R7140.

Wang HF (2001). Multi-agent co-ordination for the secondary voltage control in power system contingencies. IEE Generation, Transmission Distribution, 148:61-66.

5

Frame structure for Mobile multi-hop relay (MMR) WiMAX networks

F. E. Ismael[1] , S. K. Syed-Yusof[1], M. Abbas[2], N. Fisal[1] and N. Muazzah[1]

[1]CoE Telecommunication Technology, University of Technology, Malaysia (UTM), Johor, 81310 UTM Skudai, Malaysia.
[2]Head of Wireless Communication Cluster, MIMOS Berhadi, Malaysia.

Mobile Multi-hop Relay (MMR) WiMAX network uses Relay stations (RSs) to extend the cell coverage, and enhances the link quality and the throughput. In MMR WiMAX networks, the number of hops between the Subscriber stations (SSs) and the Base station (BS) is allowed to be more than two hops when Non transparent RS (NT-RS) is used. However, this requires modification to the frame structure of NT-RS to reduce the delay of relaying the data packets across multiple hops. Therefore, this paper presents a new NT-RS frame structure aimed to decrease the multi-hop relaying delay in order to improve the performance of the data transmission over MMR WiMAX network. The proposed frame structure serves the sub-ordinate SSs as well as NT-RSs in the access zones, while using the relay zones to communicate with its super-ordinate stations. The performance of the proposed frame structure is tested through a simulation work. The results showed that, the forwarding delay is reduced, and hence the link layer and the Transmission control protocol (TCP) throughput are improved significantly.

Key words: IEEE 802.16j, transmission control protocol (TCP) performance, non transparent relay stations (NT-RS), frame structure.

INTRODUCTION

The service quality near the cell boundary of single hop deployment of WiMAX network degrades due to bad channel. Therefore, a multi-hop system using RS to relay the data packets between BS and the end SSs is introduced in (Yang et al., 2009; IEEE802.16j, 2009). The RS can improve the performance at the cell boundaries as well as extending its coverage to areas where weak signals are received or no signal at all. The network utilizing this multi-hop structure is called Mobile multi-hop relay (MMR) WiMAX network. The functionality of the BS should be extended in order to support incorporation of RSs into the network. The BS that incorporates these new functions is called a Mobile multi-hop relay base station (MMR-BS) (Canton and Chahed, 2001).

The IEEE 802.16j allows the number of hops to be more than two when non-transparent RSs are used. In the case of more than two hops between the MMR-

BSand the SS, there is a need to coordinate the transmission of intermediate RSs in order to efficiently utilizing the available resources. There are two ways to approach this; Single frame (SF) structure and Multi frame (MF) structure. SF structure suffers from poor capacity of the relay zones as the number of hops increases. On the other hand, the forwarding delay of the data packets across multiple hops is relatively long, especially for large number of hops when MF structure is considered. As a result, the SSs at different hops experience different performances of data transmission. This can be seen clearly for services such as TCP traffic where the rate of injecting segments into the network depends on the rate of acknowledgement (ACK) reception. TCP is the most common transport protocol used in the internet, which provides end-to-end connection oriented and reliable data transmission service

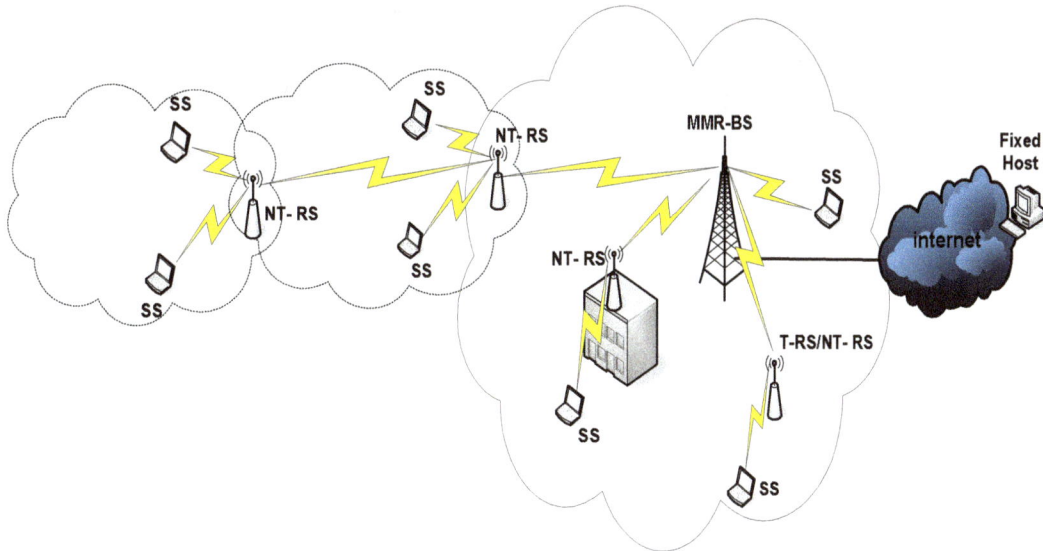

Figure 1. MMR WiMAX network architecture.

(Aweya, 2003). Due to the limitations of MF and SF structures, there is a need for a new construction of the Orthogonal frequency division multiple access (OFDMA) frame to overcome these limitations. Therefore, in this paper a new multi frame (NMF) structure for NT-RS is proposed, that is able to reduce the forwarding delay, and maintains the capacity of the MMR WiMAX network.

MOBILE MULTI-HOP RELAY WIMAX NETWORK

MMR WiMAX network architecture

The IEEE 802.16j uses various RSs to relay the data packets between the MMR-BS and end SSs. Figure 1 shows an example of MMR WiMAX network architecture consisting of one Mobile multi-hop relay base station (MMR-BS) connected to the internet through backbone network and a number of RSs to relay the data between the MMR-BS and the end SSs. The RSs used in IEEE 802.16j are backward compatible with IEEE 802.16e SSs; hence modifications are not required at the end SSs. In addition, RSs have less complexity and lower cost, as compared with the BS. Therefore, by using such RSs, an operator could deploy a network at a lower cost than using only expensive BSs to provide wide coverage (Genc et al., 2008b; Peters and Heath, 2009; Yang et al., 2009; Upase and Hunukumbure, 2008).

As shown in MMR WiMAX network model in Figure 1, the RSs can be used for coverage extension and throughput enhancement. Figure 2 shows the means of achieving these goals using different types of RSs. The coverage extension is achieved through two different scenarios; extending the coverage range of the BS and addressing coverage holes caused by shadowing as shown in Figure 1. On the other hand, enhancing system capacity will be gained by the use of multiple links with better quality (Genc et al., 2008b; Lei et al., 2008). Better link quality in terms of Signal to Noise ratio (SNR) allows the use of higher modulation level, and hence the overall throughput increases. In addition, multi-hop communication supports spatial reuse of frequency channels among RSs, which increases the overall system capacity. There are two different relaying modes of operation; transparent mode RS (T-RS) and non-transparent mode RS (NT-RS) (Genc et al., 2008b; Soldani and Dixit, 2008; Yang et al., 2009; Upase and Hunukumbure, 2008). The key difference between these two relaying modes of operation is the ability to generate and send control information to its sub-ordinate stations through the frame header. The RS operating in transparent mode cannot generate and transmit control information, but in non-transparent mode, the RSs do generate and transmit its own control information to its sub-ordinate stations.

The T-RSs (Genc et al., 2009; Upase and Hunukumbure, July 2008) only forward controls information generated by the MMR-BS, and hence they do not extend the coverage area of the BS. However, T-RS is used to enhance the system capacity in terms of throughput within the BS coverage area. T-RS has less complexity and is cheaper, as compared with the NT-RS. So, the T-RS can only operate in a centralized scheduling mode and for topology, up to two hops only. The NT-RSs (Yang et al., 2009; Sayenko et al., 2010; Upase and Hunukumbure, 2008) operate in either centralized or distributed scheduling. In the case of centralized scheduling, NT-RSs only forward the control information provided by the MMR-BS to their sub-ordinate SSs. However, when NT-RSs operate in distributed

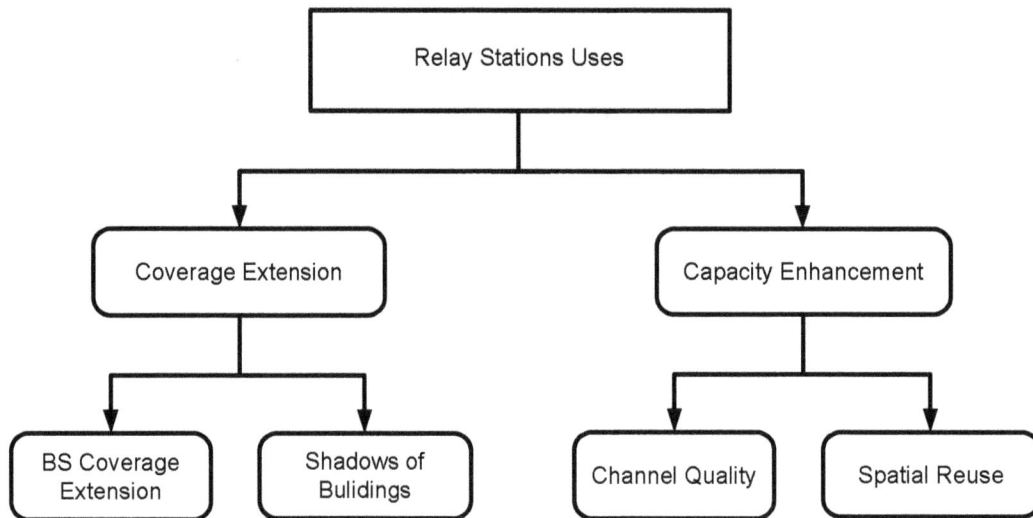

Figure 2. Relay stations functionalities.

scheduling, they generate their own control information. NT-RSs are used to provide cell coverage extension as well as capacity enhancement. They can operate in topologies larger than two hops in either a centralized or distributed scheduling mode. Despite the benefits of using of RS in the MMR WiMAX network, the resource management schemes should be modified to incorporate the RS operation. This issue is arising because in multi-hop system, there is more than one intermediate node and therefore, their transmission should be coordinated. The coordination of the intermediate nodes transmission aim to avoid interference between their transmissions and maximizes the resource utilization (Hui and Chenxi, 2009). In the following section, the frame structures and relaying modes to coordinate these transmissions will be discussed.

OFDMA frame structure and relaying modes

In IEEE 802.16e, an OFDMA frame compose of two sub-frames; a downlink (DL) sub-frame and an uplink (UL) sub-frame (IEEE802.16e-2005, 2006). The DL sub-frame is to transmit from MMR-BS to the end SS and the UL sub-frame for the reverse direction. Multi-hop functionality is defined, and hence PHY layer specifications are changed according to the multi-hop operations. The RS should receive and transmit in both UL and DL. Therefore, to accommodate the RS in IEEE 802.16j, each sub-frame of the OFDMA frame is spitted into an Access zone (AZ) and a Relay zone (RZ). The Downlink access zone (DL-AZ) is used to transmit to the end SS that is served by MMR-BS, while Uplink access zone (UL-AZ) is utilized to receive from it. Similarly, the Downlink relay zone (DL-RZ) and the Uplink relay zone (UL-RZ) are used to relay the data to and from the super-ordinate

RSs, respectively (Hoymann et al., 2006; Mach and Bestak, 2009; Sayenko et al., 2010). There are two types of relaying; transparent and non transparent, which will be discussed in the following section.

Transparent relaying

In transparent relaying, the SSs served by RSs are able to receive and decode the control information from the MMR-BS. So, the intermediate and the access RSs are not required to transmit control information by themselves. Although the SSs are inside the coverage range of the MMR-BS but by using multiple hops with the aid of RSs higher throughput is achieved (Genc et al., 2009). Thus, the goal of this type of relaying is to enhance network capacity in terms of throughput. This scheme of relaying is called transparent relaying; because the SS is unaware of the RS existence. In the transparent relaying all control information originates from the MMR-BS. The transparent relaying frame is shown in Figure 3. The control information is sent at the beginning of the DL sub-frame to all sub-ordinate stations. The control information includes; Preamble in the 1st symbol and its main usage is to enable SS to synchronize with the BS. DL-MAP determines the specified burst in the DL sub-frame allocated to each SS. UL-MAP defines the burst location and size in the coming UL sub-frame for each SS, and lastly, Frame control header (FCH) which is a special management burst that is used for the BS to advertise the system configuration. Therefore, all SSs and RSs are able to receive and decode the control information, and determine their own time to transmit or receive. The MMR-BS then proceeds in the DL sub-frame, which is utilized to send data to the sub-ordinate stations. The DL sub-frame is divided into two zones;

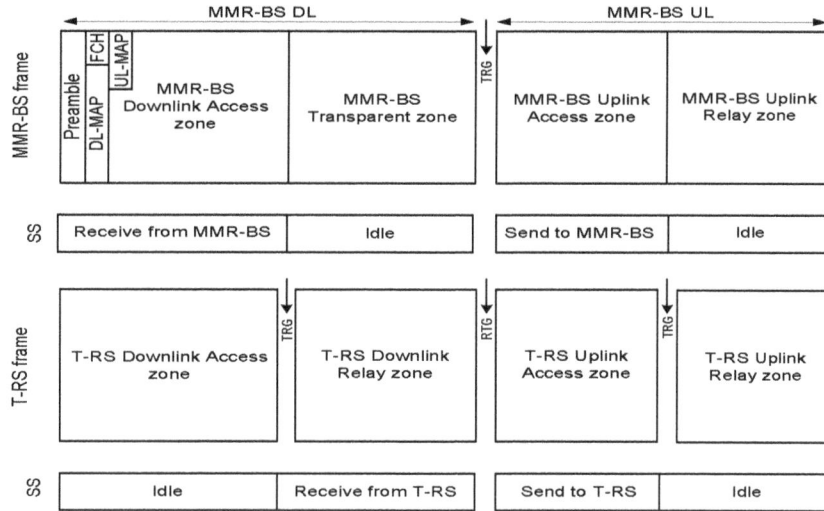

Figure 3. Transparent relaying frame structure.

DL-AZ and transparent zone. In the DL-AZ, the MMR-BS transmits to the SSs at the first hop and as well as the RSs at the first tier. While, in the transparent zone the RSs transmit to their sub-ordinate SSs. In addition, during the transparent zone MMR-BS can transmit to its subordinate SSs, or remain silent, or transmit cooperatively with the RSs. The DL sub-frame and the UL sub-frame are separated with a small transition time to switch from transmit to receive or vice versa. Similarly, the UL sub-frame is partitioned into two zones; UL-AZ and UL-RZs. In the UL-AZ, the end SSs transmit to their serving RS or MMR-BS, while in the UL-RZ, the RSs transmit to the MMR-BS (Genc et al., 2008a).

Non-transparent relaying

In non-transparent relaying, the SSs are served by RSs and cannot receive or decode the control information from the MMR-BS (Sayenko et al., 2010). So, the RSs serving these SSs must be able to generate and transmit its own control information at the beginning of each frame. The end SS considers the serving RS as its BS and is not able to deal with control information sent by MMR-BS. These RSs are called non-transparent because the SS synchronizes and receives control information from it. The SSs that are out of range of the MMR-BS are unable even to receive the control information sent by it, so the non-transparent RS sends to them its own frame and hence extend the MMR-BS coverage. The frame structure of non-transparent RS is shown in Figure 4 which consists of DL access; DL relay; UL access and UL relay zones, in which, both the MMR-BS and the RS transmits control information at the beginning of the frame. So, the SSs synchronize with the RS, which is synchronized with the MMR-BS. The DL

sub-frame starts with the AZ, which is utilized by the MMR-BS to transmit to its sub-ordinate SSs. After the DL-AZ, is the DL-RZ that begins with relay-amble (R-amble) which is control information specifically for sub-ordinate RSs. The UL sub-frame is partitioned into two parts, UL-AZ to receive from sub-ordinate SSs and UL-RZ to receive from sub-ordinate RSs. The main issue facing non-transparent relaying is how the RS and MMR-BS can transmit simultaneously in time and possibly in frequency without causing severe interference. Furthermore, non-transparent RSs are more sophisticated and thus much expensive than transparent RSs.

Three hops frame structure

In the scenario of Figure 1, the access RS transmits to the SS in the DL-AZ and expects data from its super-ordinate station in the DL-RZ. The problem is that its super-ordinate station is a RS that may also be required to receive in the DL-RZ. To address the issue of multiple hops, the frame structure of MMR-BS and the non-transparent RS are required to be adapted. There are some efforts done in the literature to address this issue. These schemes will be discussed in the following sections. There are two ways to approach this; Single frame (SF) structure with multi RZs and Multi frame (MF) structure (Koon et al., 2007; Seung-Yeon et al., 2008; Mach and Bestak, 2009; Peters and Heath, 2009; Taha et al., 2011). In SF structure with multi RZs, the number of DL-AZs and UL-RZs are determined depending on the number of intermediate NT-RS between MMR-BS and the SSs. Figure 5 shows an example of SF structure for three hop MMR WiMAX system. In the DL-RZ, the first RZ is used by MMR-BS to relay the data of the first tier

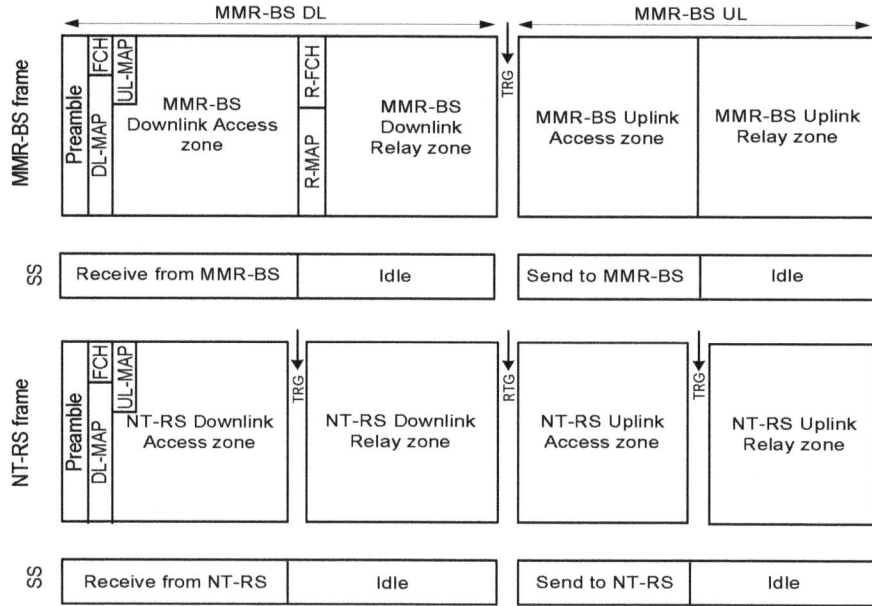

Figure 4. Non transparent relaying frame structure.

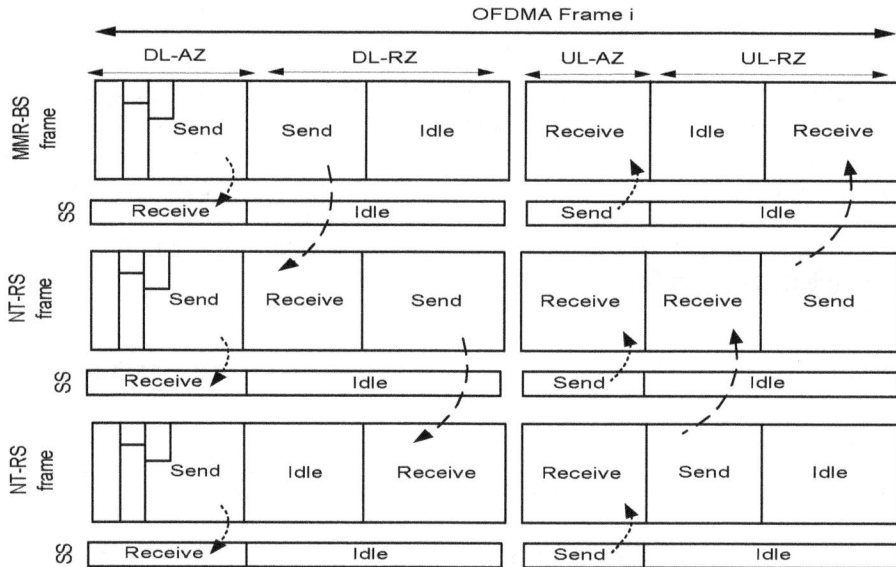

Figure 5. Single frame structure for three hops.

NT-RS. The second RZ is utilized to relay the second tier NT-RS. By the same way, they use specific RZs in the UL to send their data to their super-ordinate stations (MMR-BS or NT-RS) (Koon et al., 2007; Hui and Chenxi, 2008; Seung-Yeon et al., 2008; Taha et al., 2011). In this frame structure, the data is relayed across multiple hops in one frame duration time. However, it requires centralized resource allocation in the MMR-BS for both DL and UL, which will be very complex when large number of hops is considered. In addition, as the number of hops increases, the system capacity in terms of number of SSs and the sending data rate degrades severely. These limitations make it not suitable for multi-hop systems with large number of hops.

The other approach is to group frames together into what is called a multi-frame (Hui and Chenxi, 2008; Taha et al., 2011). In this frame structure, simultaneous transmission in the DL-AZ and UL-AZ of both NT-RSs

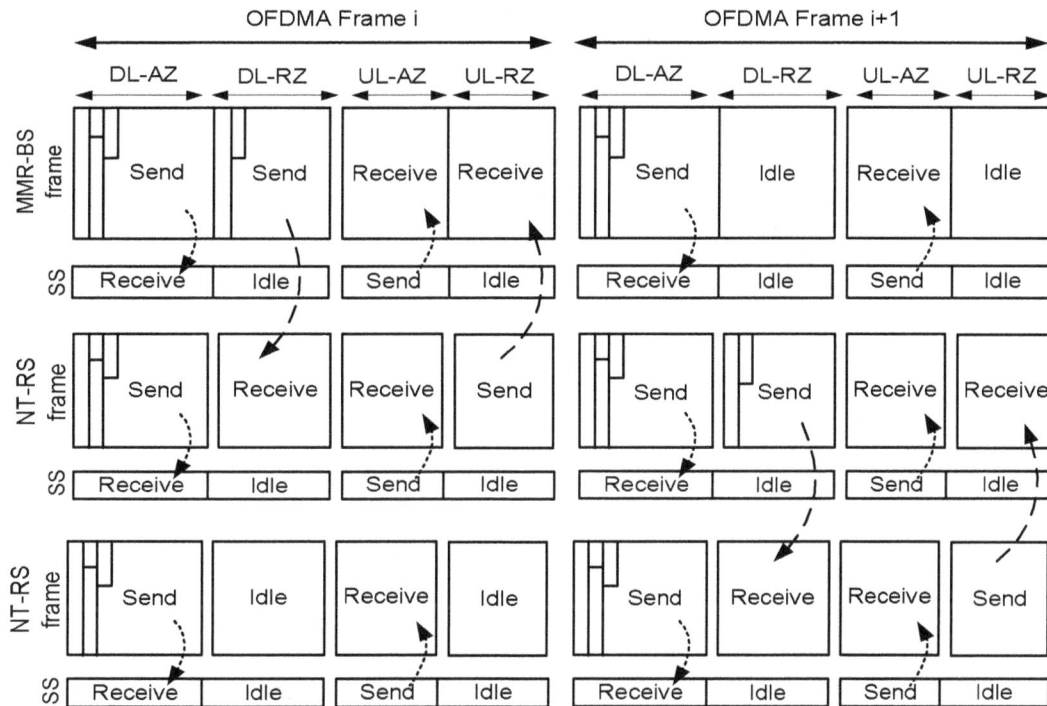

Figure 6. Multi frame structure for three hops.

and MMR-BS is allowed while in the RZs, only one station (either MMR-BS or NT-RS) is eligible to transmit and the others should be in receiving mode. The same concepts are applied for the first tier and second tier NT-RS. Therefore, RSs at odd tiers transmit in the RZs of odd-numbered frames, and RSs at even tiers transmit in the RZs of even-numbered frames. Figure 6 gives an example of MF structure for three hops MMR WiMAX system during two successive frames. The DL-RZ of the first frame (OFDMA frame i) is utilized to relay the data from MMR-BS to its sub-ordinate NT-RSs, while UL-RZ is used to relay the data from the NT-RSs in the first tier to the MMR-BS. However, the second frame (OFDMA frame i + 1) is used to relay the data between the first tier and second tier NT-RSs for the DL and the UL. This frame structure is applicable for centralized and distributed scheduling architectures.

The relay zone is used by the intermediate RSs alternatively to communicate with their super-ordinate RS and sub-ordinate RS. As result, the delay of relaying the data packets across multiple hops is increased exponentially with the hop level. Therefore, in order to enhance the performance of MMR WiMAX network, this paper proposes amendments to the multi-frame structure of NT-RS that aims to decrease the delay and maintain the capacity of MMR WiMAX network. The new multi-frame structure uses the access zones to communicate with its sub-ordinate stations and the relay zones to communicate with its super-ordinate stations. The details of the proposed frame structure will be discussed in the

following section.

The proposed new multi frame structure

NMF structure design concepts

In this section, a new multi-frame (NMF) structure for the NT-RS is proposed based on the MF structure for the time division duplex (TDD) systems. Figure 7 shows the frame structure of the MMR-BS and the frame structure of the NT-RSs at the first and second tiers. It gives the specific time for the NT-RSs to communicate with their sub-ordinates stations (SSs and NT-RSs) and super-ordinates stations (MMR-BS and NT-RSs). The proposed NMF uses the same frame structure specified by the standard (IEEE802.16j-2009, 2009) for the MMR-BS; however, the frame structure of NT-RS is customized to decrease the forwarding delay and at the same time maintain the capacity of the system in terms of number of SSs and transmission rates.

In NMF structure, the whole frame is divided into two main sub-frames, DL and UL sub-frames separated by small transmit receive guard (TRG) time to allow the MMR-BS to change the mode from transmission to receiving. Further, the downlink sub-frame is divided into three parts where the first part is for control message transmission (preamble, DL-MAP, UL-MAP and the FCH), the second one is for data transmission to the SSs that are served directly by the MMR-BS named downlink

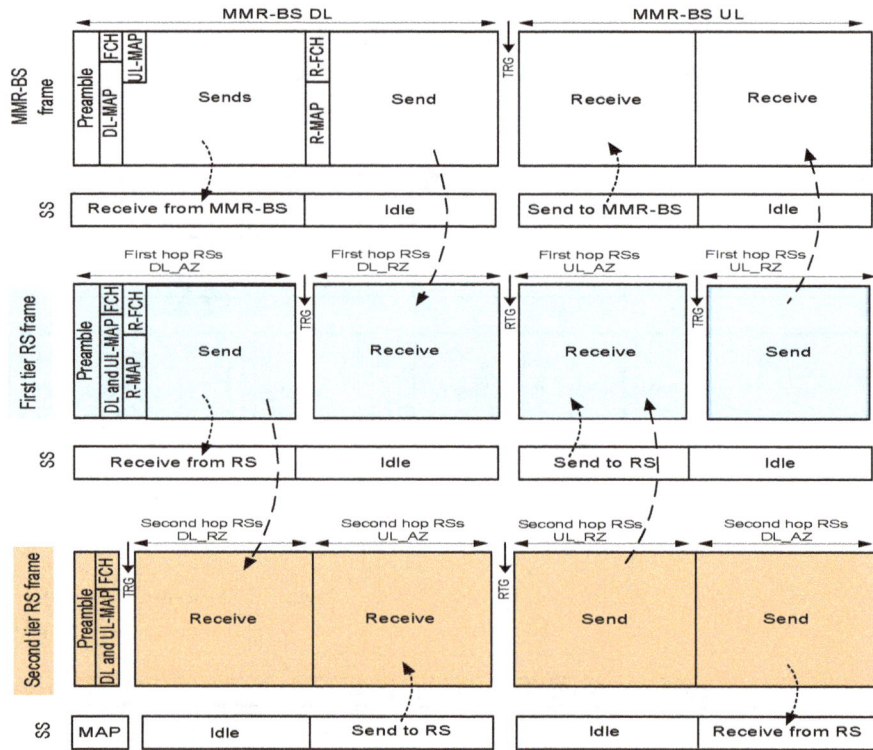

Figure 7. NMF frame structures of MMR-BS and NT-RSs.

access zone (DL-AZ) and the last part is utilized for data transmission to the first tier NT-RSs which is called downlink relay zone (DL-RZ). Also, the uplink sub-frame is partitioned into two parts, one is to receive from first hop SSs called uplink access zone (UL-AZ) and the other is to receive from the first tier NT-RSs named uplink relay zone (UL-RZ). In the proposed NMF structure, the intermediate NT-RSs utilize the access zones to communicate with their sub-ordinate stations (SSs and NT-RSs) and the relay zones to communicate with their super-ordinates station (MMR-BS or NT-RS) as shown in Figure 7. The intermediate NT-RSs receive data from their super-ordinate station in the DL-RZ and from their sub-ordinate stations in the UL-AZ. On the other hand, it transmits data to their super-ordinate station in the UL-RZ and to their sub-ordinate stations in the DL-AZ. The allocated resources to the sub-ordinate stations (SSs or NT-RSs) are determined by scheduling algorithms. There are no zone capacity limits for the SSs or the NT-RSs served by the same super-ordinates NT-RS. Therefore, it is very flexible in allocating the available resources to both SSs and NT-RSs and this allows them to utilize the available resources efficiently.

The first tier NT-RS frame structure consists of four parts separated by small guard time (TRG/RTG) to allow the NT-RS to change its mode from sending to receiving or vice versa. At the beginning of this frame structure, the control information is sent to inform the sub-ordinate

stations with their slots for transmission and reception. Next, it is followed by the downlink zone which comprises of two parts namely DL-AZ and DL-RZ. The DL-AZ is utilized for downlink data transmission to the sub-ordinate stations. The allocated slots to each SS or NT-RS are determined by the scheduling algorithm. In the DL-RZ, the first tier NT-RS receives data from its super-ordinate station, which is the MMR-BS. Consequently, the UL-AZ is used to receive data from sub-ordinate stations (SSs and NT-RSs) and the UL-RZ is utilized to transmit data to the MMR-BS. The second tier NT-RSs frame structure is constructed by the same way as that of the first tier with a few differences. The frame structure of the second tier NT-RS has only two guard times to change the mode of the NT-RS from sending to receiving or vice versa. In addition, it doesn't send control message for the sub-ordinate NT-RSs because in this paper, only three hops are considered. If it is required to extend the hops to more than three, the control messages should be added at the beginning of the last sub-frame which is used to transmit to the sub-ordinate stations. As seen from the previous discussion, the proposed NMF structure is different from MF structure in that it eliminates the superposition manner of using of the relay zones, and hence it reduces the delay of relaying the data packets across multiple hops. In the following section, the capacity of the proposed NMF frame structure will be analyzed numerically.

NMF NT-RS frame capacity

The aim of this numerical analysis is to determine the number of slots and the capacity of all access and relay zones of the proposed NMF structure. Firstly, the number of OFDMA symbols in the whole frame, downlink and uplink zones of the MMR-BS are determined by Equations (1) to (9). Equation 1 gives the sub-carrier spacing (Δf) which is determined by the sampling factor (F_s) and the Fast fourier transform (FFT) size (N_{FFT}). While, Equation (2) gives the sampling factor, where BW is the bandwidth and n is a factor depends on the value of the bandwidth, where its value is normally (8/7) or (28/25) if the bandwidth is multiple of 1.25, 1.5, 2, 2.75 MHz (Andrews et al., 2007; Ergen, 2009).

$$\Delta f = \frac{F_s}{N_{FFT}} \tag{1}$$

$$F_s = floor\left(\frac{n*BW}{8000}\right) \tag{2}$$

The useful OFDMA symbol time (T_b) is given by the Equation (3) and the cyclic guard time (T_g) in Equation (4), where G is the ratio of the cyclic guard time to the useful OFDMA symbol time, and which it takes one of the values (1/32, 1/16, 1/8 and 1/4) (Andrews et al., 2007; Ergen, 2009).

$$T_b = \frac{1}{\Delta f} = \frac{N_{FFT}}{F_s} = \frac{N_{FFT}}{\left\lfloor\dfrac{n*BW}{8000}\right\rfloor * 8000} \tag{3}$$

$$T_g = G * T_b = \frac{G*N_{FFT}}{\left\lfloor\dfrac{n*BW}{8000}\right\rfloor * 8000} \tag{4}$$

The OFDMA symbol time is the summation of the useful OFDMA symbol time and cyclic guard time as shown in the Equations (5) and (6) (Andrews et al., 2007; Ergen, 2009).

$$T_{symb} = T_b + T_g = \frac{N_{FFT}}{\left\lfloor\dfrac{n*BW}{8000}\right\rfloor * 8000} + \frac{G*N_{FFT}}{\left\lfloor\dfrac{n*BW}{8000}\right\rfloor * 8000} \tag{5}$$

$$T_{symb} = (1 + G)\frac{N_{FFT}}{\left\lfloor\dfrac{n*BW}{8000}\right\rfloor * 8000} \tag{6}$$

The number of OFDMA symbols in an OFDMA frame or part of the frame is calculated by dividing the duration of the frame or the sub-frame by the OFDMA symbol time. So, the number of OFDMA symbols in an OFDMA frame,

downlink and uplink sub-frames for the MMR-BS are determined by the Equations (7), (8) and (9), respectively. The number of symbols in the downlink and the uplink sub-frames are determined by the downlink (DL_{Ratio})$_{MMR-BS}$ to the uplink (UL_{Ratio})$_{MMR-BS}$ ratios as shown in Equations (8) and (9).

$$N_{symb}^{frame} = \frac{(T^{frame} - TRG)}{T_{symb}} = \frac{\left(T^{frame} - TRG * \left\lfloor\dfrac{n*BW}{8000}\right\rfloor\right)*8000}{(1+G)*N_{FFT}} \tag{7}$$

$$N_{symb}^{(DL)_{MMRBS}} = \left(N_{symb}^{frame}\right)*\frac{(DL_{Ratio})_{MMRBS}}{(DL_{Ratio})_{MMRBS} + (UL_{Ratio})_{MMRBS}} \tag{8}$$

$$N_{symb}^{(UL)_{MMR-BS}} = \left(N_{symb}^{frame}\right) - N_{symb}^{(DL)_{MMR-BS}} \tag{9}$$

Next, the number of sub-carriers, channels and slots in the frame structure of the MMR-BS can be determined through Equations (10) to (15). The number of subcarriers is determined by the total bandwidth (BW) divided by the sub-carrier spacing (Δf) as in Equation (10) (Andrews et al., 2007; Ergen, 2009).

$$N_{subcarr}^{frame} = \frac{BW}{\Delta f} \tag{10}$$

The available sub-carriers are divided into two parts; some of them are used for data sending while the remaining reserved as control sub-carriers as in Equation (11) (Andrews et al., 2007; Ergen, 2009).

$$N_{subcarr}^{frame} = N_{Data,subcarr}^{frame} + N_{Cont,subcarr}^{frame} \tag{11}$$

To determine the number of channels and slots for the downlink and uplink sub-frames, Partial usage sub-carriers (PUSC) permutation is used (Andrews et al., 2007; Ergen 2009). In the downlink sub-frame, the number of channels is calculated by dividing the number of data sub-carriers by the cluster size which contains 32 sub-carriers; 24 of them for data and 8 for pilot signals as in Equation (12).

$$N_{channel}^{(DL)_{MMR-BS}} = \frac{N_{Data,subcarr}^{frame}}{Cluster_{size}} \tag{12}$$

To calculate the number of slots in the downlink of the MMR-BS, the calculated number of channels is multiplied by the number of OFDMA symbols in the downlink sub-frame and the total is divided by the number of symbols per slot which is 2 for the downlink PUSC permutation as shown in Equation (13) (IEEE802.16e-2005, 2006; Andrews et al., 2007).

$$N_{slots}^{(DL)_{MMR-BS}} = \frac{N_{symb}^{(DL)_{MMR-BS}} * N_{channel}^{(DL)_{MMR-BS}}}{N_{symb}^{slot}} \tag{13}$$

For the uplink sub-frame, the number of channels is calculated by dividing the number of data sub-carriers by the size of the tile, which is a group of 4 tiles each one contains 6 sub-carriers (4*6); 18 of them for data and 6 for pilot signals as in Equation (14) (Andrews et al., 2007; Ergen, 2009).

$$N_{channel}^{(UL)_{MMR-BS}} = \frac{N_{Data,subcarr}^{frame}}{Tile_{size}} \tag{14}$$

To determine the number of slots in the uplink sub-frame of the MMR-BS, the calculated number of channels is multiplied by the number of OFDMA symbols in the downlink sub-frame and the total is divided by the number of symbols per slot which is 3 for the uplink PUSC permutation as shown in Equation (15) (Andrews et al., 2007; Ergen, 2009).

$$N_{slots}^{(UL)_{MMR-BS}} = \frac{N_{symb}^{(UL)_{MMR-BS}} * N_{channel}^{(UL)_{MMR-BS}}}{N_{symb}^{slot}} \tag{15}$$

The Equations (16) to (19) gives the number of slots in the downlink access zone, downlink relay zone, uplink access zone and the uplink relay zone for the MMR-BS, respectively. The number of slots in the downlink access zone of the MMR-BS is calculated by multiplying the number of slots in the downlink sub-frame of the MMR-BS by the ratio of the downlink access zone to the summation of the downlink and uplink ratios as in Equation (16). The numbers of slots in the downlink relay zone of the MMR-BS is calculated by excluding the number of slots in the access zone of the MMR-BS from the total number of slots in the downlink sub-frame of the MMR-BS, as in Equation (17).

$$N_{slots}^{(DL\_AZ)_{MMR-BS}} = \frac{N_{slots}^{(DL)_{MMR-BS}} * (DL\_AZ)_{Ratio}}{(DL\_AZ)_{Ratio} + (DL\_RZ)_{Ratio}} \tag{16}$$

$$N_{slots}^{(DL\_RZ)_{MMRBS}} = N_{slots}^{(DL)_{MMRBS}} - N_{slots}^{(DL\_AZ)_{MMRBS}} \tag{17}$$

The same procedure is performed to calculate the number of slots in the uplink access zone and uplink relay zone of the MMR-BS as in Equations (18) and (19).

$$N_{slots}^{(UL\_AZ)_{MMR-BS}} = \frac{N_{slots}^{(UL)_{MMR-BS}} * (UL\_AZ)_{Ratio}}{(UL\_AZ)_{Ratio} + (UL\_RZ)_{Ratio}} \tag{18}$$

$$N_{slots}^{(UL\_RZ)_{MMR-BS}} = N_{slots}^{(UL)_{MMR-BS}} - N_{slots}^{(UL\_AZ)_{MMR-BS}} \tag{19}$$

As mentioned before, the first tier NT-RSs are directly connected to the MMR-BS, they use their access zones to communicate with their sub-ordinate stations (SSs and NT-RS) and also their relay zones to communicate with the MMR-BS as in the proposed frame structure shown in Figure 7. When the MMR-BS decided to allocate resources to the SSs served by the first tier, NT-RS should consider these capacities to avoid packets dropping in the queues of the intermediate NT-RSs due to limited storage. Equations (20) and (21) determine the maximum number of slots of the DL-AZ and UL-AZ of the first tier NT-RSs as a function of the MMR-BS zones.

$$N_{slots,Max}^{(DL\_AZ)_{NT-RS_1^n}} = N_{slots}^{(DL\_AZ)_{MMR-BS}} \tag{20}$$

$$N_{slots,Max}^{(UL\_AZ)_{NT-RS_1^n}} = N_{slots}^{(UL\_AZ)_{MMR-BS}} \tag{21}$$

Equations (22) and (23) determine the maximum number of slots of the DL-RZ and UL-RZ of the first tier NT-RSs as function of MMR-BS zones, which are used to relay the data to sub-ordinate and super-ordinate stations, respectively.

$$N_{slots,Max}^{(DL\_RZ)_{NT-RS_1^n}} = N_{slots}^{(DL\_RZ)_{MMR-BS}} \tag{22}$$

$$N_{slots,Max}^{(UL\_RZ)_{NT-RS_1^n}} = N_{slots}^{(UL\_RZ)_{MMR-BS}} \tag{23}$$

The second tier NT-RSs are those served by the first tier NT-RSs, and they use the access zones to communicate with their sub-ordinate stations (SSs) and the relay zones to communicate with their super-ordinate NT-RSs as in the proposed frame structure shown in Figure 7. Equations (24) and (25) determine the maximum number of slots of the DL-AZ and UL-AZ of the second tier NT-RSs as a function of the first tier NT-RS slots, which are utilized to communicate with the sub-ordinate stations.

$$N_{slots,Max}^{(DL\_AZ)_{NT-RS_2^n}} = N_{slots,Max}^{(UL\_RZ)_{NT-RS_1^n}} \tag{24}$$

$$N_{slots,Max}^{(UL\_AZ)_{NT-RS_2^n}} = N_{slots,Max}^{(DL\_RZ)_{NT-RS_1^n}} \tag{25}$$

Table 1. Simulation parameters.

Parameter	Value
Physical Layer	OFDMA
Operating frequency	3.5 GHz
Bandwidth	20 MHz
FFT size	2048
OFDMA frame duration	20 ms
Duplex mode	TDD
TRG/RTG gap	50 µs
Cyclic prefix length	1/8
OFDMA symbols	198
DL:UL ratio	6:4
DL-AZ:DL-RZ ratio	1:1
UL-AZ:UL-RZ ratio	1:1
Fragmentation/packing	on
CRC/ARQ	on
Propagation model	Two ray ground

Meanwhile, Equations (26) and (27) determine the maximum number of slots of the DL-RZ and UL-RZ of the second tier NT-RSs as a function of the first tier NT-RS slots, which are utilized to relay data to the sub-ordinate and super-ordinate stations, respectively.

$$N_{slots,Max}^{(DL\_RZ)_{NT-RS_2^n}} = N_{slots,Max}^{(DL\_AZ)_{NT-RS_1^n}} \qquad (26)$$

$$N_{slots,Max}^{(UL\_RZ)_{NT-RS_2^n}} = N_{slots,Max}^{(UL\_AZ)_{NT-RS_1^n}} \qquad (27)$$

The slot capacity from the information theory side of view is calculated by utilizing Shannon's theorem. Equation (28) gives the maximum capacity of one sub-carrier over one OFDMA symbol, while Equation (29) gives the capacity of one slot composed of number of sub-carriers over number of OFDMA symbols.

$$C_{subcarr\_Max,symb} = \frac{BW_T}{N_{subcarr}^{frame} * N_{symb}^{frame}} \log_2(1 + \frac{E_s}{N_0}) \qquad (28)$$

$$C_{slot\_Max} = \frac{BW_T}{N_{subcarr}^{frame} * N_{symb}^{frame}} * N_{subcarr}^{slot} * N_{symb}^{slot} * \log(1 + \frac{E_s}{N_0}) \qquad (29)$$

Equations (30) and (31) give the maximum theoretical channel capacity as defined by Shannon's theorem for the whole OFDMA frame and sub zones of the MMR-BS, where * in Equation (31) represents DL-AZ, UL-AZ, DL-RZ or UL-RZ. It is calculated by summing the individual capacities of all the slots in the OFDMA frame or the

specific access or relay zone.

$$C_{frame}^{Max} = \sum_{i=1}^{N_{slots}^{frame}} C_{slot_i} \qquad (30)$$

$$C_{(*)MMR-BS}^{Max} = \sum_{i=1}^{N_{slots}^{(*)MMR-BS}} C_{slot_i} \qquad (31)$$

The more practical capacity is determined by the modulation scheme and the coding rate. Equation (32) gives the amount of data bits that can be sent in one slot, where MOD_{slot} is a factor that represents the number of bits can be sent in one symbol which determined by the modulation scheme and CR_{slot} is the channel coding rate.

$$R_{slot} = \frac{BW_T}{N_{subcarr}^{frame} * N_{symb}^{frame}} * N_{subcarr}^{slot} * N_{symb}^{slot} * MOD_{slot} * CR_{slot} \qquad (32)$$

Equation (33) gives the practical capacity of the downlink and uplink access and relay zones of the MMR-BS, where * represents DL-AZ, UL-AZ, DL-RZ or UL-RZ. It is calculated by summing the individual capacities of all the slots in the OFDMA frame or the specific access or relay zone.

$$R_{(*)MMR-BS} = \sum_{i=1}^{N_{slots}^{(*)MMR\_BS}} R_{slot_i} \qquad (33)$$

RESULTS AND DISCUSSION

In this section, the performance evaluation of the proposed NMF structure is conducted, which is composed of two parts. The first part discusses the capacity of NMF structure in terms of the number of SSs that can be admitted and the maximum transmission rates at each zone, as compared to MF and SF structures. In the second part, the effect of the proposed NMF on the relaying delay is analyzed, and hence the improvements in link layer and TCP traffic are discussed. The system model considered in this paper is three hops MMR WiMAX networks as shown in Figure 1. Simulations in ns-2 network simulator considering the system parameters of MMR WiMAX network are used in order to perform the experiments. The main features of the MAC layer and physical layer of the MMR WiMAX network are implemented in the simulation software. The simulation is performed with the simulation parameters shown in Table 1. The simulation scenarios evaluate the effect of the proposed NMF structure on TCP performance. It is assumed that the TCP segment size is 1024 bytes for all the simulations. In addition, it is assumed that all the sending hosts are always having data to be sent to the SSs in the different hops and there are sufficient resources to forward them. The NT-RS uses the next frame to relay the data received in the current frame and the link layer frame error percentage is 10%.

Capacity evaluation of NMF

The purpose of this performance evaluation part is to show how the proposed NMF structure can utilize the available relay link slots flexibly and efficiently. The network model shown in Figure 1 is considered. The simulations are conducted using the simulation parameters as stated in Table 1. The ratios of downlink and uplink access and relay zones as shown in Table 1; are 6:4 for downlink and uplink, and 1:1 for the access and relay zones. In addition, the downlink and uplink relay zones of the SF structure are divided equally between the second and third hops. The SSs are accepted with their maximum sustained rate until the capacity of the access zone is reached. Then, the excess slots above the minimum reserved rate are determined and hence more SSs are admitted with lower rates. The process of admitting new SSs is stopped when the transmission rate of each SS becomes equal to the minimum reserved rate or the available slots are insufficient to provide the new SSs with the required data rates. Different simulation scenarios and various SSs distribution are conducted. In each scenario, the maximum number of SSs supported in the access and relay zones, as well as the maximum data rates is presented.

In the first scenario, a three hop network model with only one branch comprising of two NT-RSs is considered. Figure 8 shows a distribution of the SSs of 50% in the first hop and 25% of the SSs at the second and third hops. While, Figure 9 gives the system capacity when there are no SSs in the second tier NT-RS and all the available slots are utilized by the SSs served by the first tier NT-RS. The result indicates that the maximum number of SSs supported in the access zone is the same for all frame structures. However, in the relay zone, the SF structure can accommodate only 50% percent of that can be served by MF and NMF structures. In the second scenario, three hops network architecture with one MMR-BS and two branches with two NT-RSs in each one are considered. The SSs are distributed; 50% in the first hop, they are served directly by MMR-BS access zone, and the SSs in the relay zone are divided equally between the NT-RSs as in Figure 10. Figures 8 to 10 shows that, the link data rate is raised up as the number of admitted SSs increased. However, when it reaches its maximum capacity, the link data rate remains constant even if the number of SSs is increased. In addition, MF structure and NMF structure can accommodate the same number of

SSs in the first, second and third hops and provide typical link data rate. On the other hand, SF structure can accommodate the same number of SSs as that for MF and NMF structures at the first hop, and hence a typical link data rate is achieved for them. However, in the second and third hops, the number of SSs served by SF structure is only half of those supported by MF and NMF structures. Therefore, the link data rate is reduced by 50% as compared with the data rate of MF and NMF structures. In the last scenario, the capacity of MF, NMF and SF structures are evaluated for single branch with four hops. The same simulation parameters as in Table 1 are used. The relay zone of the SF structure is divided into three equal partitions to be used by the NT-RSs at the first, second and third tiers.

Figure 11 depicts the maximum number of SSs that can be accommodated in each hop as well as the link data rate. The SSs are distributed; 50% served by the access zone of the MMR-BS and the remaining SSs are distributed equally between the NT-RS. As shown in the Figure 11, the capacity of relay zone of the SF structure is reduced to about 33% as compared to MF and NMF structure. In all the frame structures discussed above, the link data rate in each hop is increased with the increment of the number of admitted SSs until it reaches the maximum capacity. After that, if the number of SSs is increased, the link data rate remains constant. In addition, when the numbers of the NT-RSs increases, the slots available to each NT-RS are limited, and hence the number of SSs that can be served will be relatively small. This gives a limit to the number of hops that can be used, as well as the number of NT-RSs in each tier. As shown from the results, MF structure and NMF give typical capacity for all the scenarios discussed before. However, the capacity of SF structure is reduced to $1/(n - 1)$ of the capacity of MF and NMF structures, where n is the number of hops. The poor capacity of SF structure as the number of hops increases makes it inconvenient for more than two hops.

Link layer and TCP traffic performance

The delay of relaying the data packets across multiple hops using MF structure is increased proportionally with the number of hops, which results in very poor throughput of both link layer and TCP traffics. Therefore, the forwarding delay can be reduced by using different multi-frame configuration as that proposed in NMF structure. The effect of the forwarding delay on the link layer and the TCP performances will be discussed in the next section. To evaluate the effect of the proposed NMF structure, simulation with the same parameters as shown in Table 1 is conducted. The TCP traffic is transmitted to the MMR WiMAX network from the fixed hosts in the internet. The performance of one SS at each hop is considered to observe the performance enhancement in

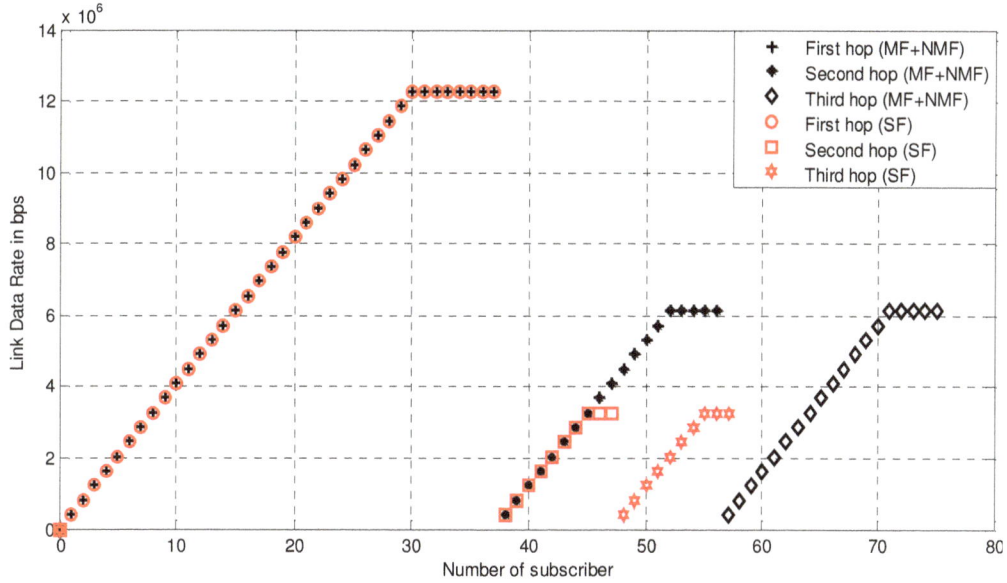

Figure 8. Access and relay zones capacity of equally distributed SSs at the relay zone.

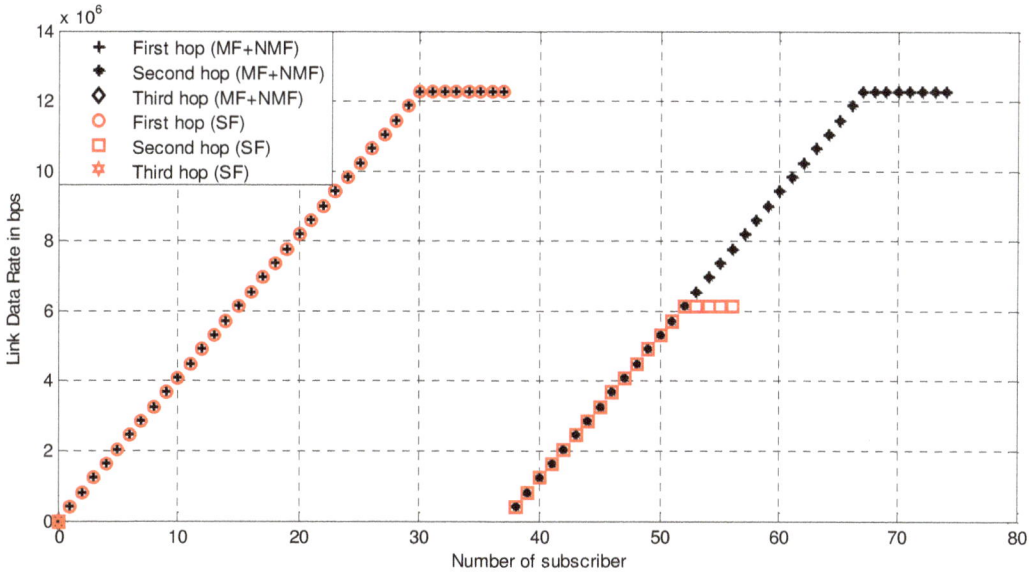

Figure 9. Access and relay zones capacity of even SSs distribution.

various hop levels. The performances of the proposed NMF structure are compared with the MF structure (Zhifeng et al., 2007). It is assumed that the NT-RS uses the next frame to relay the data received in the current frame and the link layer frame error percentage is 10%. The delays that are induced due to relaying data packets a cross n hops using MF and NMF structures are compared in Figure 12. The results in Figure 12 indicates that for MF structure, as the number of hops is increased, the link layer forwarding delay is rapidly increased with

the rate of $(2*n - 1)$, while in the proposed NMF structure, the delay is slightly increased with the rate of n, where n is the number of hops between MMR-BS and the end SSs. Figure 12 shows that the proposed NMF structure gives 33 and 40% reduction in the delay for the second and third hops, respectively. Due to shorter forwarding delay induced in the link layer from the NMF structure, as compared with the MF structure, better link layer throughput is obtained for NMF structure as shown in Figure 13. The SSs at the first hop gained almost the

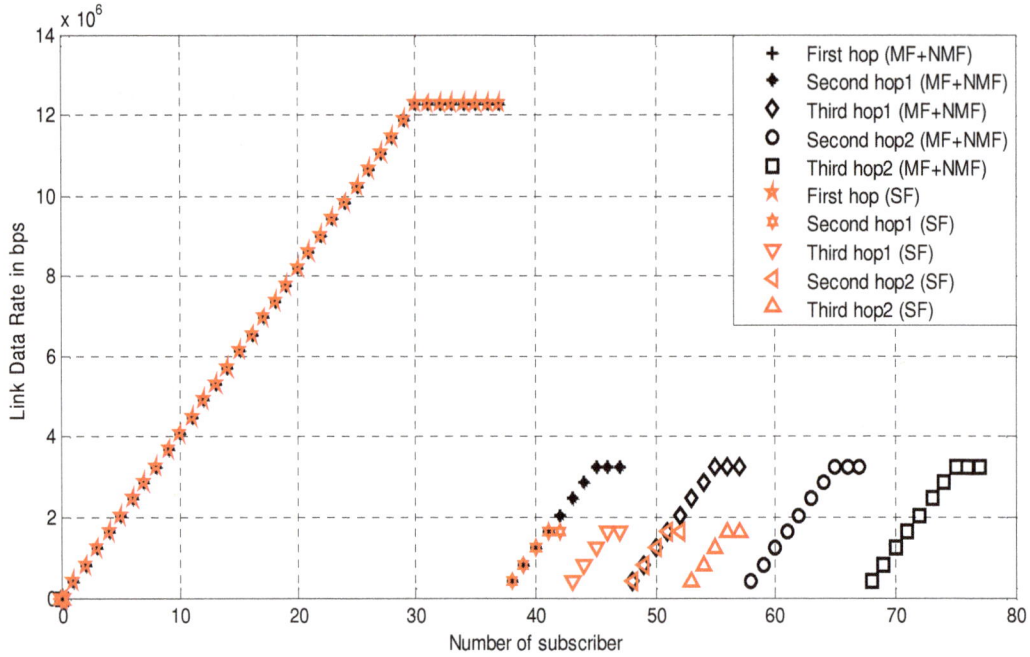

Figure 10. Access and Relay zones Capacity of three hops with two branches.

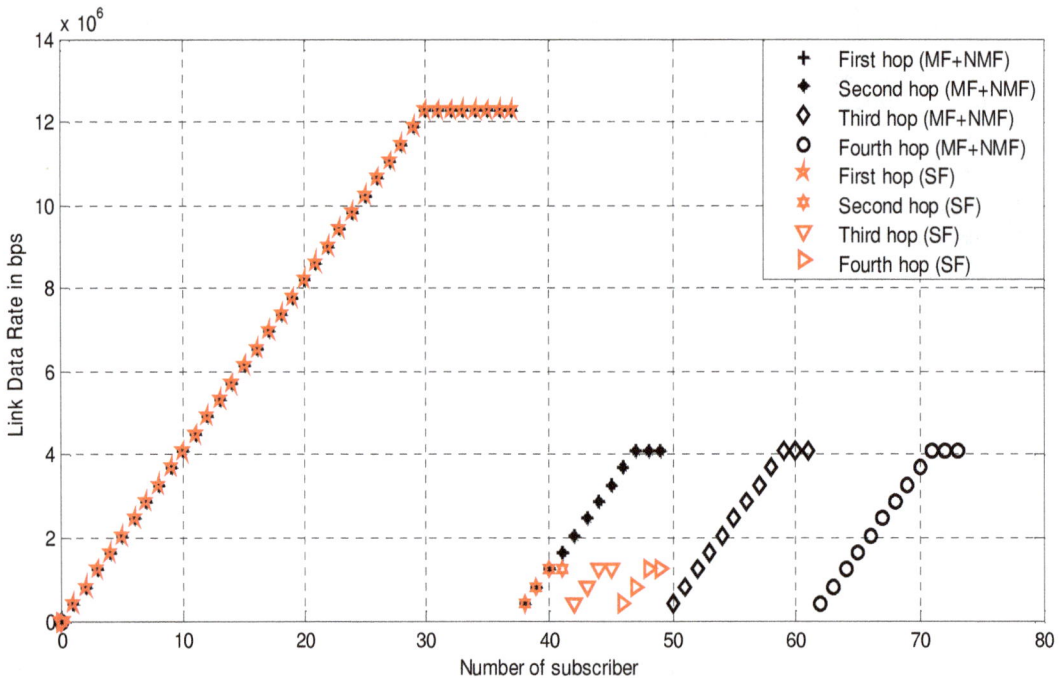

Figure 11. Access and relay zones Capacity of four hops network model.

same throughput because both frame structures experienced typical forwarding delay, while the throughput of the proposed NMF structure outperforms the MF structure by 35 and 53% at the second hop and compared with MF structure especially for the SSs at second hop and third hop. In addition, RTT is increased proportionally with the hop level for both MF and NMF structures. Due to a shorter RTT at the second and third

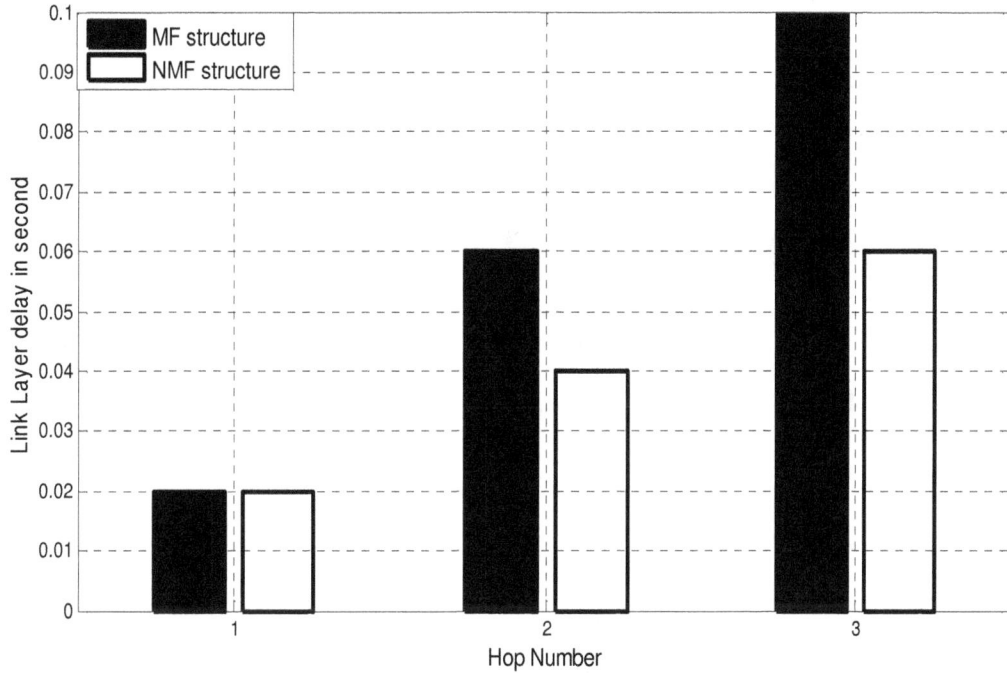

Figure 12. Link layer delays of different hops.

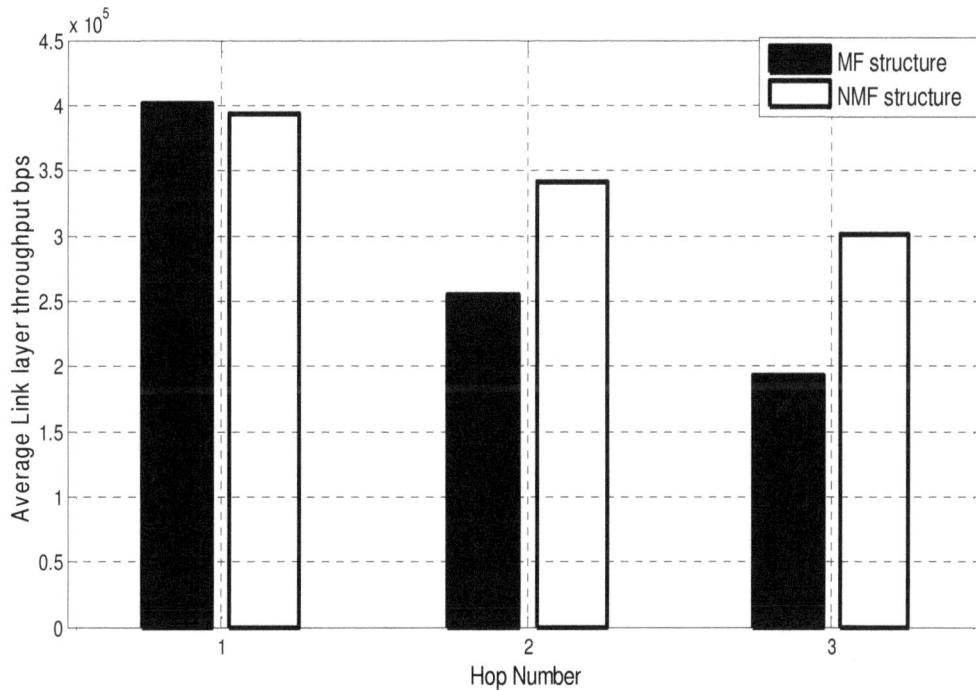

Figure 13. Link layer throughput comparisons of different hops.

hop levels, the congestion windows (CWNDs) of the SSs using the proposed NMF structure are higher than those applying MF structure (Figure 14a to c). The development of the TCP CWND of the SSs at first, second and third hops are shown in the Figure 15a to c, respectively. In Figure 15a, the SSs at the first hop that is applying the MF and NMF structure have almost the same CWND due to typical RTT. In Figure 15b, the CWND of the NMF

(a)

(b)

(c)

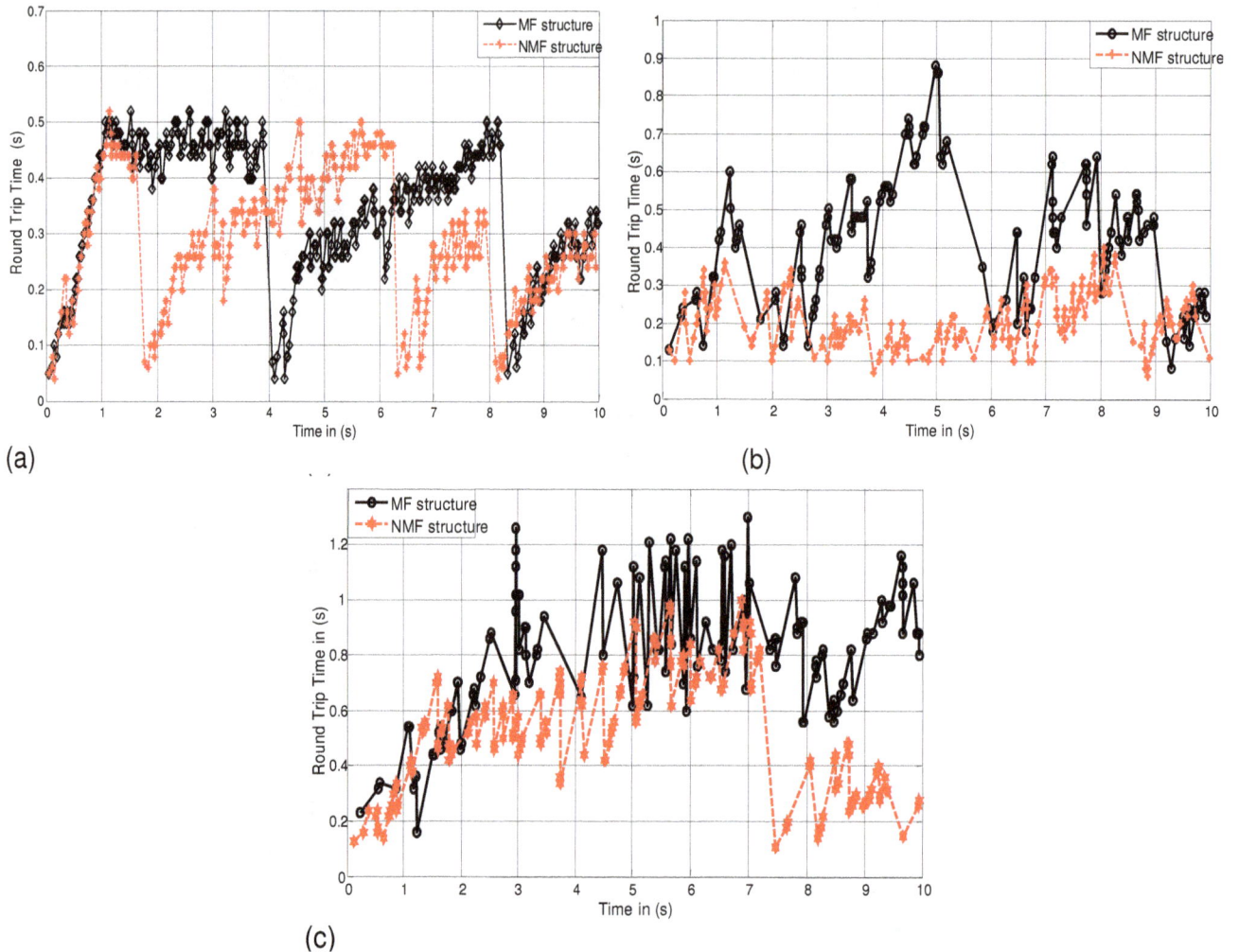

Figure 14. (a) RTT comparisons of the first hop; (b) RTT comparisons of the second hop; (c) RTT comparisons of the third hop.

structure at second hop is increased up to 20 packets before the timeout occurs, while that for the MF structure remains below 12 packets for most of the simulation time. In Figure 15c, the CWND of the NMF structure at third hopgrows from 8 up to 20 packets. However, the CWND of the MF structure lies between 6 to 10 packets. In general, as shown in Figure 15a to c, the rate of the CWND development is affected by the value of the RTT.

Therefore, the CWND of the NMF structure grow rapidly to higher values as compared with those using MF structure. In addition, the CWND of the SS at the first hop is developed faster, and it becomes slower as the number of hops increases. The reason for this is the slower TCP ACK reception rate due to longer RTT as the number of hops increases.

Higher congestion window allows the TCP host to send more TCP packets without waiting for the TCP ACKs, which results in higher TCP throughput and as shown in Figure 16. Figure 16 gives the TCP throughput comparison of the SSs at different hops. The throughputs

of the SSs at the second and third hops are increased by 30 to 40% when the proposed NMF structure is applied. However, the first hop SS gained almost the same performance for the MF and NMF structures.

Conclusion

This paper presented a new multi-frame (NMF) structure to enhance the performance of MMR WiMAX networks. The results showed that the proposed NMF structure is able to support flexible SSs distribution at different hops which are the same as that of MF structure. However, the relay zone capacity of NMF is 100% greater than SF structure when three hops architecture is considered. On the other hand, the forwarding delay is reduced with 33 and 40% for the second hop and third hop correspondingly as compared to MF structure. This allows the link layer throughput to be improved by 35% for the second hop and 53% for the third hop.

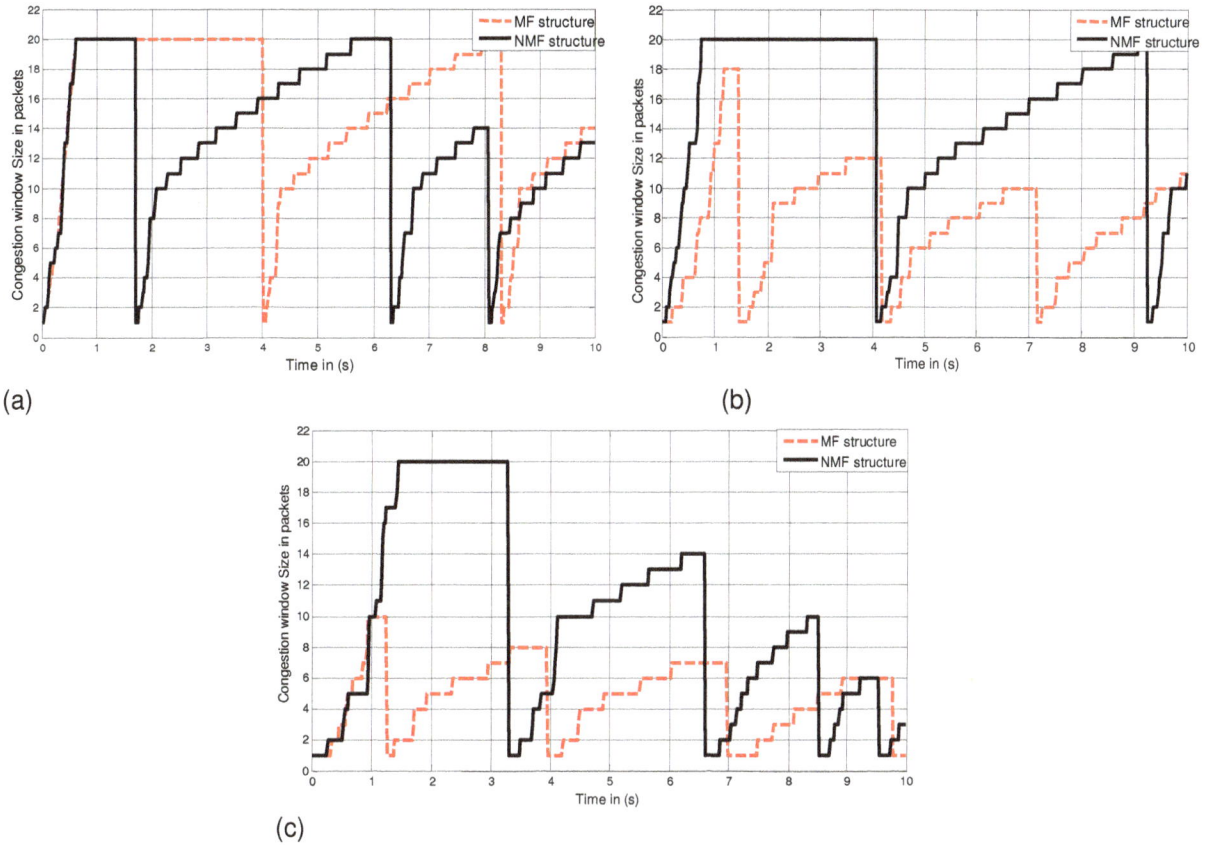

Figure 15. (a) TCP congestion window development of the first hop; (b) TCP congestion window development of second hop; (c) TCP congestion window development of third hop.

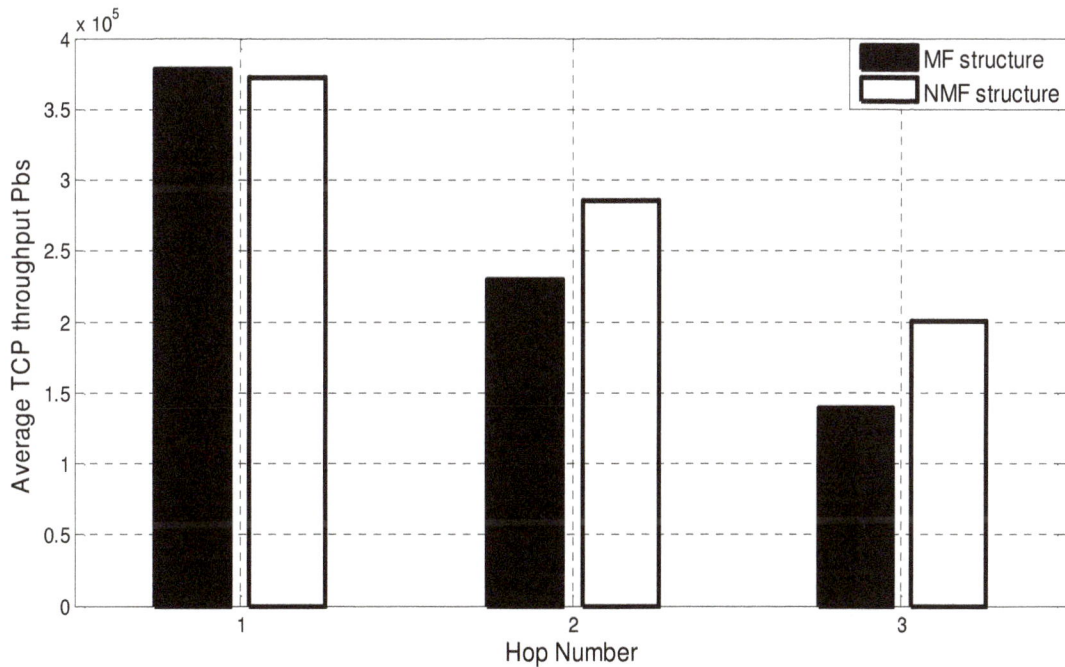

Figure 16. A comparison of average TCP throughput for different hops.

Consequently, the TCP throughput is also increased with 30 and 40% for the SSs at second hop and third hop.

ACKNOWLEDGEMENTS

This work is fully funded by RMC, UTM (Research Management Centre, UTM) and by MIMOS COE. I heartily grateful to Assoc. Prof. Dr. Sharifah Kamilah, Prof. Dr. Mazlan Abbas, Head of Wireless CommunicationCluster of MIMOS Berhad and Prof. Dr. Norsheila Fisal, Director of MIMOS for their constant supervision.

REFERENCES

Andrews JG, Arunabha G, Rias M (2007). Fundamentals of WiMAX: Understanding Broadband Wireless Networking Upper Saddle River, NJ: Prentice Hall. http://tocs.ulb.tu-darmstadt.de/187530629.pdf.

Aweya J (2003). Transmission Control Protocol. Wiley Encyclopedia of Telecommunications, John Wiley & Sons, Inc. DOI: 10.1002/0471219282.eot202.

Canton AF, Chahed T (2001). End-to-end reliability in UMTS: TCP over ARQ. IEEE Global Telecommunications Conference, 2001, GLOBECOM '01.

Ergen M (2009). WiMAX Physical Layer for Mobile Broadband, Springer US. pp. 271-307.

Genc V, Murphy S, Murphy J (2008a). Performance analysis of transparent relays in 802.16j MMR networks. 6th International Symposium on Modeling and Optimization in Mobile, Ad Hoc, and Wireless Networks and Workshops, WiOPT, 2008.

Genc V, Murphy S, Murphy J (2008b). IEEE 802.16J relay-based wireless access networks: an overview. Wireless Commun. IEEE 15(5):56-63.

Genc V, Murphy S, Murphy J (2009). Analysis of Transparent Mode IEEE 802.16j System Performance with Varying Numbers of Relays and Associated Transmit Power. IEEE Wireless Communications and Networking Conference, WCNC 2009.

Hoymann C, Klagges K, Schinnenburg M (2006). Multihop Communication in Relay Enhanced IEEE 802.16 Networks. IEEE 17th International Symposium on Personal, Indoor and Mobile Radio Communications.

Hui Z, Chenxi Z (2008). System-Level Modeling and Performance Evaluation of Multi-Hop 802.16j Systems. International Wireless Communications and Mobile Computing Conference, IWCMC '08.

Hui Z, Chenxi Z (2009). Resource Allocation in 802.16j Multi-Hop Relay Systems with the User Resource Fairness Constraint. IEEE Wireless Communications and Networking Conference, WCNC, 2009.

IEEE802.16e-2005 (2006). Local and Metropolitan Networks Part 16: Air Interface for Fixed and Mobile Broadband Wireless Access Systems, Amendment 2: Physical and Medium Access Control Layers for Combined Fixed and Mobile Operation in Licensed Bands and Corrigendum 1.

IEEE802.16j-2009 (2009). IEEE standard for Local and Metropolitan Area Networks – Part 16: Air Interface for Fixed Broadband Wireless Access Systems– Multihop Relay Specification.

Koon HT, Zhifeng T, Jinyuan Z, Anfei L (2007). Adaptive Frame structure for Mobile Multihop Relay (MMR) Networks. 6th International Conference on Information, Communications and Signal Processing.

Lei X, Fuja TE, Costello DJ (2008). An analysis of mobile relaying for coverage extension. IEEE International Symposium on Information Theory, ISIT 2008.

Mach P, Bestak R (2009). Optimization of Frame Structure for WiMAX Multi-hop Networks Wireless and Mobile Networking, Springer Boston 308:106-116.

Peters SW, Heath RW (2009). The future of WiMAX: Multihop relaying with IEEE 802.16j. Commun. Mag. IEEE 47(1):104-111.

Sayenko A, Alanen O, Martikainen H (2010). Analysis of the Non-Transparent In-Band Relays in the IEEE 802.16 Multi-Hop System. IEEE Wireless Communications and Networking Conference (WCNC).

Seung-Yeon K, Se-Jin K, Seung-wan R, Hyoung-Woo L, Choong-ho C (2008). Performance analysis of single-frame mode and multi-frame mode in IEEE 802.16j MMR system. IEEE 19th International Symposium on Personal, Indoor and Mobile Radio Communications, PIMRC 2008.

Soldani D, Dixit S (2008). Wireless relays for broadband access [radio communications series]. Commun. Mag. IEEE 46(3):58-66.

Taha A-EM, Pandeli K, Hossam H, Najah AA (2011). Evaluating frame structure design in WiMAX relay networks. Concurrency Comput. Pract. Exper. 25:608-625.

Upase B, Hunukumbure M (2008). Dimensioning and cost analysis of Multihop relay enabled WiMAX networks. FUJITSU Sci. Tech. J. 44(3):303-317.

Yang Y, Hu H, Xu J, Mao G (2009). Relay technologies for WiMax and LTE-advanced mobile systems. Commun. Mag. IEEE 47(10):100-105.

Zhifeng T, Anfei L, Koon HT, Jinyun Z (2007). Frame Structure Design for IEEE 802.16j Mobile Multihop Relay (MMR) Networks. IEEE Global Telecommunications Conference, GLOBECOM '07.

6

A modified surface wave particle motion discrimination process

Yusuf Arif Kutlu[1] and Nilgün Sayıl[2]

[1]Department of Geophysics, Faculty of Engineering and Architecture, Nevsehir University, 50300, Nevsehir-Turkey.
[2]Department of Geophysics, Faculty of Engineering, Karadeniz Technical University, 61080, Trabzon-Turkey.

The difference between polarization properties of surface waves and noise provides us with a simple way of discriminating the fundamental mode surface waves on three-component seismograms. In this process, vertical, radial and transverse component amplitudes at each frequency are weighted according to a theoretical three-dimensional particle motion pattern for a selected window length (WL) and moving interval (MI). For the epicentral distances closer than about 2200 km, the weighted functions of a formerly proposed approach are not compatible with the angular distributions of polarization properties of surface waves. It means that the former weighted functions are not perfectly able to weight surface waves amplitudes of some epicentral distances. In order to solve this problem, a modification in this study is implemented by analyzing compatibility of the weighted functions with the angular distribution of polarization properties of synthetic surface waves. The former and new processes are tested on three component synthetic seismograms and on some digital broadband records. As a result, the new filtering process (NFP) is shown to be more flexible and stable. It can be used to discriminate the fundamental mode Love and Rayleigh waves on three-component seismograms at all ranges of epicentral distances.

Key words: Former filtering process (FFP), surface wave particle motion discrimination process, new filtering process (NFP), modified surface wave particle motion discrimination process.

INTRODUCTION

Since 1960s, polarization analysis techniques for various types of data sets have been implemented to describe the signal content of time series in geophysics (Flinn, 1965; Samson, 1977; Samson and Olson, 1980; Holcomb, 1980; Kanasewich, 1981; Vidale, 1986; Plesinger et al., 1986; Jurkevics, 1988; Shieh and Herrmann, 1990; Perelberg and Hornbostel, 1994; Lilly and Park, 1995; Selby, 2001). And also several methods based on the polarization properties of different types of waves have been proposed to discriminate between signal and noise in time or frequency domain (Shimsoni and Smith, 1964; Simons, 1968; Montalbetti and Kanasewich, 1970; Blandford, 1977, 1982; Morozov and

Smithson, 1996; Patane and Ferrari, 1997; Chael, 1997; Du et al., 2000; De Meersman et al., 2006; Pinnegar, 2006; Amoroso et al., 2012).

On long period seismograms, in order to isolate the fundamental mode Love and Rayleigh waves from the effects of microseismic noise, a surface wave particle motion discrimination process was designed by Simons (1968) using the differences of polarization properties between surface waves and microseismic noise. The weighted functions of this former filtering process (FFP) are not compatible with the angular distributions of polarization properties of the fundamental mode surface waves at epicentral distances closer than about 2200 km due to time-frequency resolution loss depending mainly on the selection of window length (WL) and interval moving process. This case causes deformation of isolated fundamental mode surface waves on three-component seismograms. The aim of this study is to devise new weighted functions to solve this problem by adjusting the weighted functions according to the synthetic polarization parameters. In this respect, we made use of the compatibility between the angular distribution of polarization properties of the fundamental mode surface waves and the weighted functions of FFP. FFP and the newly proposed filtering process (NFP) are tested on three-component synthetic seismograms and also on digital broadband recordings with all ranges of epicentral distances.

SURFACE WAVE PARTICLE MOTION DISCRIMINATION PROCESS

This process is performed in the frequency domain because of the dispersive character of surface waves. Discrete Fourier transforms of vertical, radial and transverse component ground motions are calculated for a selected WL and then the interval is moved. Amplitude coefficients at each frequency are weighted according to three-dimensional theoretical particle motion pattern of Love and Rayleigh waves. The weights or adjustments do not apply to the original phase values. The weighted segments for each windowing are transformed back to the time domain, and the filtered signal is obtained after averaging over the overlapping amplitudes.

The components of the ground motion with length $N.\Delta t$ (Δt sampling rate) are derived from Discrete Fourier coefficients according to the following equations:

$$A_i(\eta f) = \left[a_i^2(\eta f) + b_i^2(\eta f)\right]^{1/2}, \qquad \eta = 0, 1, 2, ..., N/2 \tag{1a}$$

$$\Phi_i(\eta f) = \arctan\left[\frac{b_i(\eta f)}{a_i(\eta f)}\right], \qquad \eta = 0, 1, 2, ..., N/2 \tag{1b}$$

Where i = Z, R, T define vertical, radial and transverse component ground motions, respectively. $A_Z(\eta f)$, $A_R(\eta f)$, $A_T(\eta f)$ are vertical, radial and transverse amplitudes at each frequency as shown in Figure 1.

In addition, $\beta(\eta f)$ is the apparent horizontal azimuth, $\psi(\eta f)$ is the angle between major eccentricity of the particle motion ellipse and vertical component, and $\alpha(\eta f)$ is the phase difference between vertical and radial components in Figure 1, which are calculated from the following equations:

$$\beta(\eta f) = \arctan\left[\frac{A_T(\eta f)}{A_R(\eta f)}\right] \tag{2a}$$

$$\psi(\eta f) = \arctan\left[\frac{\sqrt{A_R^2(\eta f) + A_T^2(\eta f)}}{A_Z(\eta f)}\right] \tag{2b}$$

$$\alpha(\eta f) = \phi_R(\eta f) - \phi_Z(\eta f) \tag{2c}$$

The spectral amplitudes at each frequency can be weighted using the functions $\beta(\eta f)$, $\psi(\eta f)$ and $\alpha(\eta f)$ as follows (Simons, 1968):

$$A'_Z(\eta f) = A_Z(\eta f).\cos^M\left[\beta(\eta f)\right]\cos^K\left[\psi(\eta f) - \theta\right]\sin^N\left[\alpha(\eta f)\right] \tag{3a}$$

$$A'_R(\eta f) = A_R(\eta f).\cos^M\left[\beta(\eta f)\right]\cos^K\left[\psi(\eta f) - \theta\right]\sin^N\left[\alpha(\eta f)\right] \tag{3b}$$

$$A'_T(\eta f) = A_T(\eta f).\sin^M\left[\beta(\eta f)\right]\sin^K\left[\psi(\eta f)\right] \tag{3c}$$

Herein $A'_Z(\eta f)$, $A'_R(\eta f)$ and $A'_T(\eta f)$ are the weighted vertical, radial and transverse components of the ground motion, respectively. The functions $\beta(\eta f)$, $\psi(\eta f)$ and $\alpha(\eta f)$ vary in the range from 0 to 1. The exponents M, K and N are empirically obtained as 8, 8 and 4. The angle θ can be set to 0.21π (37.8°) corresponding to the theoretical horizontal/vertical displacement ratio for the fundamental mode Rayleigh waves in the Gutenberg Earth model.

The modification of FFP

Depending on WL and MI, the weighted functions of FFP may not be compatible with the angular distribution of polarization parameters calculated from synthetic seismograms at each frequency for particularly smaller epicentral distances. This case of FFP causes weak detection of surface waves on three-component seismograms. Our objective is to find new weighted functions to overcome this problem by adjusting the weighted functions to the polarization parameters (Equations 2a, b and c). In order to investigate the compatibility issue, the synthetic surface waves were computed using two sets of parameters given in Tables 1 and 2.

We subsequently calculated the synthetic polarization parameters at each frequency for the specified time sections (Sections 1, 2, and 3 represent pure Love particle motion, pure Rayleigh particle motion and mixed particle motion, respectively in Figure 2) and obtained respective angular distributions in Figure 3. Also, to derive the synthetic polarization parameters of full length Love wave particle motion, we combined Section 1 with the right sides of Section 3 around 45° for β and around 60° for ψ, respectively in Figure 3. In a similar way, for full length Rayleigh wave particle motion, we combined Section 2 with the left sides of Section 3 around 45° for β and around 60° for ψ, respectively in Figure 3.

According to the results in Figure 3, the first situation is completely associated with pure Love particle motion (Section 1 in Figures 2 and 3). There are dominating periods on the transverse component and the amplitudes on the other components are very small. The second situation is associated with pure Rayleigh particle motion (Section 2 in Figures 2 and 3). The ground motion on the horizontal plane is perfectly radial and the amplitude on the tangential component is rapidly decreasing. In both cases, it is shown that the weighted functions of FFP are perfectly compatible with the angular distribution of polarization parameters in Figure 3. This means that FFP is able to weight Love wave amplitudes in Section 1 of Figure 2 and Rayleigh wave amplitudes in Section 2 of Figure 2.

In contrast, the weighted functions of FFP are not very compatible with the angular distribution of polarization parameters around 45° (for β) and around 70° (for ψ) in Section 3 of Figure 3. This means that the weighted functions of FFP are not exactly able to weight the fundamental mode surface wave amplitudes around 45° (for β) and around 70° (for ψ) because of interactions between particle motions. In addition, for the full length Love wave (Sections 1 and 3 in Figure 3) and for the full length Rayleigh wave (Sections 2 and 3 in Figure 3), the weighted functions of FFP are not compatible with the angular distribution of polarization parameters around 45° (for β) and around 70° (for ψ). For this reason, fundamental mode Love wave and initial parts of fundamental mode Rayleigh waves are deformed after FFP processing.

Hence, the weighted functions of FFP need be rearranged for better performance around 45° (for β) and around 70° (for ψ). In the proposed filtering process (NFP), the weighted functions used in

Table 1. The first set of parameters used in computing synthetic seismograms.

Crustal model						
Layer number	Thickness (km)	P-velocity (km s^{-1})	S-velocity (km s^{-1})	Density (g/cm^3)	Quality factors	
					Q_α	Q_β
1	1.50	4.30	2.60	2.10	300	150
2	3.00	5.25	3.15	2.40	250	125
3	4.50	5.80	3.60	2.60	200	100
4	7.50	6.20	3.90	2.80	150	75
5	15.0	6.70	4.20	3.00	100	50
6	∞	8.10	4.70	3.40	50	25

Table 2. The second set of parameters used in computing synthetic seismograms.

Source parameters	
Rake	180°
Dip	90°
Strike	90°
Azimuth	180°
Focal depth	10 km
Epicentral distance	1200 km

FFP are made perfectly compatible with the angular distribution of polarization parameters. Also, the weighted functions of NFP are able to strengthen the fundamental mode surface wave amplitudes. FFP is modified by the following new equations:

$$A_Z'(\eta f) = A_Z(\eta f).WF_1^{M_1}[\beta(\eta f)].\cos^K[\psi(\eta f)-\theta].\sin^N[\alpha(\eta f)] \quad (4a)$$

$$A_R'(\eta f) = A_R(\eta f).WF_1^{M_1}[\beta(\eta f)].\cos^K[\psi(\eta f)-\theta].\sin^N[\alpha(\eta f)] \quad (4b)$$

$$A_T'(\eta f) = A_T(\eta f).WF_2^{M_2}[\beta(\eta f)].\sin^K[\psi(\eta f)+\theta/2] \quad (4c)$$

Herein

$$WF_1\ [\beta(\eta f)] = \begin{cases} 0.95+(+0.05.\cos^4(2.\beta(\eta f))) & \beta(\eta f)\le \pi/4 \\ 0.95+(-0.95.\cos^4(2.\beta(\eta f))) & \beta(\eta f) > \pi/4 \end{cases}$$

$$WF_2\ [\beta(\eta f)] = \begin{cases} 0.95+(-0.95.\cos^4(2.\beta(\eta f))) & \beta(\eta f)\le \pi/4 \\ 0.95+(+0.05.\cos^4(2.\beta(\eta f))) & \beta(\eta f) > \pi/4 \end{cases}$$

The weighted functions of NFP (WF_1 [β(ηf)] and WF_2 [β(ηf)]) were entirely derived from orthogonal functions of sine and cosine by trial and error to be compatible with the angular distribution of synthetic polarization parameters in Figure 4. The coefficients M_1 = 14 and M_2 = 14 are determined empirically. Also, the weighted function sin[(ψ(ηf)] in FFP is modified as sin[(ψ(ηf)+θ/2)] to be compatible with the polarization parameters. In a similar way, the coefficient K = 8 and horizontal/vertical displacement ratio θ/2 = 18.9 for the function sin[(ψ(ηf)+θ/2)] are determined empirically. The weighted functions sin[α(ηf)] and cos[ψ(ηf)-θ] in FFP are kept in NFP. The weighted functions of FFP and NFP are given in Figure 5.

COMPARISON OF FORMER AND NEW FILTERING PROCESSES

The filtering process is performed in several steps. All amplitudes in each component are normalized. For a selected WL and MI, Fast Fourier transforms of seismograms are calculated. In frequency domain, amplitude coefficients in each frequency are weighted. The weighted amplitudes are then transformed back to time domain. From arithmetical averages of overlapping amplitudes, filtered signal is obtained. To alleviate the windowing effect, Butterworth low-pass filtering and rounding are applied.

In this study, 10% cos tapering is applied to the edges of the WL. Addition of the lengths of 10% cos tapering to the maximum period (MP) of the seismogram gives the WL. Depending on WL, MI is determined after trial and error. This procedure can be formulated as follows:

WL - 2(10% cos tapering of WL) = MP

WL - 0.2WL = MP

WL = MP/0.8 (5a)

Ratio = WL/MI

MI = WL/Ratio (5b)

The ratio of WL/MI was given by Simons (1968) as 16. But, synthetic tests for all ranges of epicentral distances show that ratio 1.25 fits better than 16 to provide satisfactory time-frequency resolution and do not deform wave forms as shown in Figure 6.

The synthetic seismograms in Figure 2 were again utilized in Figure 7 to test FFP against NFP. Both filtering processes are not good at weighting the high frequency content of the wave forms due to the effect of windowing, low-pass filtering and rounding errors. FFP appears to fail on the initial part of the fundamental mode Rayleigh wave and on the main part of the fundamental mode Love wave. In general, NFP is better than FFP to recover the fundamental mode signals.

Table 3. Focal parameters of events used in this study.

S/N	Date (mo/d/yr)	Time (h:min:s)	Latitude (N°)	Longitude (E°)	Depth (km)	Magnitude (Ms)
1	12/26/2003	01:56:56	29.01	58.27	33.0	6.7
2	01/23/2005	22:36:08	35.95	29.71	31.9	5.4

Soruce: Incorporated Research Institutions For Seismology (IRIS).

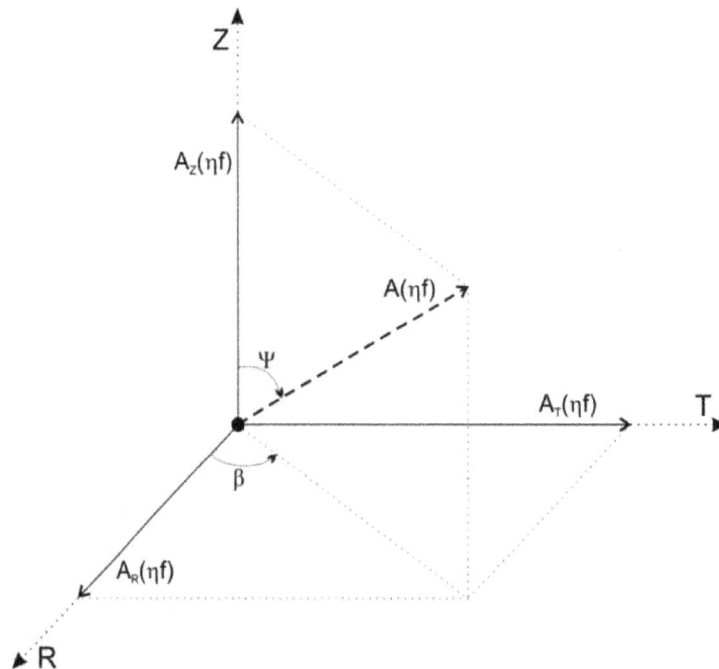

Figure 1. Components of the ground motion.

Both filtering processes were also applied on two three-component broadband digital seismograms recorded at TBZ station of Karadeniz Technical University, Department of Geophysics, Trabzon, Turkey given in Table 3.

The original traces (unfiltered), filtered traces (using FFP) and filtered traces (using NFP) for Event 1 in Table 3 with an epicentral distance of 2137 km are shown in Figure 8. Both filtering processes are successful to weight all parts of the fundamental mode Rayleigh waves in each component (from arrow 1 to the end). NFP has a noticeable achievement here (from arrow 2 to the end), whereas FFP fails in the initial part of the fundamental mode Love wave (Arrow 3).

The original traces (unfiltered), filtered traces (using FFP) and filtered traces (using NFP) for Event 2 in Table 3 with an epicentral distance of 1178 km are shown in Figure 9. NFP is more successful than FFP in terms of weighting the initial part of the fundamental mode Love and Rayleigh waves in each component (from arrow 1 to the end for Rayleigh waves, arrow 3 for Love wave).

Herein, it is clear that when the epicentral distance gets smaller, FFP fails on the initial part of Love and Rayleigh waves in each component. NFP is able to weight all parts of fundamental Love and Rayleigh waves to achieve significant improvement over FFP (Figures 7, 8, and 9).

DISCUSSION

All amplitudes are normalized to avoid complications from amplitude interactions between components. When normalization is not employed, both processes may result abnormal amplitudes because both FFP and NFP are extremely sensitive to amplitude ratios between components. On the other hand, normalization may adversely affect the result of the filtering process. To avoid amplitude interactions between components, relatively high quality data sets should be utilized. In this study, a fixed WL and MI are used for the whole traces. Because surface waves are dispersive, changing WL and MI can be devised instead of fixing them.

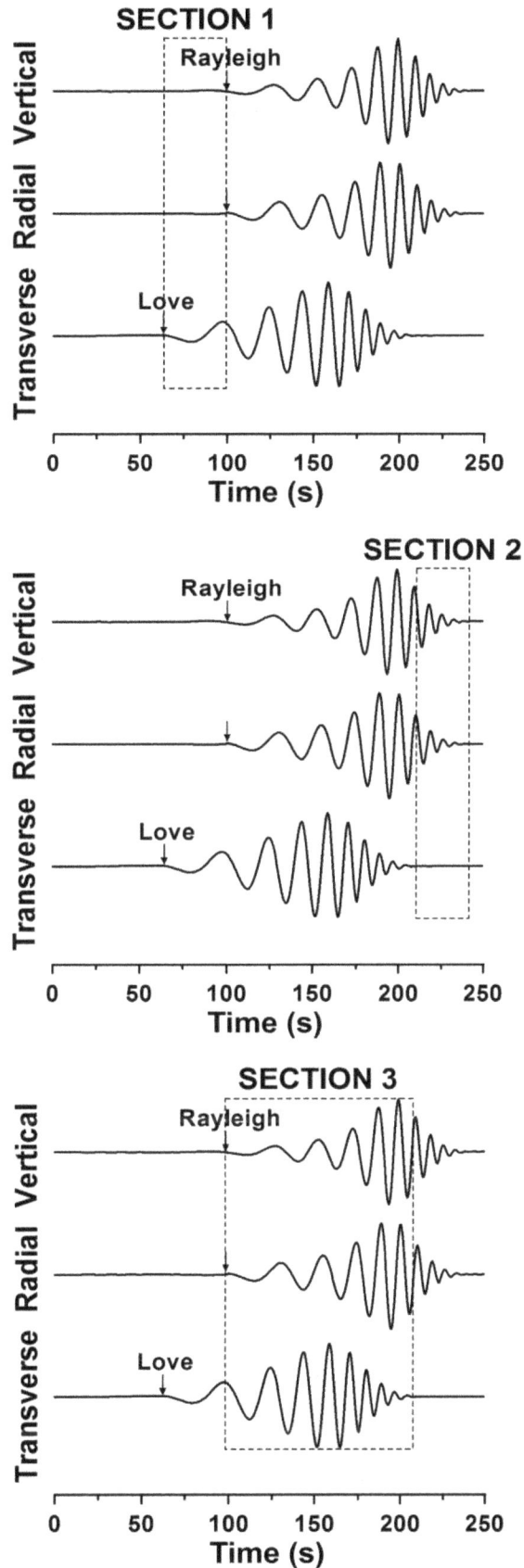

Figure 2. Time sections used to analyze synthetic polarization parameters.

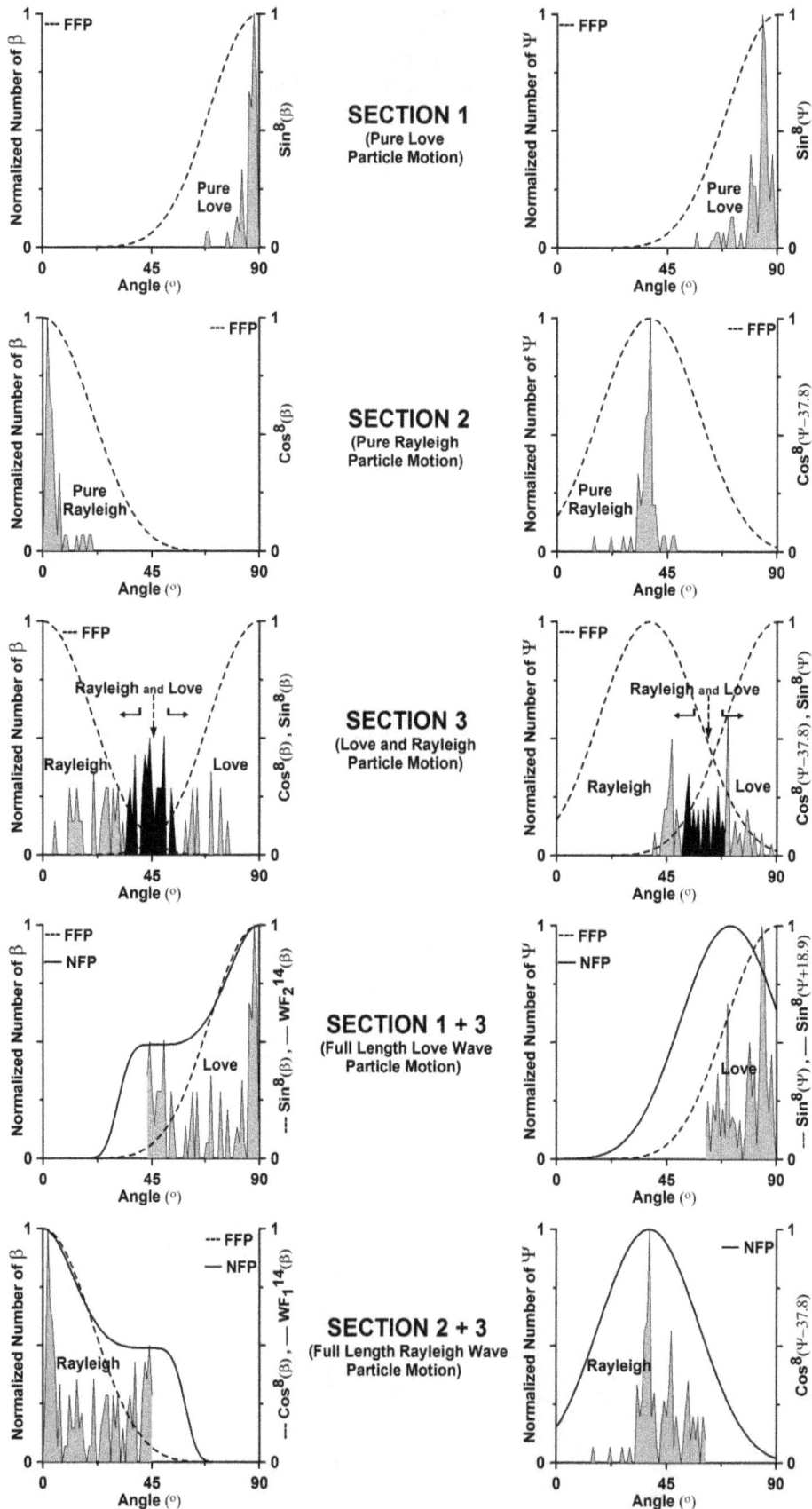

Figure 3. Investigation of compatibility between angular distribution of polarization properties and weighted functions of FFP and NFP in each section.

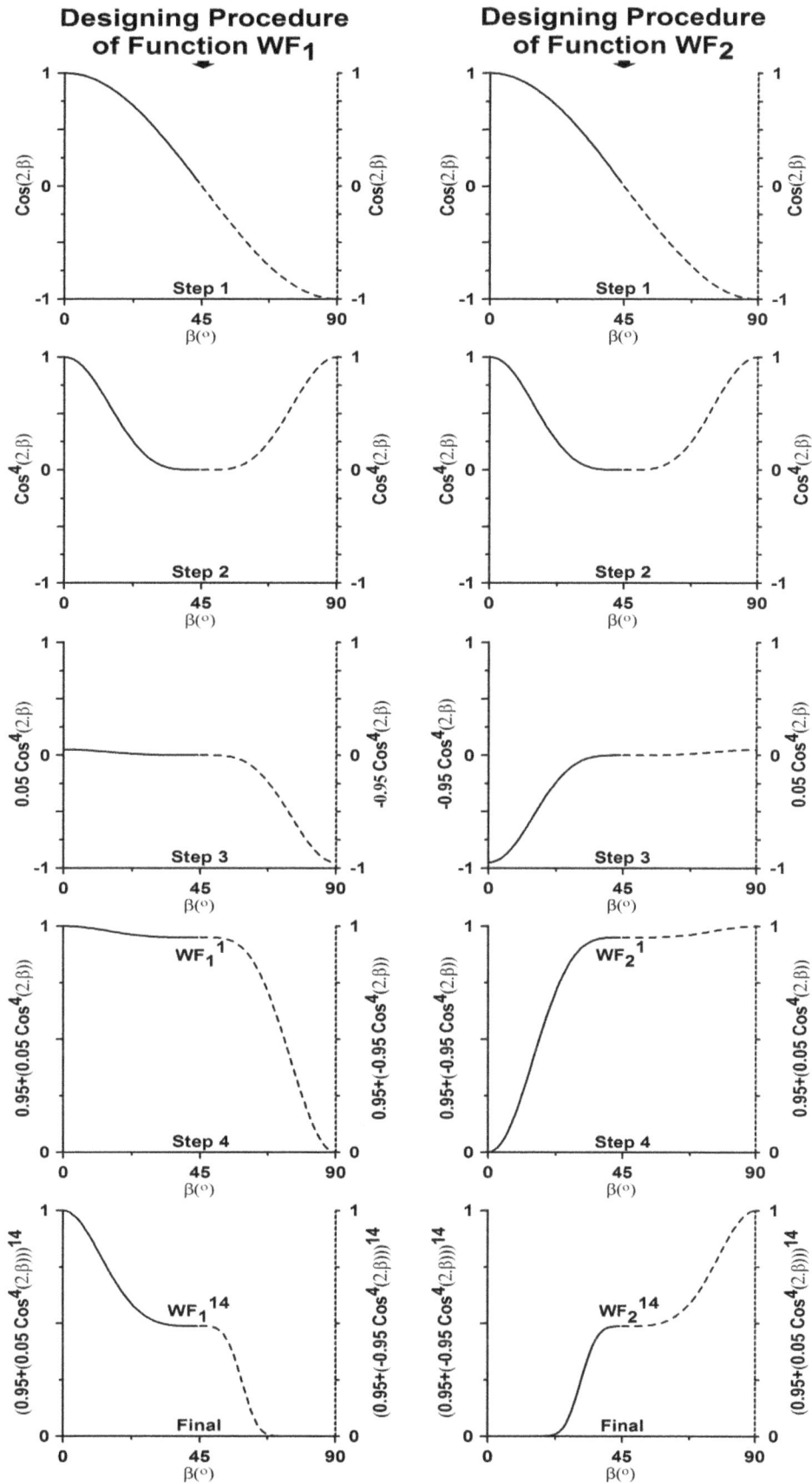

Figure 4. Designing procedures of functions WF_1 and WF_2 step by step.

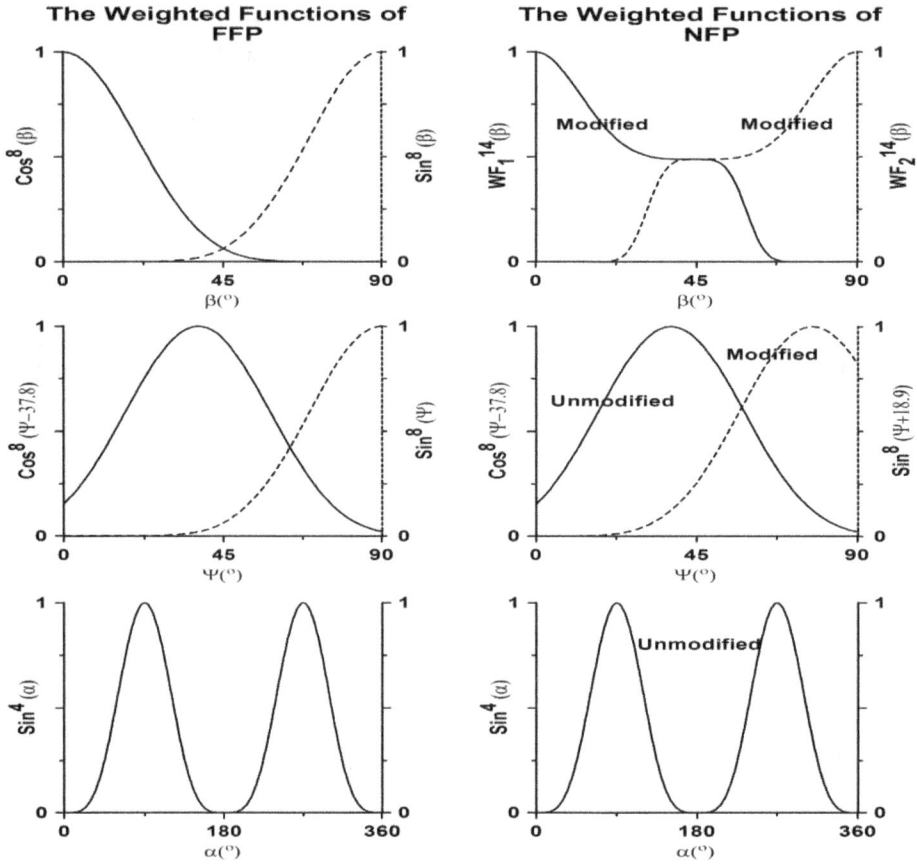

Figure 5. Weighted functions of FFP and NFP.

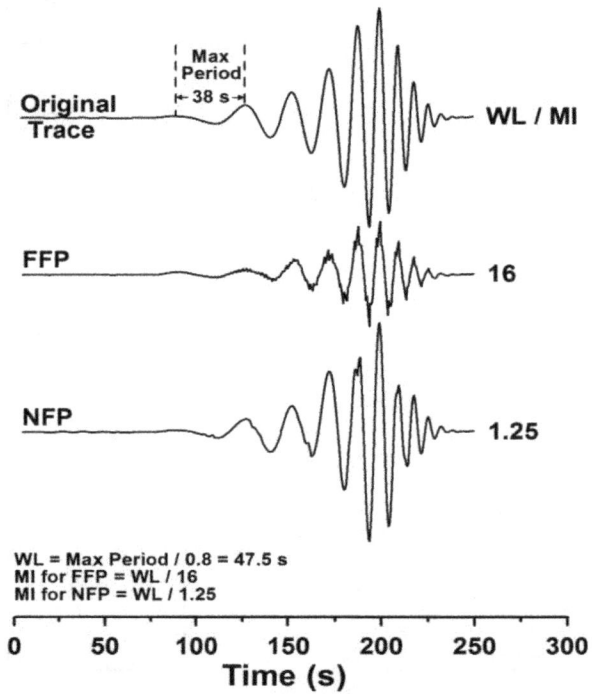

WL = Max Period / 0.8 = 47.5 s
MI for FFP = WL / 16
MI for NFP = WL / 1.25

Figure 6. The effect of windowing process on seismograms.

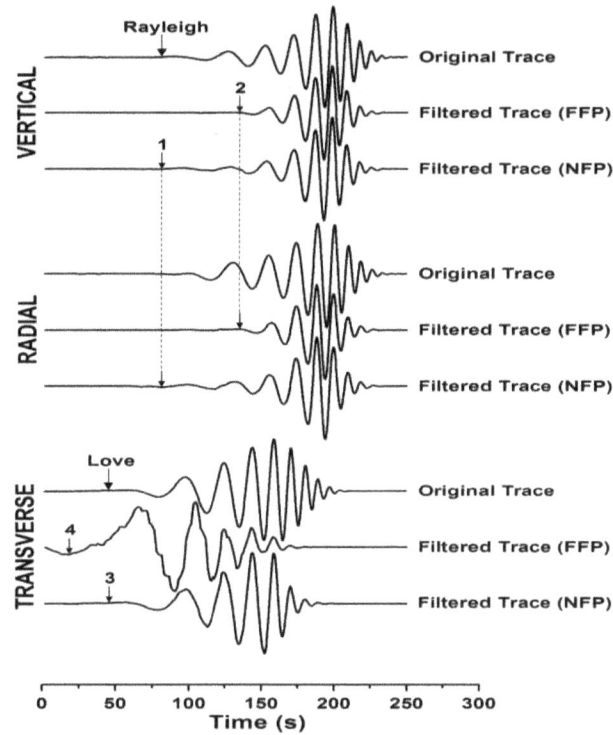

Figure 7. Original and filtered traces of synthetic seismograms.

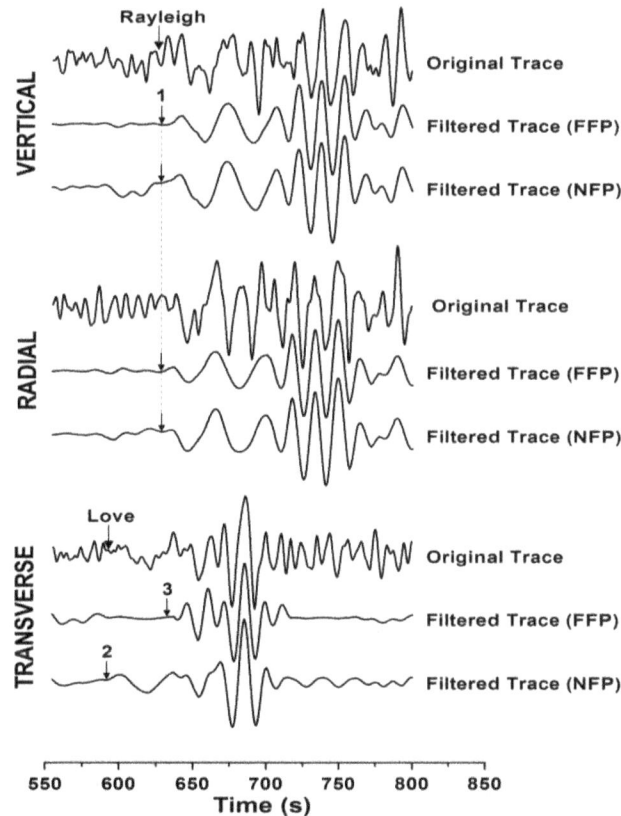

Figure 8. Original and filtered traces of Event 1 in Table 3.

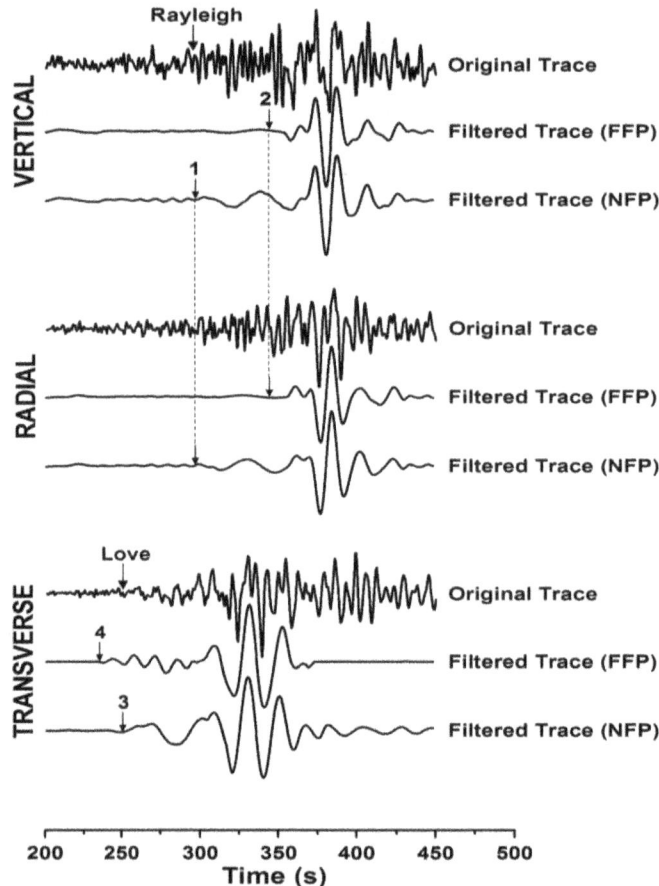

Figure 9. Filtered and unfiltered traces of Event 2 in Table 3.

FFP was modified to provide a better fit for the angular distribution of the synthetic polarization parameters derived from different time intervals. NFP given in Equations 4a, b and c is good for all epicentral distance ranges including epicentral distances smaller than ~20°. As a result, NFP is found better than FFP about weighting the fundamental mode Love and Rayleigh waves, and NFP can be safely used at all ranges of epicentral distances.

ACKNOWLEDGEMENT

We are very grateful to the Department of Geophysics, Karadeniz Technical University for the two broadband digital records used in this study.

REFERENCES

Amoroso O, Maercklin N, Zollo A (2012). S-wave identification by polarization filtering and waveform coherence analyses. Bull. Seism. Soc. Am. 102(2):854-861.

Blandford RR (1977). Discrimination between earthquakes and underground explosions. Ann. Rev. Earth. Planet. Sci. 5:111-122.

Blandford RR (1982). Seismic event discrimination. Bull. Seism. Soc. Am. 72(6):69-87.

Chael EP (1997). An automated Rayleigh-wave detection algorithm. Bull. Seism. Soc. Am. 87(1):157-163.

De Meersman K, van der Baan A, Kendall JM (2006). Signal extraction and automated polarization analysis of multicomponent array data. Bull. Seism. Soc. Am. 96(6):2415-2430.

Du Z, Foulger GR, Weijian M (2000). Noise reduction for broad-band three-component seismograms using data-adaptive polarization filters. Geophys. J. Int. 141:820-828.

Flinn EA (1965). Signal analysis using rectilinearity and direction of particle motion. Proc. IEEE. 53:1874-1876.

Holcomb LG (1980). Microseisms: A twenty-six-second spectral line in long-period earth motion. Bull. Seism. Soc. Am. 70(4):1055-1070.

Jurkevics A (1988). Polarization analysis of 3-component array data. Bull. Seism. Soc. Am. 78(5):1725-1743.

Kanasewich ER (1981). Time sequence analysis in geophysics, The University of Alberta Press, Edmonton.

Lilly JM, Park J (1995). Multiwavelet spectral and polarization analysis of seismic records. Geophys. J. Int. 122:1001-1021.

Montalbetti JF, Kanasewich ER (1970). Enhancement of teleseismic body phases with a polarization filter. Geophys. J. R. Astr. Soc. 21:119-129.

Morozov IB, Smithson SB (1996). Instantaneous polarization attributes and directional filtering. Geophysics 61:872-881.

Patane D, Ferrari F (1997). Seismpol_A visual-basic computer program for interactive and automatic earthquake waveform analysis. Comput. Geosci. 23(9):1005-1012.

Perelberg AI, Hornbostel SC (1994). Applications of seismic polarization analysis. Geophysics 59:119-130.

Pinnegar CR (2006). Polarization analysis and polarization filtering of three-component signals with the time-frequency S transform. Geophys. J. Int. 165(2):596-606.

Plesinger A, Hellweg M, Seidl D (1986). Interactive high-resolution polarization analysis of broad-band seismograms. 59:129-139.

Samson JC (1977). Matrix and Stokes vector representations of detectors for polarized waveforms: theory, with some applications to teleseismic waves. Geophys. J. Int. 51(3):583-603.

Samson JC, Olson JV (1980). Some comments on the descriptions of the polarization states of waves. Geophys. J. R. Astr. Soc. 61:115-129.

Selby ND (2001). Association of Rayleigh waves using backazimuth measurements: Application to test ban verification. Bull. Seism. Soc. Am. 91(3):580-593.

Shieh CF, Herrmann RB (1990). Ground roll: Rejection using polarization filters. Geophysics 55(9):1216-1222.

Shimsoni M, Smith SW (1964). Seismic signal enhancement three-component detectors. Geophysics 24:664-671.

Simons RS (1968). A surface wave particle motion discrimination process. Bull. Seism. Soc. Am. 58:629-637.

Vidale JE (1986). Complex polarization analysis of particle motion. Bull. seism. Soc. Am. 76(5):1393-1405.

7

Improvement of electromagnetic wave (EMW) shielding through inclusion of electrolytic manganese dioxide in cement and tile-based composites with application for indoor wireless communication systems

Johann Christiaan Pretorius and B. T. Maharaj

Department of Electrical Electronic and Computer Engineering, University of Pretoria, Pretoria, South Africa.

The electromagnetic wave absorption characteristics of composite cement-based building material have attracted much interest in recent times. Researchers have mainly focused on the 2 GHz to 12 GHz frequency range while the authors have investigated the mobile and WiFi frequency bands. The determination of characteristics such as reflection loss, absorption, attenuation and shielding effectiveness are crucial in the evaluation and development of these materials for the building industry. The authors have determined the characteristics by measuring the S_{11} and S_{21} parameters of the composite cement-based material in the Global System for Mobile (GSM) and WiFi frequency bands. MnZn-ferrite and electrolytic manganese dioxide in powder form is used as absorber material to increase the permeability of the cement-based material to improve absorption and attenuation capabilities to create a cost-effective practical electromagnetic wave absorber. The results achieved show the uniqueness of electrolytic manganese dioxide as filler in composite cement based material for electromagnetic wave shielding effectiveness improvement. Shielding of 8 dB in the GSM850 and GSM900 frequency bands and 5 dB in the GSM1800 and GSM1900 frequency bands were measured.

Key words: Electromagnetic wave absorption measurement, attenuation, magnetic composites in building material, reflection loss, shielding effectiveness, transmission loss.

INTRODUCTION

The electromagnetic wave (EMW) absorption capability of composite cement-based material has attracted much attention from researchers in recent times. Researchers have adopted various methods to determine the absorption characteristics of such materials.

Absorption is an indication of how much of the EMW energy enters the material. Attenuation indicates how much of the absorbed energy is converted into other forms of energy by the material. The most common method used to determine absorption, is to measure the reflection loss (S11) at the material by placing a conductive back plate behind the device under test (DUT). This is done practically with a vector network analyzer (VNA) and two horn antennae in an anechoic chamber (Guan et al., 2006, 2007). This experimental setup actually measures the total EMW energy attenuated by the material.

Shielding effectiveness (SE) is a combination of reflection loss, attenuation and multiple internal reflections and attenuations (Schulz et al., 1988). SE can be measured by placing the DUT between two horn antennae and measuring the transmission loss (S21) through the material (Park et al., 2006). This method gives an indication of the total shielding effect, but the actual attenuation of the EMW in the material is not available.

In literature and applications, different composite cement-based materials are used to absorb and attenuate EMW energy. Expanded polystyrene is added to cement-based EMW absorber to improve its absorbing properties (Guan et al., 2005, 2007). This method shows that the attenuation is mainly due to multiple internal reflections and scattering. SE of 6 to 16 dB in the 8 to 18 GHz frequency range is measured. Carbon fillers (Guan et al., 2006) in the form of graphite and carbon black are used for shielding and absorption respectively. SE of 5 to 15 dB is achieved in the 2 to 8 GHz frequency range with carbon black.

Ferrite and stainless steel powder can be used as fillers in wood building material for EMW absorption in indoor applications (Oka et al., 2009). Absorption of above 10 dB is measured at frequencies above 2 GHz. However, the cost of these fillings is high and processing is complicated. Research has mainly been done in the 2 GHz to 12 GHz (X-band) frequency range.

In this research, the authors report on research findings to determine the exact absorption and attenuation capabilities of cement-based building material in mobile communications and indoor wireless communication systems. Electrolytic manganese dioxide (EMD) and MnZn-ferrite (CHY13K) are used as novel magnetic powder fillers in cement-based building material, and the enhanced performance of EMD when compared to CHY13K is shown. The measurement techniques in this research determine the absorption, attenuation and SE capabilities of the material. The absorption is determined by using the gating function of a VNA measuring in the time domain. The results of the combination of the two measurement methods enable one to determine the reflection, absorption and attenuation capabilities.

MATERIALS AND METHODS

Preparation of samples

The composite cement-based samples were prepared in square tile format with 300 mm side lengths and a thickness of 20 mm. The mixture for plaster cement is typically one part cement and six parts sifted river sand. The newly developed mixture used is one part cement, x parts river sand and y parts filler powder with x + y = 6. Table 1 shows the detail of the prepared samples. The ferrimagnetic powder CHY13K is $MnZnFe_2O_4$ with an initial relative complex permeability of 13000 H/m. The magnetic powder EMD is MnO_4 produced for alkaline and lithium cylindrical and flat battery cells. These two magnetic powders were chosen as magnetic fillers because of their high complex permeability.

The prepared samples were naturally cured for at least 12 weeks to reduce the moisture content, as moisture increases the conductivity of the material, hence increasing reflection loss and SE (Kharkovsky et al., 2002).

Measurement theory

When a propagating EMW with unit amplitude is perpendicular incident (normal incidence) on an absorbing material with a conductive back plate, the impedance η normal to the material is given by

$$\eta = \eta_1 \frac{\eta_2 + j\eta_1 \tanh \gamma d}{\eta_1 + j\eta_2 \tanh \gamma d} \tag{1}$$

with η_1 the complex intrinsic impedance of the material, η_2 the impedance of material at distance d in the material, and γ the propagation constant (Schulz et al., 1988; Nie et al., 2005; Kim et al., 2004).

The complex intrinsic impedance of the material is

$$\eta_1 = \sqrt{\frac{j\omega\mu}{\sigma + j\omega\varepsilon}} \tag{2}$$

where $\varepsilon = \varepsilon_0 \varepsilon_r$ is the complex permittivity, $\mu = \mu_0 \mu_r$ is the complex permeability of the material and $\omega = 2\pi f$ is the frequency in radians per second.

The free space permittivity is $\varepsilon_0 = 8.854 \times 10^{-12}$ F/m, $\mu_0 = 4\pi \times 10^{-7}$ H/m is the free space permeability while ε_r and μ_r are the relative complex permittivity and permeability, respectively of the material (Baoyi et al., 2012). The propagation constant γ in the material can be expressed as:

$$\gamma = j\omega\sqrt{\mu\varepsilon}\sqrt{1 \quad j\frac{\upsilon}{\omega\varepsilon}} = \alpha \quad j\beta \tag{3}$$

where α denotes the attenuation constant in nepers per meter and β the phase constant in radians per second of the material. The impedances and propagation constant in Equations (1) to (3) are dependent on the frequency.

The transmission loss T, resulting from both the front and back interfaces and with the successive internal reflections neglected because of very high internal attenuation *(A > 15 dB)*, is the product of the transmission coefficients across the two interfaces. Hence:

$$T = \frac{4\eta_1 \eta_0}{(\eta_1 + \eta_0)^2}. \tag{4}$$

The wave impedance is $\eta_0 = 12($ for free space. With the internal attenuation less than 15 dB, successive internal reflections and attenuations cannot be neglected and the net transmission loss T becomes:

$$T = \left[\left(\frac{4\eta_1 \eta_0}{(\eta_1 + \eta_0)^2} \right) \Big/ \left(1 - \frac{(\eta_0 - \eta_1)^2}{(\eta_0 + \eta_1)^2} e^{-2\gamma D} \right) \right] e^{-\gamma D} \tag{5}$$

By definition, the total SE is (Mishra et al., 2013; Hutagalung et al., 2012):

$$
\begin{aligned}
SE &= A + R + B \\
&= -20 \log |T| \\
&= 20 \log |e^{\gamma D}| \\
&\quad -20 \log \left| \frac{4\eta_1 \eta_0}{(\eta_1 + \eta_0)^2} \right| \\
&\quad +20 \log \left| 1 - \frac{(\eta_0 - \eta_1)^2}{(\eta_0 + \eta_1)^2} e^{-2\gamma D} \right|
\end{aligned}
\tag{6}
$$

Equation (6) is the complete formula for SE of a single shield with A

Table 1. Manufactured absorber test samples.

0EMD	1 × cement, 6 × sand
1EMD	1 × cement, 5 × sand, 1 × EMD
2EMD	1 × cement, 4 × sand, 2 × EMD
3EMD	1 × cement, 3 × sand, 3 × EMD
4EMD	1 × cement, 2 × sand, 4 × EMD
4.5EMD	1 × cement, 1.5 × sand, 4.5 × EMD
5EMD	1 × cement, 1 × sand, 5 × EMD
3CHY13K	1 × cement, 3 × sand, 3 × CHY13K
4CHY13K	1 × cement, 2 × sand, 4 × CHY13K
5CHY13K	1 × cement, 1 × sand, 5 × CHY13K

the attenuation, R the reflection loss and B the correction term due to successive reflections and attenuations internal to the material.

Measuring the transmission coefficient S_{21} between the two antennae with the DUT in the propagation path of the EMW, will reveal the SE of the material. The difference between the measured S_{11} as the reflection coefficient Γ, and S_{21} the SE, results in the attenuation. The absorption capability of the sample is determined from $(1-\Gamma)$.

Measurement setup

The measurement setup used to measure total SE is shown in Figure 1. The setup consisted of a HP37269D VNA, and two ultra broadband open horn antennae (Bantsis et al., 2012). The transmit antenna (TX) used is a 500 MHz to 4 GHz double-ridged horn antenna, part number 470523 from Saab-Grintek Technologies. The receive antenna (RX) used is a model 3115 double-ridged waveguide horn antenna manufactured by ETS-Lindgren with a frequency range of 750 MHz to 18 GHz. The S_{21} parameter was measured with the VNA in an anechoic chamber to determine the transmission coefficient of the test samples.

With the setup for S_{11} measurements, only the TX antenna and the VNA was used, while S_{11} parameters of a conductive plate with the same dimensions as the absorber material was measured and used as a reference. The absorption and reflection loss of the material is then determined without a conducting back plate by measuring S_{11} and S_{21} of the absorbing material (Sagnard and El Zein, 2005).

The measurements were done in an anechoic chamber lined with Eccosorb AN absorption material operational for a 600 MHz - 40 GHz frequency range. The chamber is designed to reflect less than -20 dB of normal incident energy. The test setups were calibrated to ensure minimum VSWR and path loss. Traditional time domain reflectometry (TDR) is achieved by launching an impulse or step into the test device and observing the response in time with a wideband receiver. The transform used by a VNA resembles TDR; however, the VNA makes swept frequency response measurements and mathematically transforms the data into a TDR-like display. This is useful to determine impedance mismatches which are the main cause of reflection loss. The time domain gating function on the VNA allows selective removal of reflection or transmission responses. The data can then be converted to the frequency domain for analysis.

Measurements were done in the following frequency bands: GSM850: 824–894 MHz; GSM900: 890-960 MHz; GSM1800: 1.71-1.88 GHz; GSM1900: 1.86-1.99 GHz; WiFi: 2.4-2.484 GHz (Figure 2).

RESULTS AND DISCUSSION

The various samples in Table 1 were investigated and the 5EMD was analyzed in this research due to its superior performance. Figures 2 to 4 show the measured results. The total loss indicates the transmission loss determined by measuring S_{21}. The reflection loss and attenuation were determined from the S_{11} measurement. The transmission loss was calculated using Equation (6) and compared with measured results from 800 MHz to 2.8 GHz in Figure 5. The multiple internal reflections and attenuations were included in the measured attenuation.

The 5EMD sample had low reflection loss and very good absorption in the GSM850 and GSM900 mobile communication bands as indicated in Figure 2. For the GSM1800 and GSM1900 frequency bands the attenuation is lower at the higher frequency bands as shown in Figure 3. For the WiFi frequency band as shown in Figure 4, the attenuation was also lower when compared to Figure 2. Hence, this resulted in a lower SE.

Figure 5 shows good correlation between the measured results and calculated data. In Figure 6 the superior SE of EMD over MnZn-ferrite is clearly visible. Figure 6 also shows the increase of SE with the addition of EMD to the cement based composite.

Measuring the S11 parameter to determine the absorption and reflection capabilities and S_{21} parameter to determine the transmission loss of the material enabled the authors to determine the attenuation of the EMW in the material. The major advantage in using such a measurement technique is that one is able to distinguish between reflection loss, attenuation, absorption and transmission loss and this enables the researcher to manipulate these characteristics to improve the performance by adjusting the composite building material for EMW absorption and attenuation. The measured results in Figure 2 show that EMD can be used effectively in cement-based material to obtain a typical SE of 8dB. With the measured results shown in Figure 6, the authors also demonstrate in this research that the SE of plaster cement can be improved by adding EMD powder to its matrix to create a composite cement-based building

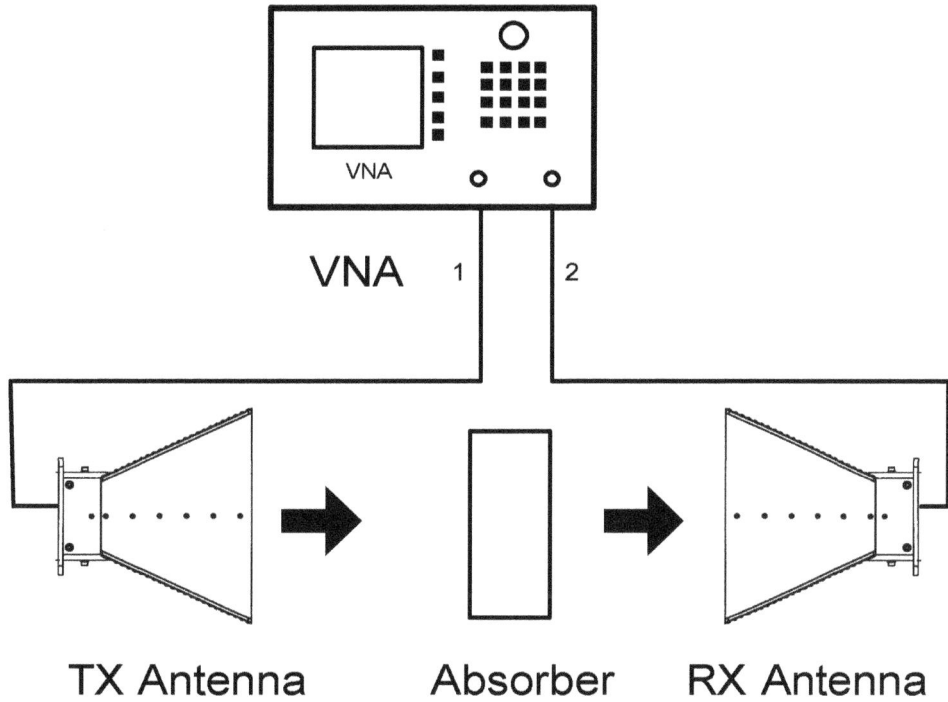

Figure 1. Measurement setup for determination of transmission loss (S_{21}).

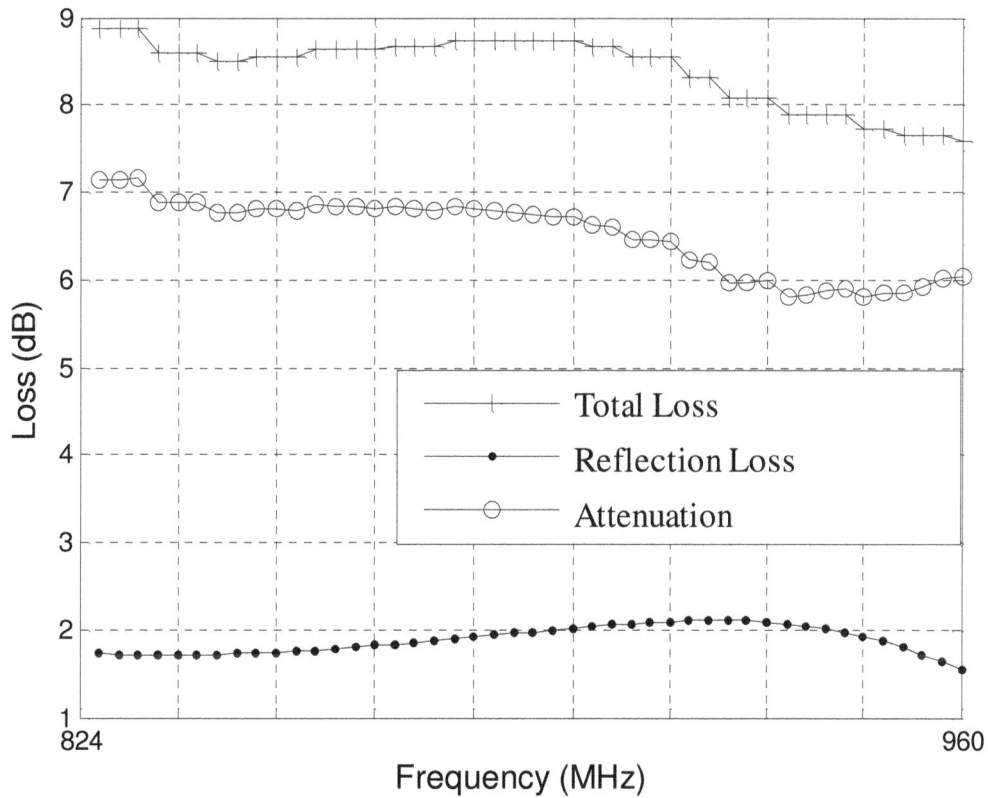

Figure 2. Measured results in the GSM850 and GSM900 frequency bands of the total loss (SE), attenuation and reflection loss.

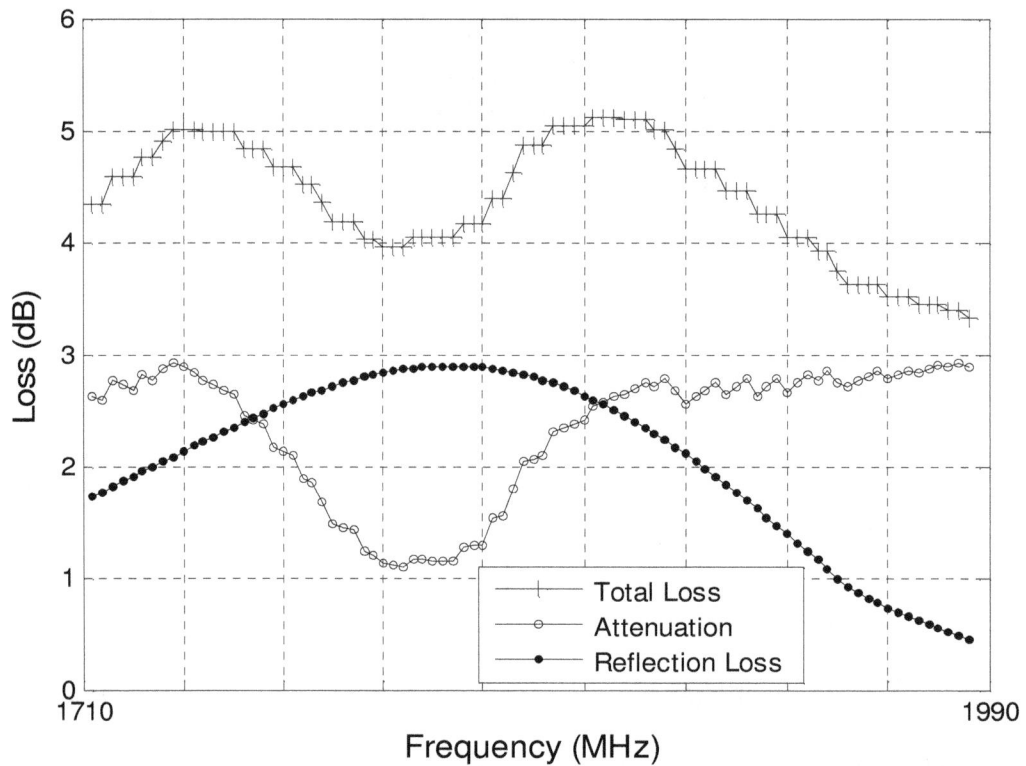

Figure 3. Measured results in the GSM1800 and GSM1900 frequency bands of the total loss (SE), attenuation and reflection loss.

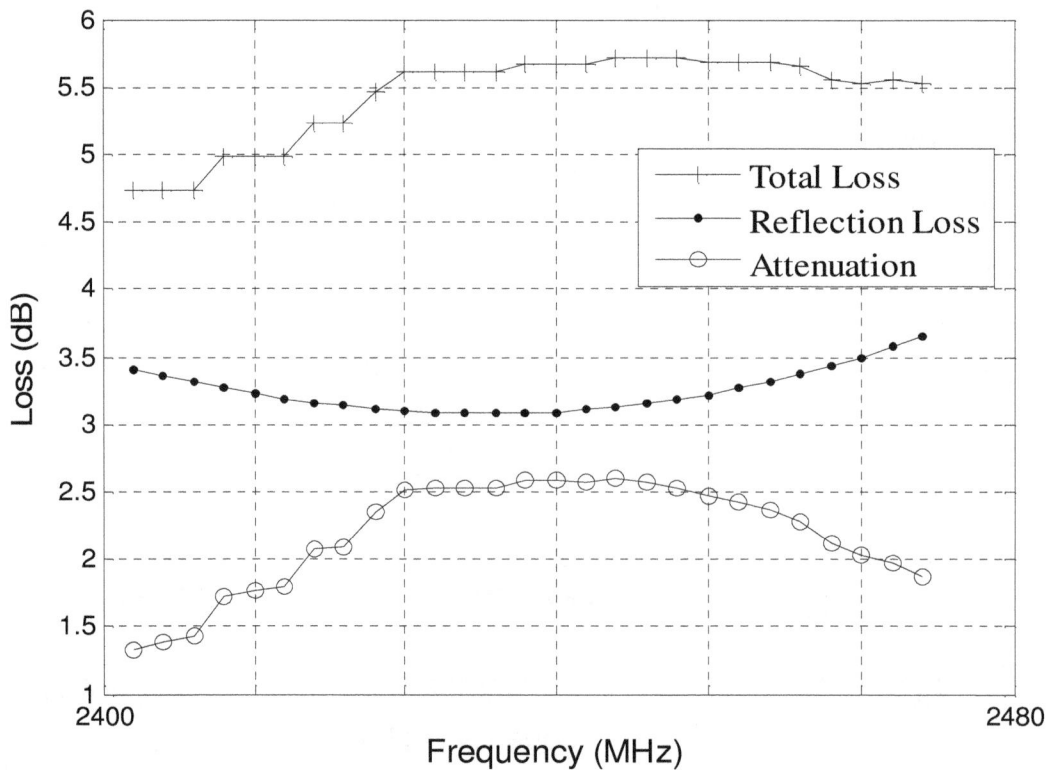

Figure 4. Measured results in the WiFi frequency band of the total loss (SE), attenuation and reflection loss.

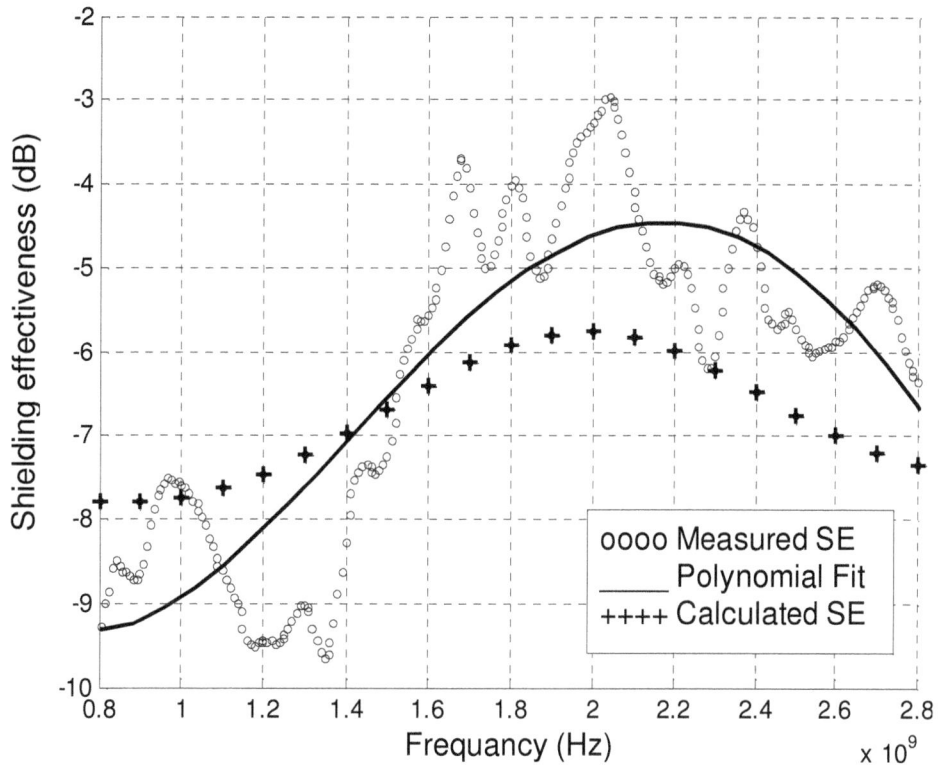

Figure 5. Comparison of measured results and calculated data of the transmission loss from 800 MHz to 2.8 GHz.

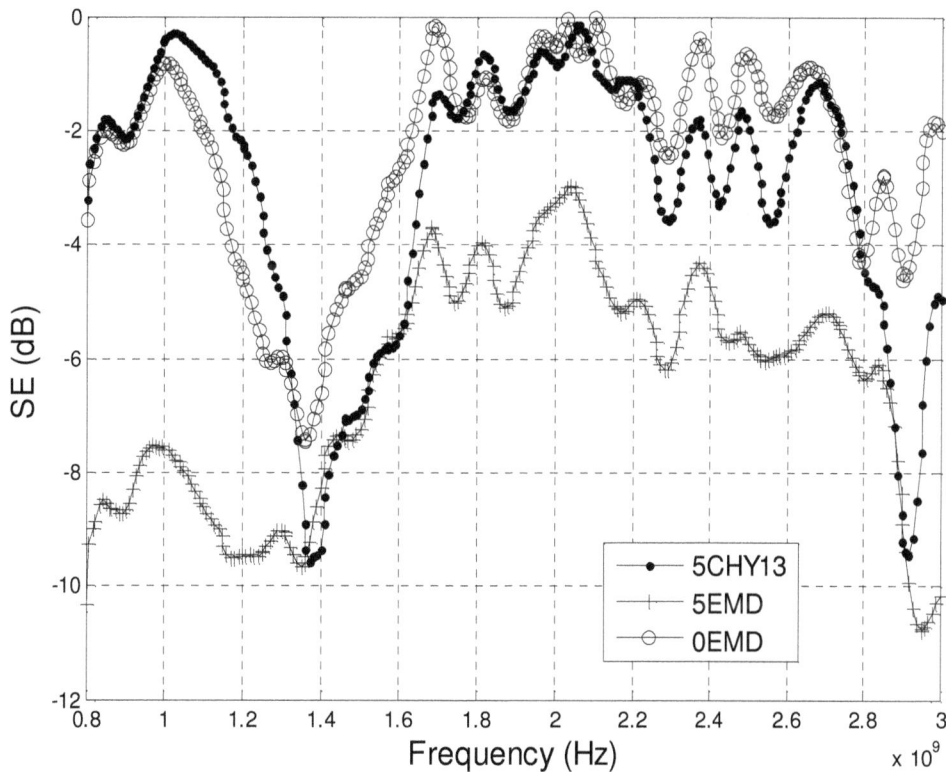

Figure 6. SE comparison of 0EMD, 5EMD and 5CHY13 samples.

material absorber. This new absorber material showed superior SE performance compared to the commonly used MnZn-ferrite.

Conclusion

It is believed that this composite cement-based material which the authors investigated and characterized in this research can practically be applied in a plaster cement form or as pre-manufactured tiles to shield priority indoor premises effectively from outside electromagnetic interference.

ACKNOWLEDGMENT

The authors wish to thank Saab-Grintek Technologies in Centurion, South Africa for the availability of their anechoic chamber and test equipment.

ABBREVIATIONS

EMW, Electromagnetic wave; **VNA,** vector network analyser; **SE,** shielding effectiveness; **DUT,** device under test; **EMD,** electrolytic manganese dioxide; **TX,** transmit; **RX,** receive; **TDR,** time domain reflectometry.

REFERENCES

Bantsis G, Mavridou S, Sikalidis C, Betsiou M, Oikonomou N, Yioultsis T (2012). Comparison of low cost shielding-absorbing cement paste building materials in X-band frequency range using a variety of wastes. Ceram. Int. Elsevier 3:3683-3692.

Baoyi L, Yuping D, Shunhua L (2012). The electromagnetic characteristics of fly ash and absorbing properties of cement-based composites using fly ash as cement replacement. Const. Build. Mater. Elsevier 27:184-188.

Guan H, Liu S, Duan Y (2005). Expanded polystyrene as an admixture in cement-based composite for electromagnetic absorbing. J. Mater. Eng. Perform. pp. 68-72.

Guan H, Liu S, Duan Y, Cheng J (2006). Cement based electromagnetic shielding and absorbing building material. Cement Concrete Compos. Elsevier 28:468-474.

Guan H, Liu S, Duan Y, Zhao Y (2007). Investigation of the electromagnetic characteristics of cement based composites filled with EPS. Cement Concrete Compos. Elsevier 29:49-54.

Hutagalung SD, Sahrol NH, Ahmad ZA, Ain MF, Othman M (2012). Effect of MnO_2 additive on the dielectric and electromagnetic interference shielding properties of sintered cement-based ceramics. Ceram. Int. Elsevier 38:671-678.

Kharkovsky SN, Akay MF, Hasar UC, Atis CD (2002). Measurement and monitoring of microwave reflection and transmission properties of cement-based specimens. IEEE Trans. Instr. Meas. 51(6):1210-1218.

Kim HM, Kim K, Lee CY, Joo J, Cho SJ, Yoon HS, Pejakovic DA, Yoo JW, Epstein AJ (2004). Electrical conductivity and electromagnetic interferance shielding of multiwalled carbon nanotube composites containing Fe catalyst. Appl. Phys. Lett. 84(4):589-591.

Mishra M, Singh AP, Dhawan SK (2013). Expanded graphite-nanoferrite-fly ash composites for shielding of electromagnetic pollution. J. Alloys Compd. Elsevier 557:244-251.

Nie Y, He H, Feng Z, Xiong B (2005). Absorbing properties of the magnetic composite electromagnetic wave absorber. Proc. IEEE Int. Symp. Microwave, Antenna, Propagation and EMC Technologies for Wireless Communications, Beijing, China, 8-12 August, 2005, pp. 724-727.

Oka H, Tanaka K, Osada H, Kubota K, Dawson FP (2009). Study of electromagnetic wave absorption characteristics and component parameters of laminated-type magnetic wood with stainless steel and ferrite powder for use as building materials. J. Appl. Phys. 105:1-3.

Park KY, Lee SE, Kim CG, Han JH (2006). Fabrication and electromagnetic characteristics of electromagnetic wave absorbing sandwich structures. Compos. Sci. Technol. Elsevier 66:576-584.

Sagnard F, El Zein G (2005). Characterization of building materials for propagation modeling: Frequency and time response. Int. J. Electron. Comm. Elsevier 59:337-347.

Schulz RB, Plantz VC, Brush DR (1988). Shielding theory and practice. IEEE Trans. Electromagn. Compatibility 30(3):187-200.

8

Performance analysis of ultra wideband media access control protocol providing two class traffic

Mohamed E. Wahed*, Mohamed K. Hussein, Mohamed H. Mousa and
Mohamed Abdel Hameed

Faculty of Computers and Informatics, Suez Canal University, Ismailia, Egypt.

Ultra wideband (UWB) technology based primarily on the impulse radio paradigm has a huge potential for revolutionizing the world of digital communications especially wireless communications. Multiple packets can be assembled into a single frame at the medium access control (MAC) layer, which can significantly improve the throughput performance. This paper presents a detailed performance analysis for UWB multi-band MAC protocol. The MAC protocol is presented as two different clases. The first class involves all bands as data bands, while in the second calss one of its band is used for traffic control. The analysis is based on using multi-server queuing model and a detailed simulation. Each data band is represented as a server in the queuing model. The analysis is complemented with extensive simulations. In the experimental results, the protcol performance is presented as mean packet size, average packet delay, and blocking probability. A number of important conclusions were outlined from this study. First, the increase of transition rate between two classes has noticeable effect on the average number of packets in the queue and in the system, while the increase of the utilization rate U has a remarkable same effect. Next, the absence of one data band effects on the average number of packets decreases as the transmission rate decreases. Finally, the use of a multi-band approach provides an inherent flexibility in operation to coexist with other wireless networks.

Key words: Ultra wideband (UWB), wireless communications, medium access control (MAC), ad hoc networks, multi-server queuing model.

INTRODUCTION

Ultra wideband (UWB) is the cutting edge wireless short-range technology which has been the focus of interest in both academia and industry for applications in wireless communications (Karapistoli, 2012; Federal Communications Commission, 2002; Cuomo et al., 2002; Ghavami et al., 2004). UWB technology has many benefits owing to its UWB nature, which include high data rate, less path loss and better immunity to multipath propagation, availability of low-cost transceivers, low transmit power and low interference (Agusut and Ha, 2004; Merz et al., 2005). Further, the UWB spectrum can be partitioned into multiple comparatively narrow frequency bands that are mutually orthogonal and can be used simultaneously. This will lead to the use the available spectrum more efficiently (Chaudhry et al., 1992; Di Benedetto

et al., 2005).

There are two types of UWB wireless communication approaches, the single-band approach and the multi-band approach. In the single-band approach, the communication is based on using a single band with time-hopping as the basic means of access (Federal Communications Commission, 2002). However, in the multi-band approach the available frequency bandwidth is divided into B bands. $B - 1$ of these bands are used for data transmissions and are referred to as data bands. The remaining band is used for request control packets only (Broustis et al., 2006, 2007).

There has been mutliple research works on the design of the medium access control (MAC) protocol for multi-band UWB-based wireless networks that support ad hoc communications. For example, (Broustis et al., 2006, 2007), a multi-band MAC protocol for use with UWB-based ad hoc networks has been proposed. The proposed protocol design is based on separation of

*Corresponding author. E-mail: mohamed_badawy@ci.suez.edu.eg.

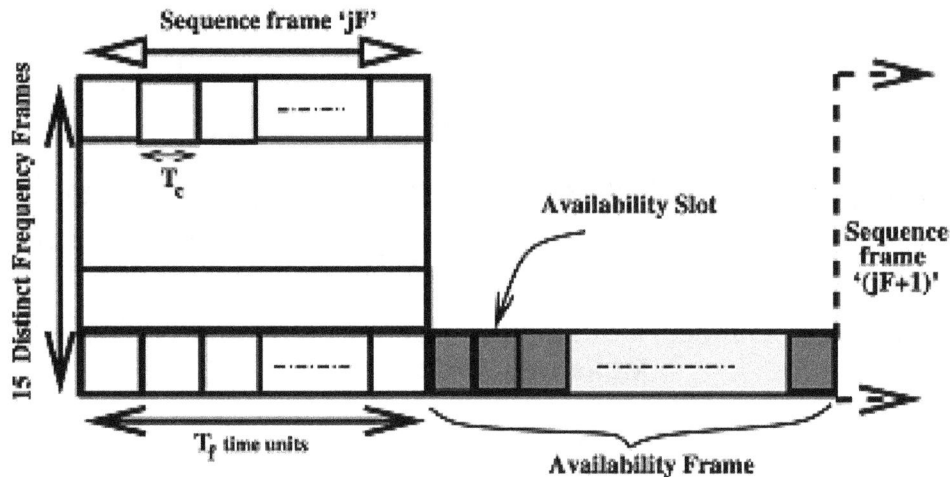

Figure 1. The frame structure multi-band MAC protocol.

control and data onto different bands. Simply, each data exchange begins with a rendezvous transaction on a control channel, where one node sends an explicit request message to the other, which signals its preparation to continue through a simple partial-response message. Subsequently, the two nodes switch their attention to the data bands and select one of them for the actual data exchange using first come, first serve (FCFS) discpline. This separation has two main advantages. First, since all nodes share a common unreserved channel only for short control messages, the contention on the shared channel is limited. Second, once a pair of nodes agrees to communicate on a data band, the communication can be continuous (no need for the use of time hopping sequences), and thus, it is highly efficient. The presented result indicate that the throughput of the proposed scheme is significantly higher compared to a single-band approach that combats delay spread by increasing the spacing between pulse transmissions.

This study aims to provide a detailed performance analysis for UWB multi-band MAC protocol. This is performed by analyzing the performance UWB multi-band MAC protocol in terms of utilizing the multiple bands. The analysis is based on using multi-server queuing model. This queuing model has been firstly explored by Chaudhry et al. (1992), Jurdak et al. (2005), Shoukry et al. (1994), and Shoukry et al. (1995). In the presented model, the data bands B are used instead of servers C in the multi-server queuing model (M[x]/M/B; B-1/FCFS), which in turn are responsible for the data transmission between users in a wireless ad hoc network. Finally, the effectivness of the presented model is evaluated using extensive simulations.

THE MULTI-BAND MAC PROTOCOL OVERVIEW

Here, a brief overview is given on the basic concepts and the operation of the multi-band MAC protocol. The key idea is to have a communicating pair of nodes exchange data over a private band as opposed to a single common band. In the multiple bands, the available frequency bandwidth is divided into B bands. B-1 of these bands are used for data transmissions, called data bands. The remaining band is used for request control packets only, called the Request Band or Req-Band, and is usually assigned to the first band. The multi-band MAC protocol is designed based on the physical separation of the available UWB bandwidth of 7.5 GHz into multiple bands, each of which spans 500 MHz of the spectrum (Broustis et al., 2007).

Frame structure

Figure 1 shows the frame structure of the multi-band MAC protocol. Across all the bands, time is broken into superframes, which are separated by smaller availability frames. All data and control communication takes place during superframes. The availability frame is used to indicate whether each band will be busy or not in the next superframe. The availability frames alleviate the possibility of collisions of data transmissions in the superframes. Further, each superframe consists of F sequence frames, each of which in turn consists of Tf/Tc chip-times. The availability frame is sandwiched between the last sequence frame of the jth superframe and the first sequence frame of the $(j+1)^{st}$ superframe (Broustis et al., 2007).

Operations of the MAC protocol

The protocol implementation at each node can be represented by a finite state machine, as shown in Figure 2. The available bandwidth is divided into B bands as follows:

1) One band for request and information about the state of both sender and reciever (control band).
2) The rest bands for data transmissions and acknowledgements (data bands).

The band availability maps into the following frames:

3) Superframes: Transmission of all control and data packets.
4) Availability frames: Declare intention to keep using a band.

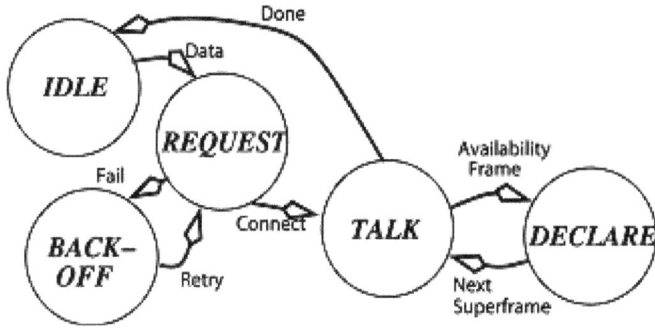

Figure 2. Depiction of the multi-band MAC protocol operations.

MULTI-SERVER QUEUING MODELING

Here, we study a multi-server queuing model ($M[x]/M/C; C-1/FCFS$) to discuss multi-band MAC protocol so we specify data bands B instead of servers C. As a result, the multi-server queuing model will be in the form ($M[x]/M/B; B-1/FCFS$), which in turn is responsible for transmission of data between users in wireless ad hoc networks using UWB technique.

Model description

In the multi-server queuing model, packets arrive in batches according to compound Poisson process with mean batch arrival rate λ (Chaudhry and Grassmann, 1979). Let $N(t)$ be the number of batches of packets which have arrived by time t where $\{N(t), t \geq 0\}$ can be modeled as a Poisson process. Next, we fit a discrete distribution to the sizes of the successive batches where the batch sizes must be positive integers. Thus, the total number of packets to arrive by time t denoted by $Z(t)$ is given by:

$$Z(t) = \sum_{i=1}^{N(t)} x_i \quad for \quad t \geq 0 \tag{1}$$

Where x_i is the number of packets in the ith batch. The x_i's are assumed to be independent identically distributed random variables which are also independent of $\{N(t), t \geq 0\}$, then the stochastic process $\{Z(t), t \geq 0\}$ is said to be a compound Poisson process. In this case, the arrival batch size X has a positive Poisson distribution where no batch size equals zero, so we will exclude the value zero from the Poisson distribution that is given as:

$$P(x; \theta) = \theta^x e^{-\theta}/x!, x \geq 0, \theta > 0 \tag{2}$$

Now, let $\{a_x\}$ represent a probability sequence that governs batch size (that is, an arriving batch has size x with probability a_x). This probability a_x is defined as:

$$a_x = P(x; \theta)/[1-P(0;\theta)] = \frac{\theta^x e^{-\theta}/x!}{1-e^{-\theta}} \tag{3}$$

Where θ is a parameter of batch size. The average batch size m in this case will be given by:

$$E(X) = \theta/(1-e^{-\theta}) = m \tag{4}$$

Packets are served according to exponential service time distribution with parameter $\mu > 0$. The B and (B-1) in the model notation above are the number of parallel data bands. Service is on a FCFS basis. In addition, the queue length in our model is assumed to have no limit. As indicated by the notation ($M[x]/M/B; B-1/FCFS$), batches of packets arrive at random times with mean packet arrival rate λ. We investigate specifically a distribution of the batch size, namely a positive Poisson batch sizes. The queuing system under study has homogeneous parallel data bands where service time has exponential distribution with mean $1/\mu$. Our system alternates between two types of classes of system operation, this is due to the type of information that senders desire to communicate. In one class-1 of operation all B data bands are available and in the other class-0 only ($B-1$) data bands are available. The possibility of collisions due to multiple new senders choose the same band. This means that only one data band is allowed to collide randomly according to a discrete uniform distribution that assigns one of the data bands to be out of service where the band collision has equal chance over all data bands in the system. The mean time that the system operates with B data bands and ($B-1$) data bands is $1/\alpha$ and $1/\beta$, respectively.

The queue discipline for batches of packets is FCFS, while the service discipline within the batches is based on randomly choosing one of the packets mentioned earlier. If a batch of r packets arrives, while the system has n requested senders and $B-r \geq 1$ and $r \geq B-n$, then $B-n$ are served immediately. The $r-B+n$ delayed senders must wait for d departures, $d = 1,2,.... , r-B+n$. On the other hand, if the packet arrives while the system is in state n, $n \geq B$, all the r packets are delayed. Thus, of these r packets who arrive when $n = B+k$, $k = 0, 1, 2...$, the packets under consideration must wait for d departures, $d = k+1, k+2, ..., k+r$. The conditional probability of the packets waiting for d departures before its service commences, given the state of the system n just before the arrival of the packet's batch, is given by,

$$\sum_{r=B-n+d}^{\infty} \frac{r a_r}{m} \frac{1}{r} = \frac{1}{m} \sum_{r=B-n+d}^{\infty} a_r, \quad d = 1, 2,...0 \leq n \leq d +B-1 \tag{5}$$

Where the batch size X is a random variable with distribution given by:

$$a_r = P(X = r), r \geq 1$$

and X has mean m, $0 \leq m = \sum_{r=1}^{\infty} r a_r < \infty$, and variance σ^2, $0 < \sigma^2 < \infty$

Packets of certain batch are served randomly. If X is the size of a batch, then packet i, $i = 1,2,...,X$ has the same chance of joining service. In order to generate an equal chance to all packets in a given batch of size X, either a uniform assignment of the order in which they may be served is used or using a predefined permutation sequence. This routine helps to organize every packet in the batch.

Modeling assumptions and notations

Table 1 shows the different notation which are used in the proposed model. The proposed model has many assumptions. First of all, the system at any instant of time is in one of two classes of operation Class-0 or Class-1. The requested packets are mobile and independent from each other to form an ad hoc network. Further, the queue has B identical exponential bands in parallel with each band having a service rate μ. Groups of packets arrive at a multi-band in accordance with Poisson process with parameter λ.

Table 1. Summary of parameters.

Parameter	Description
B	Number of data bands
N	The number of batches
λ	Mean batch arrival rate
μ	Constant service rate of a band
m	The average batch size
L_Q	The average number of data packets in the queue
W_Q	The waiting time per data packet in the queue
W_S	The waiting time per data packet in the system
L_S	The average number of data packets in the system
Class-1	All B data bands are available for serving the packets in this class
Class-0	All (B-1) data bands are available for serving the packets in this class
$P_{0,i}$	The probability of having no data packets to transmit in the system when the system is in Class-i, i = 0,1
β	Conversion rate from Class-0 to Class-1
α	Conversion rate from Class-1 to class-0
U	The utilization of traffic
θ	The parameter of the batch size distribution
T_C	Chip-time (time spacing between pulses)
P_B	The blocking probability that defines the probability that all data bands are busy.
Pe	Bit error probability (the expectation value of the BER (Bit Error Rate) which can be considered as an approximate estimate of the bit error probability)
CBR	Constant bit rate (the form of a technique which is used for the purpose of measuring the rate at which the encoding of the data packets takes place).

Packets are transmitted in accordance of batches to exchange data over the ad hoc network. The duration of the superframe (batch) is set to 11,200 Tc, Tc = 6 × 10 to 8 s. In the transmission process, we use both CBR = 0.04 s and Poisson traffic with arrival rate = $1/\lambda$, the service time is exponential with the same mean $1/\mu$. The times that system operates with B data bands and (B-1) data bands has exponential distribution with mean $1/\alpha$ and $1/\beta$, respectively. The batch size X follows a positive Poisson distribution with parameter θ. Two or more packet transmissions sent at the same time on the same band will collide. All packets which collide will then initiate back-off timers, where they remain for a random delay before returning to the request state. The overall simulation time is 10000 s. Packets are served according to FCFS discipline, while nodes are served randomly. The utilization traffic for the system is given by $U = m\lambda / B\mu$ and the condition for existence of the steady state is $U < 1$.

The analysis approach

The simulation steps can be described in the following steps:

1) Generate 1000 random variables (indicated as B bands),
2) Generate 1000 interarrival times for 1000 different batches from exponential distribution with mean $1/\lambda$,
3) Generate a random batch size for each batch in step 2 using positive Poisson distribution with parameter θ,
4) To specify the service times for each packet in the successive batches, generate random variables from exponential distribution with mean $1/\mu$,

5) Specify the intervals of time that the system operates with B data bands by generating random variables from exponential distribution with mean $1/\alpha$,
6) To specify the intervals of time that the system operates with (B-1) data bands, generate random variables from exponential distribution with mean $1/\beta$.
7) To specify which data band will be collide, generate random variables from discrete uniform distribution with parameter B. Where the integers 1, 2, 3,...,B occur with equal probability,
8) Calculate the event time of back-offs and retrying for each band based on the intervals of time during which the system works with B-1 and B data bands, respectively,
9) Calculate the traffic rate U from equation $U = m\lambda/B\mu$,
10) Find the number of all packets that enter the system as a total of sizes of batches N that arrive to the system,
11) Convert the transfer rate from packets per second into bits per second where packet length = 8 bits/s,
12) Find the next arrival time of a batch and packet, respectively,
13) Specify the class at which an arriving batch will find the system as follows:

a) If the arrival time of a batch is greater than the event time of back-off and less than the event time of retrying, the system will be in Class-0.
b) If the arrival time of a batch is greater than event time of retrying and less than the event time of back-off, the system will be in Class-1.
14) An arriving packet will immediately start service if one of data bands is free or wait until any data band becomes free,
15) Calculate the departure time of a packet from system as the total of arrival time plus service time,

Table 2. Average number of packets in the queue L_Q.

U	Transition rate between two classes (Class-0 and Class-1)				
	0.000	**0.250**	**0.500**	**0.750**	**1.000**
0.100	0.067	0.069	0.078	0.084	0.093
0.200	0.165	0.337	0.547	0.859	0.928
0.300	1.014	1.676	2.309	2.840	3.415
0.400	4.095	4.822	5.163	6.175	6.954
0.500	7.471	7.954	8.457	9.047	9.885
0.600	10.598	11.201	11.891	12.576	13.356
0.700	14.021	16.896	18.835	22.012	25.869
0.800	28.170	32.549	44.705	62.976	87.874
0.900	92.825	130.172	200.513	310.145	408.679

Mean size of batch = 5.896; Number of data bands = 5.

16) Calculate the waiting time of a packet in the queue,
17) Calculate the waiting time of a packet in the system,
18) Calculate the busy time for each data band,
19) Calculate the departure time of a batch,
20) Calculate the total of waiting times of packets in queue,
21) Calculate the total of waiting times of packets in system,
22) Repeat Steps 17 to 21 until all packets in a given batch are served,
23) Define the different time intervals during which system operates in the two classes,
24) Repeat the steps from 13 to 26, 1000 times,
25) The probability that the system is in Class-i, i = 0,1, while there is no any packet in the system $P_{0,i}$ from equation,

$$P_{0,i} = \frac{\Sigma \text{ The number of batches that arrive to find the system is idle in Class-i, i = 0,1}}{\text{The number of all batches}}$$

26) Calculate the probability that no packets are in system P_0 from equation,

$$P_0 = \frac{\Sigma \text{ The number of batches that arrive to find the system is idle}}{\text{The number of all batches}}$$

Where, $P_0 = P_{0,0} + P_{0,1}$
27) Calculate the blocking probability P_B, that is, the probability that any packet in an arriving batch   must wait from equation,

$$P_B = \frac{\text{The number of waiting packets in the queue}}{\text{The number of all packets}}$$

28) Calculate the average waiting time in queue W_Q from equation,

$$W_Q = \frac{\Sigma \text{ The waiting time per packet in the queue}}{\text{The number of all packets}}$$

29) Calculate the average waiting time in system W_S from equation,

$$W_S = \frac{\Sigma \text{ The waiting time per packet in the system}}{\text{The number of all packets}}$$

30) Calculate the average number of packets in the queue L_Q from equation,

$$L_Q = \frac{\Sigma \text{ The waiting time per packet in the queue}}{T_S}$$

31) Calculate the average number of packets in the system L_S from equation,

$$L_S = \frac{\Sigma \text{ The waiting time per packet in the system}}{T_S}$$

where T_S is the overall simulation time we observed in the system

which equals 10000 s.

SIMULATION RESULTS

The simulation program was tested extensively for values of (U, B, X) where $0.1 \leq U \leq 0.9$, $1 \leq B \leq 100$ and batch size $X \leq 100$. Tables 2 to 7 contain two variables. The first variable represents the traffic utilization S, and the second variable represents transition rate between two classes which is equal $\alpha/(\alpha+\beta)$, while the other values inside each table represent performance measures (output data). These measures are Ls, L_Q, W_Q, W_S, P_0, P_B in Tables 2 to 7, respectively. Based on the approach explained previously, the input data contains the number of data bands B, the mean time that the system organizes with B data bands denotes by $1/\alpha$, the mean time that the system organizes with (B-1) data bands denotes by $1/\beta$, the number of batches N, the parameter size of batches θ, mean service time $1/\mu$ and the mean inter-arrival time $1/\lambda$. Different performance measures are calculated. These measures include:

i) The expected number of packets in the queue and in the system will be L_Q and L_S, respectively.
ii) The expected waiting time per packet in the queue and in the system are W_Q and W_S, respectively.
iii) The idle probability of having zero packets in the

Table 3. Average number of packets in the system L_S.

U	Transition rate between two classes (Class-0 and Class-1)				
	0.000	0.250	0.500	0.750	1.000
0.100	0.185	0.437	0.647	0.959	1.328
0.200	1.854	2.586	3.079	3.840	4.415
0.300	4.914	5.776	6.109	6.940	7.315
0.400	7.995	8.722	9.563	9.975	10.554
0.500	10.971	11.654	12.057	12.977	13.885
0.600	14.598	17.241	20.801	23.546	28.156
0.700	30.021	35.996	40.635	46.012	52.869
0.800	58.270	67.569	74.305	84.076	92.814
0.900	110.025	162.176	220.463	340.185	418.669

Table 4. Average waiting time of packets in the queue W_Q.

U	Transition rate between two classes (Class-0 and Class-1)				
	0.000	0.250	0.500	0.750	1.000
0.100	0.185	0.237	0.367	0.480	0.598
0.200	1.274	1.986	2.779	3.340	3.995
0.300	4.714	5.276	6.309	7.240	8.315
0.400	8.959	9.772	10.534	11.375	12.254
0.500	12.071	15.624	18.657	22.977	25.885
0.600	24.598	32.211	40.401	48.546	53.156
0.700	43.021	58.596	78.335	98.012	118.869
0.800	88.250	120.579	144.335	179.076	214.514
0.900	157.025	223.162	350.463	470.185	578.669

Mean size of batch = 5.896; Number of data bands = 5.

Table 5. Average waiting time of packets in the system W_S.

U	Transition rate between two classes (Class-0 and Class-1)				
	0.000	0.250	0.500	0.750	1.000
0.100	1.185	1.337	1.767	2.280	2.998
0.200	3.284	3.956	4.779	5.340	5.995
0.300	6.724	7.276	8.309	9.040	9.908
0.400	11.959	13.772	15.594	17.395	19.254
0.500	22.571	25.694	30.617	35.907	40.625
0.600	40.998	44.251	48.401	55.576	63.196
0.700	63.441	80.526	95.335	108.012	138.869
0.800	122.210	142.529	164.352	189.456	224.564
0.900	197.725	255.162	360.463	490.185	598.669

Mean size of batch = 5.896; Number of data bands = 5.

system P_0.
iv) The probability that all bands are busy, blocking probability P_B.

We note that the traffic utilization U changes with variation in the value of $1/\mu$. So the value of transition rate between two classes (Class-0 and Class -1) is varied due to the changes in values $1/\alpha$ and $1/\beta$. Results shown in Tables 2 to 7 are obtained by these input data. Figures 3 to 8 are the explanations for Tables 2 to 7, respectively. These graphs show the effect of performance measures on the system resulting for different traffic utilization U and transition rate between two classes. For example, Figure 3 denotes the relation among traffic utilization

Table 6. The idle probability of having zero packets in the system P_0.

U	Transition rate between two classes (Class-0 and Class-1)				
	0.000	0.250	0.500	0.750	1.000
0.100	0.838	0.844	0.851	0.825	0.816
0.200	0.775	0.787	0.797	0.762	0.755
0.300	0.651	0.662	0.685	0.632	0.619
0.400	0.537	0.566	0.596	0.515	0.459
0.500	0.368	0.392	0.432	0.349	0.299
0.600	0.249	0.264	0.293	0.228	0.187
0.700	0.165	0.188	0.205	0.135	0.106
0.800	0.076	0.109	0.148	0.049	0.034
0.900	0.0246	0.0514	0.0877	0.0211	0.0032

Mean size of batch = 5.896; Number of data bands = 5.

Table 7. The probability that all bands are busy, blocking probability P_B.

U	Transition rate between two classes (Class-0 and Class-1)				
	0.000	0.250	0.500	0.750	1.000
0.795	0.772	0.755	0.737	0.717	0.100
0.855	0.812	0.795	0.778	0.757	0.200
0.889	0.862	0.831	0.792	0.785	0.300
0.918	0.895	0.876	0.846	0.826	0.400
0.944	0.919	0.898	0.872	0.849	0.500
0.972	0.958	0.939	0.911	0.893	0.600
0.982	0.971	0.945	0.927	0.912	0.700
0.991	0.977	0.969	0.959	0.948	0.800
0.998	0.989	0.983	0.971	0.956	0.900

Mean size of batch = 5.896; Number of data bands = 5.

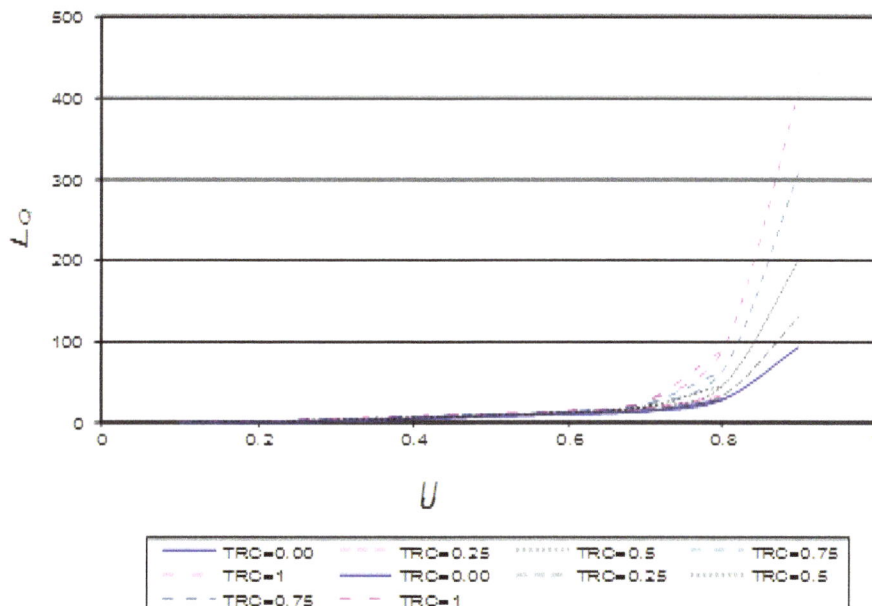

Figure 3. L_Q versus U where (m = 5.896, B = 5).

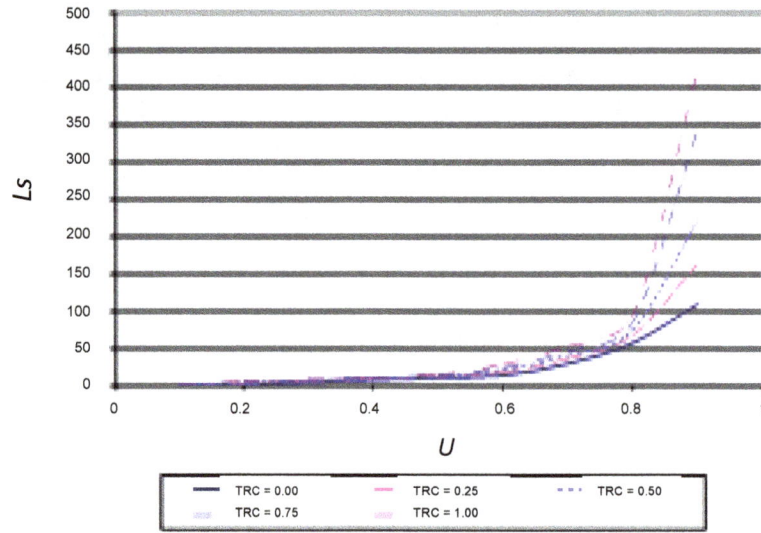

Figure 4. L_S versus U where (m = 5.896, B = 5).

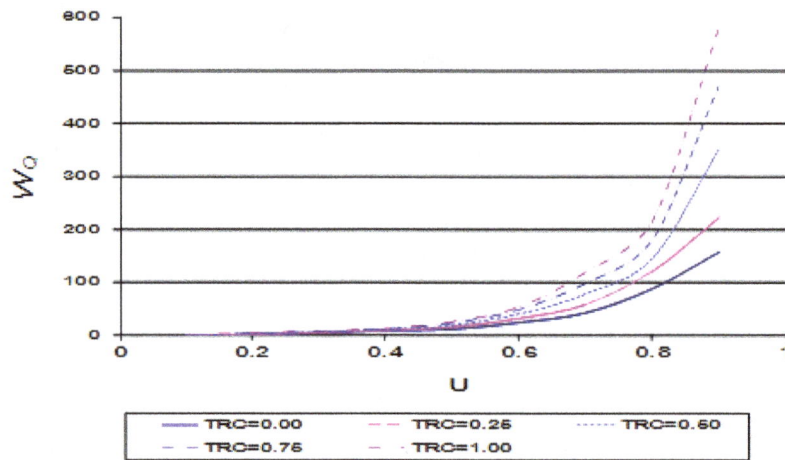

Figure 5. W_Q versus U where (m = 5.896, B = 5).

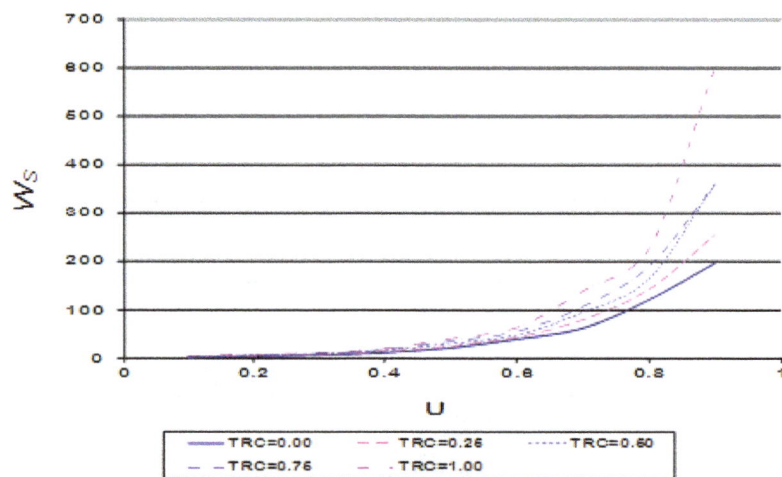

Figure 6. W_S versus U where (m = 5.896, B = 5).

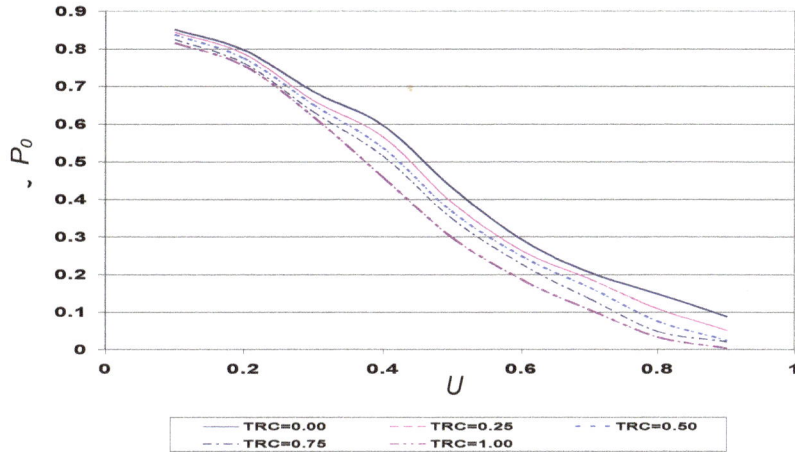

Figure 7. P_0 versus U where (m = 5.896, B = 5).

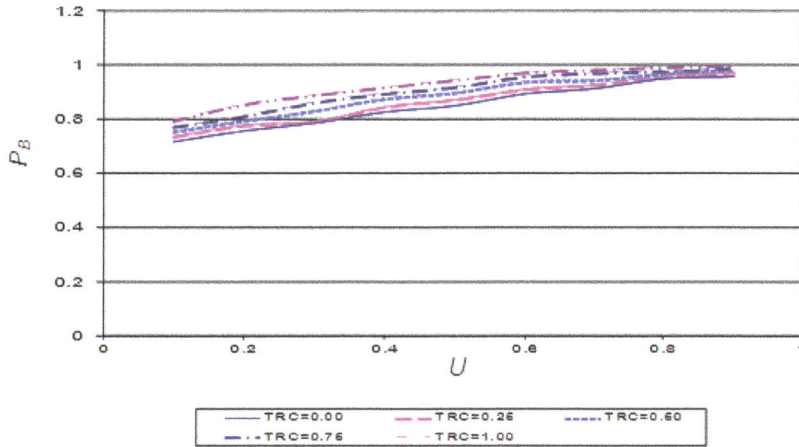

Figure 8. P_B versus U where (m = 5.896, B = 5).

U and the average number of packets in the queue L_Q with the increasing of transition rate between two classes TRC which equals $\alpha/(\alpha+\beta)$.

Tables 2 and 3 show that the increases of transition rate between two classes (Class-0 and Class-1) has noticeable effect on the average number of packets in the queue and in the system, while the increase of the utilization rate U has a remarkable effect on the average number of packets in the queue and in the system. Further, the average number of packets in the queue and in the system increase as the average batch size increases. So, the increase of TRC affects both L_Q and L_S when the average batch size increases. The most effect on the average number of packets in the system and in the queue is happening when the utilization rate U is very close to unity. The absence of one data band affects the average number of packets as utilization rate is near the unity (heavy load traffic), which means that transmission rate decreases.

Tables 4 and 5 show that the averages waiting time in the queue as well as in the system is highly influenced by the absence of one of the bands for both low and high utilization rate of data bands. Table 6 shows the probability of having no packets in the system P_0 when a batch arrives. This means that the percentage of time the system is idle is not affected by the absence of one of the data bands; while P_0 decreases as TRC increases for the utilization rate, and P_0 decreases noticeably as TRC increases for the high utilization rate of bands. Moreover, P_0 decreases as the utilization rate of data bands increases.

Table 7 shows that the blocking probability P_B is highly influenced by the increase of the utilization rate of data bands, while P_B increases as the utilization rate increases. So, the blocking probability has remarkable effect resulting from absence of one of the data band, where it increases as the TRC increases.

Finally, the performance measures are changed in

response to the changes of the operating parameters. We documented the behavior of the system when one of the data bands temporarily leaves the system with useful graphical representation to give an opportunity to notice the system behavior over the traffic utilization and the transition rate between two classes (Class-0 and Class-1). Further, the use of a multi-band approach provides an inherent flexibility in operation to coexist with other wireless networks. The approach we present is conjoint with the UWB physical layer and takes into account the regulations imposed by the FCC. One main requirement for real-time service is delay guarantee, as packets with a large delay may be considered useless and discarded.

Conclusion

In this study, we have presented a detailed performance analysis for UWB multi-band MAC protocol. The analysis is based on using multi-server queuing model and a detailed simulation. Each data band is represented as a server in the queuing model. A number of important conclusions were outlined from this study. First, the increase of transition rate between two classes (Class-0 and Class-1) has noticeable effect on the average number of packets in the queue and in the system, while the increase of the utilization rate U has a remarkable same effect. Next, the absence of one data band effects on the average number of packets decreases the transmission rate decreases. Finally, the use of a multi-band approach provides an inherent flexibility in operation to coexist with other wireless networks

REFERENCES

Agusut NJ, Ha DS (2004). An Efficient UWB Radio Architecture for Busy Signal MAC protocols. In IEEE SECON pp. 325-334.

Broustis I, Krishnamurthy SV, Faloutsos M, Molle M, Foerster JR (2007). Multiband media access control protocol in impulse-based UWB ad hoc networks. IEEE Trans. Mobile Comput. 6(4):351-366.

Broustis I, Molle M, Krishnamurthy S, Faloutsos M, Foerster J (2006). A new binary conflict resolution based MAC protocol for impulse-based UWB ad hoc networks. Wireless Commun. Mobile Comput. 6(7):933-949.

Chaudhry ML, Grassmann WK (1979). Further results for the queuing system $M^{[x]}$/M/C. J. Opl. Res. Soc.30:755-763.

Chaudhry ML, Templeton JGC, Medhi J (1992). Computational results of multiserver bulk arrival queues with constant service time $M^{[x]}$/D/C. J. Opns. Res. 40:229-238.

Cuomo F, Martello C, Baiocchi A, Fabrizio C (2002). Radio Resource Sharing for Ad Hoc Networking with UWB. In IEEE JSAC 20(9):1722-1732.

Di Benedetto MG, De Nardis L, Junk M, Giancola G (2005). (UWB)2: Uncoordinated, Wireless, Baseborn Medium Access for UWB Communication Networks. Mobile Networks Appl. 10:663-674.

Federal Communications Commission (FCC) (2002). First Report and Order in The Matter of Revision of Part 15 of the Commission's Rules Regarding Ultra wideband Transmission Systems. ET-Docket 98-153, FCC 02-48.

Ghavami M, Michael LB, Kohno R (2004). Ultra Wideband Signals and Systems in Communication Engineering. John Wiley and sons: Chichester.   pp.   125-160. http://onlinelibrary.wiley.com/book/10.1002/0470867531.

Jurdak R, Baldi P, Lopez CV (2005). U-MAC: A Proactive and Adaptive UWB Medium Access Control Protocol. In Wireless Commun. Mobile Comput. 5(5):551-566.

Karapistoli E, Gragopoulos I, Tsetsinas I, Pavlidou F (2012). Location-aided medium access control for low data rate UWB wireless sensor networks. Wireless Commun. Mobile Comput.12(11):956-968.

Merz R, Widmer J, Le Budec J, Radunovic B (2005). A Joint PHY/MAC Architecture for Low-Radiated Power TH-UWB Wireless Ad Hoc Networks. In Wireless Commun. Mobile Comput. 5(5):567-580.

Shoukry EM, Gharraph MK, Hassan NA (1994). Multi-server bi-level queuing system with heterogeneous servers. J. Egypt. Stat. 39:208-221.

Shoukry EM, Gharraph MK, Hassan NA (1995). Multi-server bi-level queuing system with Erlagian service time and server breakdowns. J. Adv. Model. Anal. 27:47-64.

9

Mobile agent based clustering and maintenance using secure routing protocol for mobile ad hoc network

R. Pushpa Lakshmi and A. Vincent Antony Kumar

PSNA College of Engineering and Technology, Dindigul, Tamilnadu, India.

Routing in mobile ad hoc network is the challenging task due to the dynamic nature and resource limitations of network. Network clustering deals with partitioning network into clusters based on some rules. Clustering guarantees limited resource utilization and network scalability. In this paper, we mainly focused on using mobile agents for collecting information about cluster members and cluster maintenance. Ensuring security of mobile agent is a difficult task. In this work, the authorization of mobile agent is achieved using session key or secret key shared between mobile agent owner and node. We applied a distributed private key generation scheme, which generates secret key of cluster members based on 'n' key shares. The session key is generated based on past frequent traffic pattern that exist between the nodes. The behavior of mobile agent with respect to various security issues such as replay attack, non repudiation, denial of service and unauthorized access was discussed. The performance of the proposed scheme is evaluated under presence of varying number of malicious nodes. According to the observed results, the proposed protocol guarantees high packet delivery ratio and low delay, compared to cluster based routing protocol.

Key words: Cluster, mobile agent, distributed key generation, mobile ad hoc network, security.

INTRODUCTION

Mobile agents (MA) are commonly used in mobile ad hoc network (MANET) to collect network information and for network maintenance. MANET is wireless network with dynamic topology and without centralized control. MA is mobile software code that migrates from host to host. Due to its autonomy property, a MA can work without centralized control. MA also reacts automatically to the changes in the network environment. Due to dynamic nature, MA is suitable for MANET.

To collect information about network environment, more amounts of data need to be exchanged and processed across all nodes in the network. This increases the network traffic. MA reduces the network traffic, as it visits and collects data directly from mobile nodes. This reduces network latency (Wayne and Tom, 1999). Ensuring security of MA is difficult as the MA can be attacked by mobile node (MN) or other MA. A MA can also attack MN. One of the main challenges in using MA is to deal with security issues such as confidentiality, authentication, authorization, and non repudiation (Sarwarul Islam Rizvi et al., 2010; Marikkannu et al., 2011).

Previous work

In Pushpa Lakshmi and Vincent (2011), we proposed a secure dominating set based routing protocol which applies a fuzzy logic controller to evaluate the trust value of nodes. The network is partitioned into clusters, where the cluster heads (CH) are elected based on their trust value and probability of future contact. The trust value of each node is computed based on their packet drop rate, packet forwarded successfully, and packet forwarded with alteration. Routing is carried out through trustable nodes in the network. To increase the level of security, a composite key management technique is applied. For key revocation process, mobile agent (MA) system is used to collect revoke point values for cluster members. Initially, revoke point value of all nodes is 0. When MA executes on mobile node (MN), the MN can update the revoke point value of suspected node. Based on revoke point value, the trust level of suspected node can be classified as not trustable, fully trustable, normal, low, and avg. The paper does not cover about security issues of MA code and data, election of common leader in cluster hierarchy, and selection of 'n' key serving nodes.

Ensuring security in MA system is important because MA code and data can be updated by other MA, affected by MN at runtime, and runtime MN may be affected by MA (Michael et al., 1998; Priyanka et al., 2010). In this current paper, we mainly focus on MA code integrity and authorization, MA owner authentication, MN authentication, and MA data protection. We also present improved MA system for key deactivation process, 'n' key serving nodes selection process, and common leader election process.

Related work

Secure image mechanism proposed in (Tarig, 2009) uses a secure image controller (SIC), which creates a copy of MA. The system classifies the node as trusted and untrusted. The original MA will directly be executed on trusted nodes. Whereas the encrypted copy of MA will be executed on untrusted nodes. When MA returns, SIC decrypt the MA code and compare hash digest of returned MA with hash digest of original MA code. The hash digest will be same, if the MA code is unaltered. Else SIC recorrect all the altered portion of MA code, using its backup copy. This method withstands eavesdropping and alteration attacks. But it does not address about attacks like masquerading, unauthorized access, replay attack, and non repudiation.

The security mechanism proposed in (Giovanni, 1997) maintains log files for agent's execution process. The log file records all activities performed during MA execution process. By checking log file, the agent owner can identify whether the MA is working properly as expected or not. This method requires maintenance of large log files. The security model proposed in (Sarwarul Islam Rizvi et al., 2010) is based on threshold cryptography. Changes to the agent code can be detected by computing message integrity code (MIC). The message is digitally signed by private key. The private key is generated by combining 'n' partial key shares. To decrypt the MA code and data, the MN must have proper private key. If MN is compromised, it will not generate correct key share. This method ensures security services like confidentiality, integrity and authenticity.

Environmental key generation method (Riordan and Schneier, 1998) generates key for encryption and decryption based on environmental parameters. Agent code and data is encrypted using the generated key. Agent can decrypt the code only when specified environmental condition exists. If the required environmental condition did not exist in the node, the decryption key will not be generated correctly. The agent cannot decrypt the code without knowing correct decryption key. Mobile agent model proposed in (Yikai, 2011) uses policy based cryptography. Authorization of MN is ensured by using policies defined by agent owner. MN will decrypt MA code only when it holds the policies defined by MA owner.

Mobile agent system proposed in (Shibli et al., 2009, 2010) uses security assertion markup language (SAML) ticket to authenticate MA code. SAML tickets are issued by policy decision point (PDP) server. Change to MA code is identified by verifying SAML tickets. Data accessibility is controlled by policy tokens, which specifies access control rules for MA and runtime node. Key distribution server generates group keys based on access policies of MA and MN. Data is encrypted by group key, which only allows the authorized MN to access the agent's platform data. It uses PKCS7 signed and enveloped data type to avoid unauthorized access of data (Shibli et al., 2010).

PROPOSED WORK

In this paper, we describe three mobile agents (MA) to perform the following tasks inside a cluster:

1. For key deactivation, MA collects each cluster member's trust opinion about other members in the cluster.
2. To elect common leader at higher level in hierarchy of clusters.
3. Collect trust value details of cluster members, used for selection of key serving nodes.

The process of MA is periodically initiated by CH. MA starts from CH and randomly moves to one of the neighboring cluster member. After reaching the neighbor, MA executes its process, collect the specified details and move to the next neighbor node. The movement of MA continues until it visits all nodes mentioned in the member list. Finally, when the MA return back to the CH, the CH use the collected data for the process of key deactivation, common leader election and key serving nodes selection. Any malicious node in the network can modify MA code and its data. Malicious node can also create false MA. MA code is protected by applying one way hash function. To authenticate the node, a parallel key based scheme is proposed. The identity of CH and cluster members is validated using their secret key or session key. The CH will start a new version of MA, when old version fails to reach the CH within specified duration. The new version will be transmitted in different route.

Keys generation

Public key generation

The public key of new node is obtained by applying one way hash function on node's *ID*. If G1 is an additive group of prime numbers of order q and G2 is a multiplicative group of prime numbers of order q. The one way hash function $H : \{0,1\}^* \rightarrow G$ is defined. The public key is periodically updated based on node's trust value. The newly computed public key depends on old public key and node's current trust value and is shown in Equation (1).

$$Publickey, P_N = \begin{cases} H(ID_N), newnode \\ H(P_N \parallel trustvalue_N), existingnode \end{cases} \quad (1)$$

Private key generation

A distributed private key generation scheme is used, where the private key of node is generated by (n+1) serving nodes, based on Equation (2). 'n' serving nodes are elected by CH based on trust value of the node. The value of 'n' can be selected based on security level required by nodes, available resources and mobility of node.

$$\mathrm{Pr}\, ivatekey, S_N = S \cdot H(P_N) \quad (2)$$

Where $S = (S_1 + S_2 + S_3 + \ldots + S_N + S_{CH_{ID}})$,

$S_1, S_2, \ldots S_N$ − Secret key shares generated by 'n' serving nodes and $S_{CH_{ID}}$ − Secret key shares generated by *CH*.

The steps involved in private key generation of node N are as follows:

1. CH selects 'n' serving nodes based on trust value.

2. CH sends $KEYSHARE(P_{ID}, ID)$ to 'n' serving nodes.
3. Each serving node generate the secret key share based on Equation (3) and send $SHAREREPLY(P_{ID}, KS_{ID})$.

$$Keyshare, KS_i = h_{ID_i}^{r_i} \quad (3)$$

Where i=1 to n, CH, $h_{ID_i}^{r_i} = H(ID_i \parallel TV_i)^{r_i}$, ID_i − Identification number of node i, TV_i − Trust value of node I, r^i − Random number generated by node i
4. CH checks for correctness of key shares using Equation (4)

$$e(P_i, KS_i)^{r_i} \overset{?}{=} e(P_i^{r_i}, KS_i), \quad i= 1\ to\ 'n' \quad (4)$$

The key share is correct, if Equation (2) is true. Else it shows the malicious behavior of the node. No other node in the cluster except CH knew about public key of serving nodes. The public key of serving nodes varies based on its trust value. So, other members cannot act like serving nodes.
5. Whole secret key is generated by CH based on Equation (5)

$$S_N = \sum_{i=1}^{n} (h_{ID_N}^{r_i})$$

Secret key of node N, $\qquad\qquad (5)$

Session key generation

The concept of mining based on backtracking is applied to generate the session key. The session key is generated based on previous traffic patterns that exist between the nodes. Each node maintains a table that holds details about last few traffic carried on in the network with other nodes. Session key is generated mainly based on frequent traffic pattern (FP) that exists between the nodes and is shown in Equation (6).
Session key shared by nodes *N1* and *N2*,

$$SK_{N1N2} = H(FP \parallel ID_{N1} \parallel ID_{N2} \parallel E_{PN1PN2}(r_{N1}P_{N1} \parallel r_{N2}P_{N2})) \quad (6)$$

Where r_{N1} and r_{N2} are random numbers selected by *N1* and *N2*. The random numbers generated by nodes are securely exchanged by applying encryption using node's public key. Nodes N1 and N2 exchange their random numbers as follows.

$$N1 \rightarrow N2 : E_{P_{N2}}(r_{N1})$$

Only the nodes N1 and N2 have similar traffic pattern. N1 and N2 will share the session key only when their FP is similar. If FP is different, it indicates the malicious behavior of the node.

$$N2 \rightarrow N1 : E_{P_{N1}}(r_{N2})$$

Key deactivation

MA maintains a hash table which includes CHID encrypted using secret or session key, node ID, trust value of the node, and revoke bit vector encrypted using secret key of the node or session key shared between the node and CH. Revoke bit vector is an 'n' bit vector, where 'n' represent number of nodes in the cluster. Hash index to access table entry is computed based on $N \bmod M$. N is the node ID and M is a constant selected based on cluster size. MA also maintains a route variable and an 'n' bit visited vector. Initially, route variable is set as null and visited vector bits are set as 0.

Mobile agent parameters

Mobile agent message package contains three parts: header, code and data. The header includes fields encrypted by MA's symmetric key. MA's symmetric key is assigned by the CH. The value of the header fields are updated by the MA code during its execution in each MN. The code represents the executable MA code. The data field contains a hash table with hash entries for each cluster members.

Header fields: Agent ID – a unique number assigned by MA owner to identify each mobile agent.
Exptime – It represent agent's life time. The agent get deleted automatically when it time expires. The agent must return back to the owner before its expiry period. When it fails, the owner initiates a new copy of MA with new agent ID. The maximum number of copies the owner initiates is limited and decided by the owner.
Time stamp – Time stamp of MA represent the time elapsed from beginning of its execution.
Visited vector – It is an 'n' bit vector initially assigned as 0. 'n' represents number of member nodes in the cluster. Whenever the MA visits a MN, the bit corresponding to that MN is set as 1. MA returns back to the owner when all the nodes are visited.
Member list – In the member list, the owner specifies the IDs of all nodes to be visited by MA. The bits in visited vector are assigned to MN based on the order of ID specified in the member list. For example, the first node in member list is N1, and then the first bit in visited vector is assigned to N1. The execution of MA code is limited only to the nodes listed in member list.

Route vector – This specifies the route information. Initially, the MA owner set it as null. Whenever a mobile node is visited, the ID of the node is appended in the route vector. This route vector information will be used by owner to forward future mobile agents.
Count – Number of nodes the agent visits during its traversal, calculated using number of 1's present in visited vector. MA executes its code only when counter <= *memcnt*. MA will be deleted if counter exceeds the limit. *memcnt* represents total number of member nodes present in the cluster.
Agent rights – The owner specifies the permission granted for MA. It includes details about resource utilization and read/write access.

Data field: Hash table – Hash table contains a hash entry for each node in member list. Each entry contains CHID, UNO, code hash, ID, revoke bit vector, and trust value.
CHID - Cluster head identification number.
UNO – Unique number assigned for MA.
Code hash – Hash value for the MA code computed by the CH.
ID – Identification number of cluster member.
Revoke bit vector - 'n' bit vector, where each bit corresponds to one cluster member.
Trust value – node's trust value.

CH initialization

The CH creates an entry for each member in the hash table. Each entry includes 4 fields: CHID, node ID, revoke bit vector, and trust value. Initially revoke bit vector and trust value are assigned as 0 for all members. CH also initializes the following:

Route vector = null; visited vector = 000000(n = 6); member list = {set of cluster member ID}
CH assigns a time limit for MA and starts the execution of MA process. The structure of the hash table is shown in Figure 1, in which *Skey* is current secret key of the node/session key shared between CH and the node.

The steps involved in CH initialization are as follows:

1. Initialize MA parameters: Agent ID, exptime, count = 0, visited vector = 000000, member list = {$ID_1, ID_2, ..., ID_n$}, route vector = null.
2. Initialize hash table entries. For example, 0th index entry is initialized as $E_{SkeyCHID_0}(CHID \mid ID_0 \mid 00000010)$.
Each hash entry is encrypted by secret key of the node or session key shared between the node and CH.
3. Select the next node to traverse.
4. Move to the next selected node.

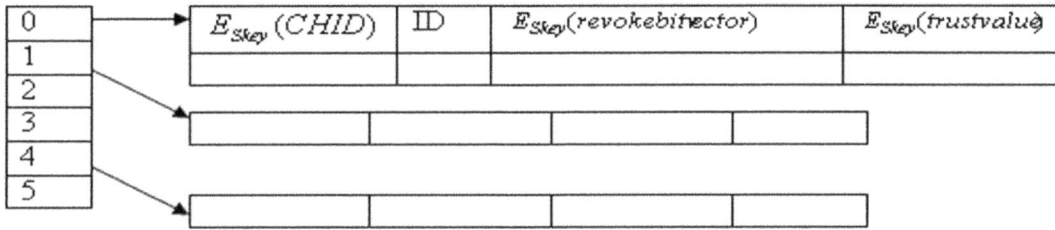

Figure 1. Data field of mobile agent. Data field contains a hash table. Each hash entry includes 4 fields. Field I: cluster head ID, agent unique number, hash value of agent code, agent rights. Field II: node ID, Field III: revoke bit vector, Field IV: trust value of node. The fields are encrypted using corresponding node's secret key or session key.

Mobile agent process

The steps involved in MN process are as follow:

Verification process at MN

1. Check the route vector. The MA code is already executed at current MN, if its ID is in route vector. In this case, MN stops the MA execution process.
2. Decrypt the code hash value using MN's secret key or session key.
3. Compute the hash value of the received MA code.
4. Compare the computed hash value with the decrypted hash value. If the hash values are equal, follow Step 4. If the hash value is invalid, the MN stops the MA execution process and does the following:

a. MN sends an error message (ERR), where PID represents previous node
ID.
b. On receiving ERR, CH sends UNTRUST (PID) to cluster members.
c. CH maintains a backup copy of MA code. It restarts the MA in different path excluding PID.
d. MN validates the owner of MA. Validation is done by decrypting the CHID specified in hash table entry.
e. Using hash index, MN identifies its corresponding hash table entry.
f. It decrypts the CHID either using the session key shared between them or using its current secret key.
g. If the agent is from malicious node, the decryption in the above step cannot take place correctly. In this case, MN terminates the MA execution process as follow:

i. MN verifies the access rights assigned for MA.
ii. It checks the life time of MA. If time expires, MN will stop the MA process.
iii. It checks the count value. If it exceeds the limit, MN stops the MA process.
iv. It checks the UNO of the MA. If UNO is invalid, MN will stop MA execution.

MA process – Header updation

1. Increment count by 1,
2. Update the visited vector. Set the bit corresponding to the current MN as 1,
3. Append the ID of current MN to route vector,
4. Encrypt all the header fields using MA's symmetric key.

MA process – Data updation

1. Set the bit value of revoke bit vector as 1, if MN suspects the trust worthiness of the node corresponding to that bit,
2. Encrypts the revoke bit vector using the session key shared between the current MN and CH or using MN's current secret key,
3. Compute MN's trust value using fuzzy logic controller. The trust value is within the range 0 and 10 ((0-1) not trusted, (2-3) low, (4-7) normal, (above 7) fully trustable).
4. Encrypts the trust value using the session key shared between the current MN and CH or using MN's current secret key,
5. Select the next node to be visited,
6. Move to the next node. Repeat from Step 1, until all nodes in member list are visited,
7. Return back to CH.

CH – Key deactivation

The steps involved in key deactivation decision making process by CH are as follows:

1. On receiving the MA before time expires, CH decrypts all revoke bit vectors. If time out occurs, CH starts a new version of MA and transmits in different route,
2. The CH identifies the node ID corresponding to the bit column having more number of ones,
3. The CH prepares a revocation list, which includes ID of the node whose key must be deactivated,
4. To ensure the security of the list, the node ID detail is encrypted using secret key or session key,

Figure 2. Key revocation list. The list is prepared by cluster head. Each node in the list contains ID of the suspected node. Each node value is encrypted using corresponding node's secret key or session key shared between node and cluster head. The list is transmitted across trustable nodes using mobile agent.

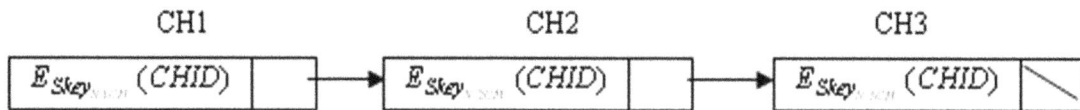

Figure 3. Common leader election. Cluster head with highest trust value is elected as leader. Elected leader ID is passed to all other cluster heads using a list. The ID is encrypted using secret key or session key of cluster head.

CH prepares a key revocation list which includes the ID of the suspected nodes, encrypted using session key or secret key. The revocation list is passed only to the trustable nodes in the cluster. To pass the revocation list, CH starts a MA in which member list includes only the ID of trustable nodes. MA follows the steps specified in MA code. On receiving the list, each node accesses its corresponding data, decrypt it and identify the ID of the suspected node. The cluster member will not consider any future messages received from the suspected node. The structure of the key revocation list is shown in Figure 2.

Selection of 'n' key serving nodes

MA – collection of trust values of members

Each node maintains information about their packet forwarding status and its neighbors. Trust value of the node is computed using fuzzy logic controller designed in our previous work (Pushpa Lakshmi and Vincent, 2011). The procedure follows the steps specified in MA code. In the computation process, MA activates the fuzzy controller to evaluate the trust value of the node. The trust value is secured by encrypting it with the secret key of the node or session key shared between the node and CH.

CH-key serving nodes selection

The steps involved in selection of 'n' key serving nodes are as follows:

1. On receiving the MA before time expires, CH decrypts the received trust value and sort the corresponding node IDs in descending order of their trust value. If MA is not

returned within the time limit, the CH issues a new version of MA and transmits it in different route,
2. CH selects top 'n' node IDs, which will act as key serving nodes in secret key generation.

Common leader election

The common leader which acts as administrator in the hierarchical cluster is elected based on trust value. The CH with maximum trust value is elected as common leader in the higher level of the hierarchical structure. The CH1 initiates a MA for common leader election. The member list includes IDs of all CHs. The procedure follows the steps specified in MA code. MA travels to the next CH, using the virtual link established between the CHs:

1. CH1 decrypts the received trust values and find the CHID with maximum trust value,
2. CH1 prepares a list that includes the elected CHID encrypted using secret key or session key. The structure of the list is shown in Figure 3.

Security issues

The MA in the network perhaps affected by malicious node or by other MA. The MA may also utilize the resource of MN or access the information of MN, without the permission of MN. The behavior of our MA is discussed based on security aspects.

Masquerading

In this attack, the malicious node will act as other node and may change the MA code and data. In the proposed model, the data in hash table is encrypted using the secret key of the node corresponding to the hash index or using the session key shared between the node and CH. The node that reads the data must first decrypt the message to validate the owner of MA. Decryption will be done only if the secret key or session key of the node is known. The secret key of the node is generated dynamically based on previous key and trust value of the node. The session key is generated based on past network traffic that exists between that node and CH. The malicious node can not identify the network traffic pattern of other node. So, the malicious node that acts as other node can not decrypt the MA data without knowing secret or session key. MA process continues only when the node is able to validate the owner. Updation to MA code can be detected using the hash value of MA code and data. Thus, the MA process ensures confidentiality, as the MA code and data are accessible only by authorized node.

Unauthorized access and alteration

Unauthorized access cannot be carried out by mobile node, as the data encrypted using secret key or session key of the node. Without permission, a node cannot access or update MA data. This ensures MA data integrity. Only the authorized MN can process MA. Even if the authorized MN later acts as malicious node, the MA data are secured as they are encrypted using corresponding node's secret key or session key. Any alteration in MA code can be detected using the code hash value. The malicious node is identified and informed to other members through UNTRUST message.

Replay attack

Time stamping is applied to avoid repeated transmission of MA. The owner of the MA attaches a timestamp which determines the lifetime of MA. The MA will be dropped when timestamp reaches expiry time. Route vector includes the ID of visited nodes. MA is transmitted only to the unvisited nodes. The current node stops the MA execution process, if it is already visited.

Non repudiation

The owner of MA, that is, CH include its CHID in the MA hash table. The CHID is used by MN for owner verification process. The CHID is encrypted by secret key or session key of the node. The secret key of each node is generated based on partial key distribution method by CH. The session key is based on past network traffic pattern that exists between the node and CH. Any node other than CH cannot generate secret or session key of a

node. So, CH cannot later deny the transmission of MA. Thus, the proposed model ensures authentication of MA and MN.

Eavesdropping

The header section of MA is encrypted using MA's symmetric key. A unique symmetric key and UNO is assigned for each MA by the CH. So, a MA cannot access the encrypted header fields of another MA. The MA data fields are encrypted by node's session or secret key. The data fields are accessible only after the validation of MA owner, MA code and MA UNO. So, data fields of a MA are secured from other MA.

Denial of service

Agent rights specify the access rights assigned for MA. If MA violates the agent rights, the MN will stop the MA execution process.

RESULTS AND DISCUSSION

The simulation results of the protocol obtained using ns-2 simulator. The performance of the proposed protocol is evaluated by comparing the results with secure cluster based routing protocol. Table 1 summarizes the simulation parameters. We used the metrics packet delivery ratio, end to end delay, and packet drop rate for performance evaluation. Packet delivery ratio is the ratio of number of data packets delivered to the destination. End to end delay refers to the delay involved in transmitting packet from source to destination, which covers queuing delay, processing delay, propagation delay, and transmission delay. Packet drop rate is the number of packets dropped by the node. The network is simulated with 40 nodes, moving in an area of size 1500 × 1500 m. The source node transmit packet of size 1024 bytes. Packets are transmitted at rate of 4 per sec.

Figure 4 shows that the packets delivered by our proposed scheme is high even in the presence of malicious nodes. In our proposed scheme, the route is selected based on the trust worthiness of the nodes. The private key and session key are generated mainly based on the trust value of the nodes. The trust value of malicious nodes will be low, and they will be removed automatically while selecting trustable shortest routing path. When the number of malicious nodes is 5 our proposed scheme was able to transmit 90 packets, whereas the secure cluster based scheme transmits only 20 packets. When the malicious nodes count increased to 20 our proposed scheme was able to deliver 700 packets, whereas secure cluster based scheme delivers only 500 packets. Our proposed scheme produce high

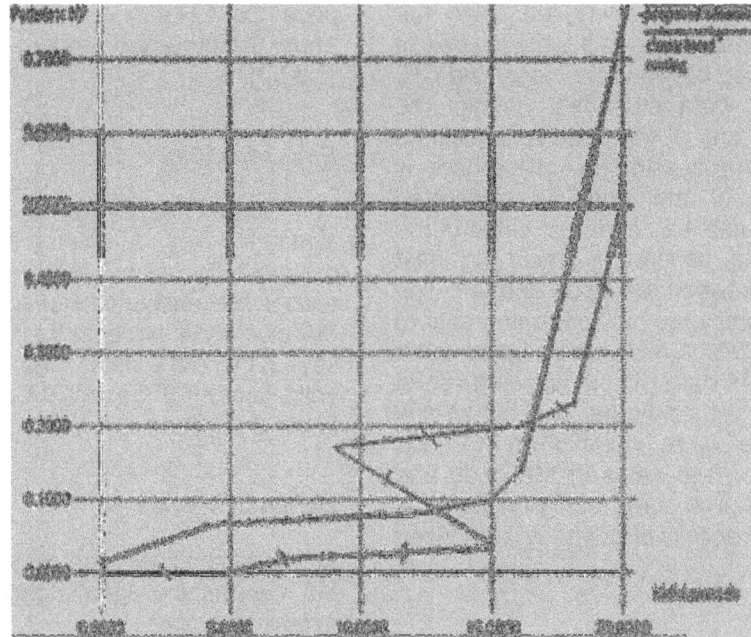

Figure 4. Packet delivered Vs Number of malicious nodes.

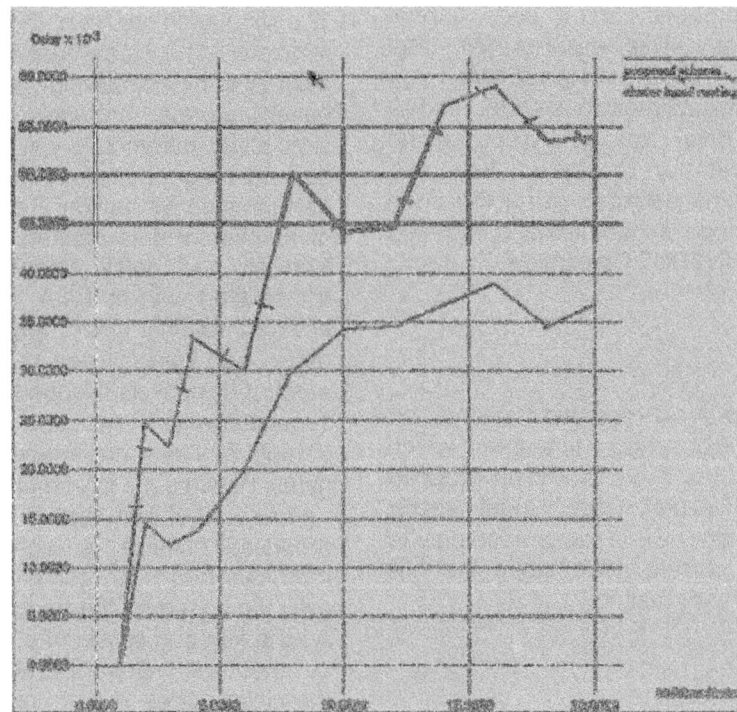

Figure 5. End to End delay Vs Number of malicious nodes.

packet delivery ratio even in presence of malicious nodes. The packet delivery ratio achieved in presence of varying number of malicious node is shown in Table 2. Figure 5 shows the end to end delay of proposed scheme and secure cluster based routing scheme. The end to end delay of our proposed scheme increases with increase in number of malicious nodes. However, the average end to end delay of our proposed scheme is less

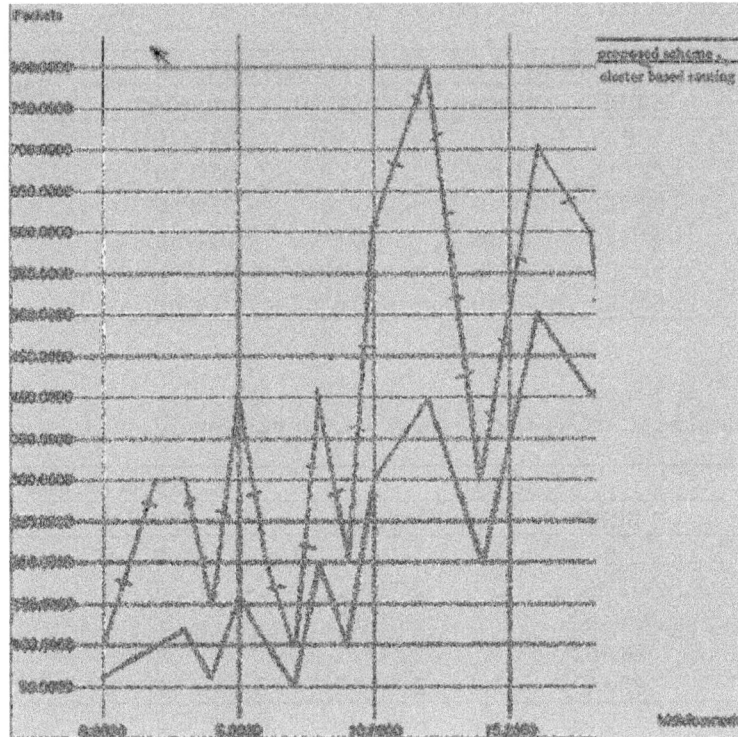

Figure 6. Packets dropped Vs Number of malicious nodes.

Table 1. Simulation parameters.

Parameter	Value
Transmitter range	250 m
Bandwidth	2 Mbps
Simulation time	20 s
Environment size	1500 m × 1500 m
Traffic type	Constant Bit Rate (CBR)
Packet rate	4 pkts/s
Packet size	1024 bytes

than delay incurred by secure cluster based routing scheme. When the number of malicious nodes in the network is 15, the end to end delay experienced by our proposed scheme is 38×10^{-3} which is less than 57×10^{-3} delay experienced by the other scheme. Our proposed scheme experienced nearly 15% less delay compared to secure cluster based routing scheme. The routing delay in presence of varying number of malicious node is shown in Table 2.

On average, the number of packets dropped by the other scheme is two times higher than the drop rate of our proposed scheme. The number of packets delivered by the proposed scheme is high even in presence of malicious nodes. Table 3 lists out routing delay and

packet delivery ratio of network in presence of varying number of nodes with respect to processing time.

Figure 6 shows the number of packets dropped by the proposed scheme and secure cluster based routing scheme. The packet drop rate of the proposed scheme is low in number of malicious nodes.

Conclusion

In this paper, we presented mobile agent based cluster maintenance mechanism using secure routing protocol for mobile ad hoc network. Securing mobile agent from mobile node, mobile agent from other mobile agent and

Table 2. Delay and packet delivery ratio of network with varying number of malicious node.

No. of malicious nodes	Delay		Packets delivered		Saving in packet delivery
	CBRP	Proposed	CBRP	Proposed	
2	0	0	0	$0.02*10^3$	$0.02*10^3$
3	$5*10^{-3}$	$5*10^{-3}$	0	$0.03*10^3$	$0.03*10^3$
5	$32*10^{-3}$	$17*10^{-3}$	0	$0.08*10^3$	$0.08*10^3$
10	$44*10^{-3}$	$34*10^{-3}$	$0.03*10^3$	$0.08*10^3$	$0.05*10^3$
15	$57*10^{-3}$	$38*10^{-3}$	$0.04*10^3$	$0.1*10^3$	$0.06*10^3$
20	$54*10^{-3}$	$37*10^{-3}$	$0.5*10^3$	$0.8*10^3$	$0.03*10^3$

Table 3. Delay and packet delivery ratio of network with respect to time.

Time	Delay		No. of packets delivered	
	CBRP	Proposed	CBRP	Proposed
0	0	0	0	0
5	$4.2*10^{-3}$	$6*10^{-3}$	20	20
10	$15*10^{-3}$	$13.5*10^{-3}$	30	170
15	$26*10^{-3}$	$22*10^{-3}$	40	300
20	$37*10^{-3}$	$32*10^{-3}$	60	440

mobile node from mobile agent is achieved using cryptography based key generation mechanism.

Authorization of mobile agent and mobile node is ensured using the secret key of mobile node or session key shared between mobile agent owner and mobile node. Mobile agent is applied for various cluster maintenance tasks such as collection of cluster member details, selection of key serving nodes, key revocation, and selection of cluster leader. The simulation results show that the proposed scheme is effective. We compared the result with cluster based routing protocol (CBRP). The proposed scheme provides high packet delivery ratio and low delay compared to CBRP, in presence of malicious nodes.

REFERENCES

Giovanni V (1997). Protecting mobile agents through tracing. Proceedings of the Third ECOOP Workshop on Operating System support for Mobile Object Systems, Finland. pp. 137-153.

Marikkannu P, Adri-Jovin JJ, Purusothaman T (2011). An Enhanced Mobile Agent Security Protocol. Eur. J. Sci. Res. 5(3):321-331.

Michael SG, Jennifer CB, David GH (1998). Mobile agent and security, IEEE Communications Magazine, July. pp. 76-85.

Priyanka D, Kamlesh D, Govil MC (2010). Security Issues in Mobile Agents. Int. J. Comput. Appl. 11(4).

Pushpa Lakshmi R, Vincent AKA (2011). A Secure Dominating set based routing and key management scheme in Mobile Ad hoc Network. WSEAS Trans. Commun. 10(10):297-307.

Riordan J, Schneier B (1998). Environmental Key Generation towards Clueless Agents Mobile Agents and Security, G. Vigna, ed., Springer-Verlag. pp. 15-24.

Sarwarul Islam Rizvi SM, Zinat S, Bo S, Md. Washiqul I (2010). Security of Mobile Agent in Ad hoc Network using Threshold Cryptography, World Academy of Science. Eng. Technol. pp. 424-427.

Shibli MA, Giambruno A, Muftic S, Lioy A (2009). MagicNET: Security System for Development, Validation and Adoption of Mobile Agents, 3rd IEEE International Conference on Network and System Security. pp. 389-396.

Shibli MA, Yousaf I, Muftic S (2010). MagicNET - Security System for Protection of Mobile Agents. 24th IEEE International Conference on Advanced Information Networking and Applications (AINA) APR 20-23, Perth, Australia. pp. 1233-1240.

Tarig MA (2009). Using Secure-Image Mechanism to Protect Mobile Agent against malicious Hosts, World Academy of Science. Eng. Technol. pp. 439-444.

Wayne J, Tom K (1999). Mobile Agent Security, National Institute of Standards and Technology, Special Publication 800-19, August, US department of commerce, Technology administration, National Institute of Standards and Technology.

Yikai W (2011). Policy-based secure agents, Master of computer science thesis, University of Wollongong, http://ro.uow.edu.au/theses/3261/.

10

On consistency and security issues in collaborative virtual environment systems

Abdulsalam Ya'u Gital[1] , Abdul Samad bn Ismail[1] and Shamala Subramaniam[2]

[1]Department of Computer Systems and Communications, Faculty of Computer Science and Information Systems, Universiti Teknologi, 81310, UTM Johor, Bahru, Malaysia.
[2]Department of Computer Technology and Network, Faculty of Computer Science and Information Technology, Universiti Putra Malaysia, 43400 Serdang, Malaysia.

This paper survey security issues in collaborative virtual environments (CVEs) systems. In CVE, multiple users work on different computers which are interconnected through different networks to interact in a shared virtual world. Due to the nature of the geographically disperse users and their connection via different networks, there are numerous security threats that denied fulfillments of most important CVE requirements which have been ignored (e.g. consistency). In this paper, we outlined the types of collaborative virtual environment applications that can be affected by security threats and attacks, it discussed some of the most important CVE systems security requirements, and then discussed the different types of security threats and attacks related to CVE systems security requirements. Finally, we describe the state of the art of CVE system security.

Key words: Collaborative virtual environments (CVE), security requirement, threats, attack.

INTRODUCTION

Currently, because of the explosive growth of the computer and communication technology, many valuable materials in military training systems and manufacturing systems in industries can be shared with each other via internet. Quite a number of researches have been done in computer applications for facilitating collaboration among multiple and distributed users, but rarely people work in the area of security in CVE systems. In CVEs, one of the main research topics is how to efficiently transmit messages to provide scalability, minimized delay, and reliability (Yong et al., 2008). CVEs need to be designed to allow groups of people from a diverse set of

organizations and locations to work together easily and securely. Security of such an environment is a crucial issue; this is because of the nature of types of data to be transmitted during collaborative activities. Among many issues in the design and implementation of collaborative virtual environment, the major ones include but not limited to security, scheduling and e-resource discovery (Signh and Signh, 2010).

This paper reviews available relevant literatures on collaborative virtual environments security issues. Most published works consider CVE requirements such as scalability, consistency, reliability, and implement series of solutions without considering security issues. Since CVE systems rely solidly on network to perform all transaction, security of these systems are important and weakness in it may lead to unsatisfactory results. All the research conducted on either improving scalability and or reliability exposes the system to many security threats due to the distributed nature of the infrastructures and did not provide security solutions. The paper further introduces collaborative virtual environment and the types of collaborative application that can be affected by security threats, followed by review of CVE security requirements. It further went on to review different types of security threats related to CVE systems. It concludes the work and proposes solutions.

COLLABORATIVE VIRTUAL ENVIRONMENTS

Collaborative virtual environment (CVE) allows participant from distant geographic location to share a common virtual environment including virtual entities and resources maintained by a group of computers which can support effective communication between the users to achieve better coordination tasks. Applications of CVEs include education, massively multiplayer online games (e.g., World of Warcraft), virtual worlds (e.g., Second Life), military training, industrial remote training, and collaborative engineering (Deng and Lau, 2012). As the number of concurrent participants is becoming larger, data exchange between the participants increases, the security of CVE systems is not guaranteed because of the location of the participants which is from different network, and data transmitted must pass through different network before it gets to destination.

There are two types of models mostly use for implementing CVE systems: Client server and Peer-to-peer. Even though client server with a single server cannot scale due to increasing number of users in such a systems (Hu et al., 2011a), but it offer the strongest security, as all important state transitions can be verified and safely stored on the server. The server accepts client input directly. The server has total control over how the CVE state is updated and can take into account any factors deemed relevant (John and Jon, 2010).

In CVE systems, all users share the same virtual space, and each of them is being represented by an entity within the virtual environment. When a user connects to the environment, moves and/or interacts with other entities, the CVE systems require the update to be transmitted in order to update its own state, and to distribute the update of state to other users (Hu et al., 2011b). With the expansion of the scale of applications and the increasing number of users, the security of the systems needs special attention for successful collaboration. Collaboration is often encouraged on the basis that it delivers greater productivity. At the heart of collaboration is the ability of the group to contribute. It is also the case that collaboration is one of a number of different ways of working together and in that sense, it is important to consider its security to protect the integrity and confidentiality of the transmitted data. While it is evident that encouraging collaboration through the use of technology has merit, it is also important to realize that successful collaboration in this day and age requires elements of technology, process and people.

There are two basic foundations of CVEs. At first, 3D virtual worlds provide the three-dimensional view and immersive environment. Second, distributed systems are necessary to offer multi-user and collaborative tools capabilities. CVEs create realistic 3D (virtual reality) displays and provide a rotational capability for views inside, above, beside, or under objects and systems in reduced, normal, or large scale. It makes the significant reduction of the time of new commercial product development and military system operational readiness, and overall development and manufacturing costs (Yong et al., 2008).

Latency which is the time interval from the time a user perform an action to the time other users will noticed the action, represents the quality of service provided to users by the system since it determines how fast changes in the virtual world are noticed to the proper client computer (Reuda et al., 2007). In this case, with the distributed nature of the users, any network that is affected by security threats such as DDoS will notice delay beyond the maximum expected for successful collaboration.

Types of collaborative applications affected by security threats

Collaborative application can be categorized according to the nature of the problem at hand. Most of these collaborative applications fall into the following six groups:

(i) Collaborative work environments (for conducting collaborative work such as military training, engineering design, visualization, documentation, etc.).
(ii) Meetings, seminars and conferences over the internet.
(iii) Simulation of face-to-face contacts where visual quality is critical (such as recruitment interviews, medical diagnoses and remote surgical operations).
(iv) Distance learning environments (for providing course materials, holding a tutorial, carrying out a team project, and conducting an examination).
(v) Networked computer games.
(vi) Leisure and entertainment (including 3D navigation and virtual embodiment) etc.

COLLABORATIVE VIRTUAL ENVIRONMENTS ARCHITECTURE

The most popular architectures used for network virtual collaborative environment design are the well known peer-to-peer architecture and client server architecture with a single server or multiple servers (Mecedonia et al., 1994). These architectures have several drawbacks that require researchers' attention, considering the current types of CVE system handling thousands simultaneously collaborating users. Many researchers contributed a lot to CVE systems in different ways and have achieved a great success for instance, Hu et al. (2011a), Yong et al. (2008), Wang (2011), Hu et al. (2011b), Morillo et al. (2010), Lin et al. (2006), Deng and Lau (2012), Chen and Chen (2006), Li (2011), Lin et al. (2008), Carlini and Ricci (2006), Kulkarni et al. (2007), Sandhu et al. (2011), Ahmed and Shirmohammadi (2008), Chen et al. (2010), Nguyen et al. (2009), Tang et al. (2010), Nguyen et al. (2011), Ta et al. (2010), Shao-Qing et al. (2003) and Hiroki and Yoshitaka (2008), but did not consider security issues which is another factor that may lead to unsatisfactory results in their findings.

Architectures based on networked servers are becoming a de-factor standard for DVE systems (Yong et al., 2008; Reuda et al., 2007). Each client in the system is attached to one of the distributed servers, when a user perform a task, the user computer controlling it sent an update message to the user computer controlling other avatars (Reuda et al., 2007). In order to maintain consistency and update view of the virtual worlds that are linked via different network, security of the link must be guaranteed and free from attack such as DDOS which are common in today's networks. This type of threats can seriously cause inconsistency that may lead to unsatisfactory result as stated previously.

Peer-to-peer architecture

In this communication architecture model, each user sends it update directly to other users. The idea is that all components in the distributed system have the same responsibilities acting both as clients and servers. There is no central server to keep status of the whole system. Each peer maintains its own copy of the virtual environment states and exchanges data directly with other peers (Bu et al., 2007; Berket et al., 2005; Khoury et al., 2007; Pan and Francis, 2004). When a program makes changes to its own database, it sends the update data out so that other programs can update their individual databases (Yong et al., 2008). This architecture has the advantages of low communication latency and fault tolerance capability, for a single client's fault will not

cause whole system to crash. Conversely, there is communication complexity with the model as each user has to adopt the filtering algorithm to reduce the consumption of network resources which causes inconsistency of the system (Hu et al., 2011a). These network resources can also be affected by security threats that may result in an inconsistency situation even with the capabilities of the peer-to-peer.

Client server architecture

The client server architecture is classified as either single server or multi server architecture.

Client server with single server

In this model, all the clients' send update to the server; the only common server collects all of the data from the different clients' machine, and sends the results back to each participating client's machine. Each participant's application communicates only with a server that is responsible for passing messages to other clients. Although this model simplifies security implementation, and has a simple data structure to store and handle the data, it is not scalable. Therefore, it is avoided due to increasing number of collaborators. All other models adopted are subjected to a lot of security threats.

Client server with multiple servers

In this model, each client sends updates to the server it is connected to, and the server transmits to other clients and the remaining servers. The management of the virtual environment relies on the several interconnected servers and each server handles a portion of the virtual environment (Hu et al., 2011a). Security of systems implemented using such a model is facing a great challenge; this is because a change or modification by any security treats may lead to a serious error.

From the above description, one notices that servers and the clients execute series of functions to keep the consistency in the virtual environment. At the server side, the server perform the function of receiving the update messages from the clients, updating the whole virtual environment and transmitting updates of the virtual environment to other clients and servers. Moreover, at the clients' side, clients must execute functions of receiving the user's input as the update message, transmitting the update message to the server and receiving the update messages from the server to keep the virtual environment up-to-date. Whenever there are security threats such as Denial of Services (DoS) and or

Distributed Denial of Service (DDoS), the computing and communication resources suffers and the entire system becomes slower and suffers a long time delay, thereby resulting in conflicting virtual world status at a given time.

CVE SYSTEM SECURITY REQUIREMENTS

Computer security can be defined from the aspect of information flow in the networked environment (Corona et al., 2009), currently security in CVE systems in an active research area. Network security is a threat, intrusion, denial of services on a network infrastructure that will analyze your work and gain information to eventually cause the network to crash or to be corrupted. Any network devices that are not being monitored are the main source of information leakage in most organizations (www.ayuverda.hubpages.com). In CVE data such as military attack preparation by group of army from different region, manufacturing system information, etc passed several threads and challenges when it comes to security. Security is a crucial issue when it comes to virtual collaboration because the participants are from geographically disperses location and is connected through different network. Each participant is expected to receive all the transaction. In this transaction, security is a big challenge for reliable and secure transmission of data from and to all participating members during virtual collaboration. The following are security requirements in collaborative virtual environment:

Authentication

This is a mechanism that a user uses to validate data during collaboration with other members of the team. Without this, attackers get access to the data and the data can be modified without the notice of the genuine users. Authentication prevents attackers from getting access to the network and the data on the network (Bullock and Benford, 1999).

Confidentiality

The data transmitted to other members of the collaborating team should only be understood by them. And the system needs to protect the channel transmitting the data so that attackers cannot get access to the data. In this case, both stored data and data in transmission should be protected from attackers (Song et al., 2005).

Integrity

This ensures that the data on transmission process is not

deleted or modified by any malicious programs or unauthorized users by the system. The system should be able to inspect viruses and backdoor programs (Salles et al., 2002).

Availability

The network should provide guaranteed services to all participating members at all the time despite attack from attackers. Users should be able to access the system whenever they want to use it because of time critical processes (Yong et al., 2008).

Non-repudiation

The source of all updates or modification should be known and identified by the system. In that case, the system should maintain the origin of the data and the information received (Yong et al., 2008). To ensure all the above security attributes, security assurance has to be put in place. The classification of computer assurance process is as shown in Figure 1.

SOME COMMON NETWORK ATTACK AND SECURITY THREATS THAT CAN AFFECT CVE SYSTEMS

CVE systems share the same internet with all other applications, therefore the different types of attack on the internet forms part of the attacked to be considered while discussing security issues in CVE systems.

Denial-of-service (DoS) and distributed-denial-of-service (DDoS) attacks

Denial of service attack can cause inconsistency that may lead to unpleasant result. A denial of service attack is a type of Internet attack that is aimed at large websites by consuming both computing and communication resources, disruption of routing information and physical network components. These attack results to slow network performance and inability to access any web site among others. A distributed denial of service attack (DDoS) occurs when multiple compromised systems or multiple attackers flood the bandwidth or resources of a targeted system with useless traffic (Gul and Hussaini, 2011).

This type of attack in a time dependent systems such as CVE can cause inconsistency thereby violating the requirement for successful collaboration. In an application like military training, group demonstration of lunching attack and or group study of map area for a mission, it

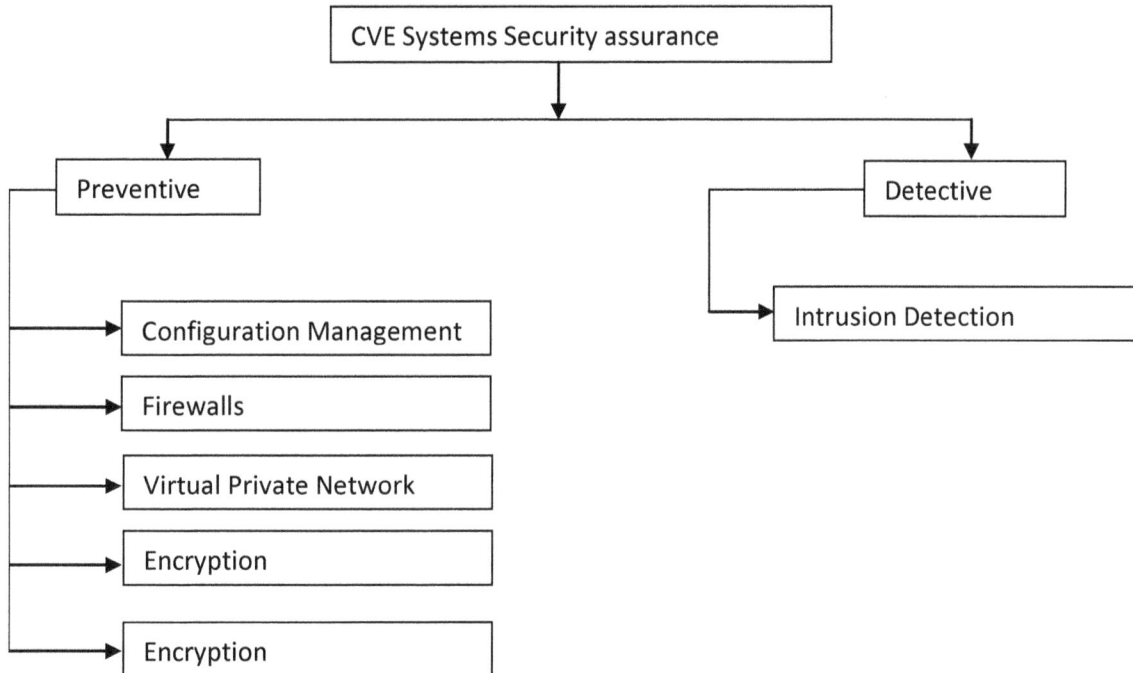

Figure 1. Classification of Collaborative Virtual Environment Security Policy.

may leave among the participant others with out-of-date plan. That may lead to failure or cause serious casualty. Many serious network security problems are caused by Distributed Denial of Service (DDoS) attacks and virus worms-spreading (Davie and Medved, 2009; Desnoyers and Shenoy, 2007). DDoS attacks always paralyze the services which network nodes can provide and occupy the network bandwidth by flooding volumes of traffic to the victims. One attack node may contribute low-rate malicious traffic but attack traffic from widely distributed attack nodes is aggregated toward to the victim (Wang and Huang, 2009; Scaforne, 2007).

Eavesdropping

This is the process of gathering users' machine information such as IP address, the operating system use by the machine and the service the machine is offering in order to launch an attack that is not likely to be noticed by the user. In general, the majority of network communications occur in an unsecured format, which allows an attacker who has gained access to data paths in your network to interpret the traffic (www.ayuverda.com). When an attacker is eavesdropping on your communications, it is referred to as sniffing or snooping. The ability of an eavesdropper to

monitor the network is generally a biggest security problem when it cone collaborative military training, manufacturing systems and in Education (On-line examination).

Sniffing

This type of attack generate similar problem or security threat in CVE systems as described in mapping eavesdropping. Packet sniffing is the interception of data packets traversing a network. A sniffer program works at the ethernet layer in combination with network interface cards (NIC) to capture all traffic traveling to and from internet host site. Further, if any of the Ethernet NIC cards are in promiscuous mode, the sniffer program will pick up all communication packets floating by anywhere near the internet host site. A sniffer placed on any backbone device, inter-network link or network aggregation point will therefore be able to monitor a whole lot of traffic.

Most of packet sniffers are passive and they listen to all data link layer frames passing by the device's network interface. There are dozens of freely available packet sniffer programs on the internet. The more sophisticated ones are the once that allow more active intrusion (www.ayuverda.com).

Spoofing

Any internet connected device necessarily sends IP datagram into the network. Such internet data packets carry the sender's IP address as well as application-layer data. If the attacker obtains control over the software running on a network device, they can then easily modify the device's protocols to place an arbitrary IP address into the data packet's source address field. This is known as IP spoofing, which makes any payload appear to come from any source. With a spoofed source IP address on a datagram, it is difficult to find the host that actually sent the datagram.

Hijacking (man-in-the-middle attack)

This is a technique that takes advantage of a weakness in the TCP/IP protocol stack, and the way headers are constructed. Hijacking occurs when someone between you and the person with whom you are communicating is actively monitoring, capturing, and controlling your communication transparently. For example, the attacker can re-route a data exchange. When computers are communicating at low levels of the network layer, the computers might not be able to determine with whom they are exchanging data. Man-in-middle attacks are like someone assuming your identity in order to read your message. The person on the other end might believe it is you, because the attacker might be actively replying as you, to keep the exchange going and gain more information (Anderson, 2007).

STATUS OF CVE SYSTEMS SECURITY

In order to come up with secured, CVE systems, it has been found necessary to evaluate the existing CVE platform base on the security requirements of the CVE systems. Other requirement can also feature for other reference and not for the purpose of this review. The CVE security requirement and other closely related requirements here serve as the evaluation criteria to show whether or not a particular CVE platform satisfies fully these requirements. Many researches to realize scalability, reliability, consistency, responsiveness, extensibility, persistency in CVE have been taking place for more than 20 years, but did not consider security aspect which is vital to any organization. Yet achieving scalability, reliability, consistency, responsiveness in most of the platform is yet to be met. The other entire requirements have effect on the security of the systems. Participant access to CVE objects and information becomes an important topic of discussion because of the

growth in the use of CVE. In virtual reality games, storefronts, classrooms, and laboratories for example, the need to control access to spaces and objects is integral to the security of activities in these virtual realms. However, limited access controls are typically available in CVEs (Wright and Madey, 2010). There is a limited number of efforts that deal with security controls in CVE systems (Wright and Madey, 2010).

CVE system such as massively multiplayer online gaming (MMPG) has experienced tremendous growth over the past decade. The number of players, game operators, game designers, and gaming companies with stake in this industry has also increased remarkably (Gupta et al., 2009). As a result, the need for security in MMPGs is becoming increasingly critical. Cheating, virtual frauds, and other security attacks are becoming increasingly widespread in the virtual world (Debbie Jiang, 2011). In 2006, it was estimated that there are more than 10 million people playing MMPGs, with the number doubling every two years (Brian, 2007). One of the most popular MMPGs is Blizzard's World of Warcraft (WoW), which reached a subscriber base of 12 million in October 2010.

Due to the architecture of MMPGs and the large number of participants, there is an inherent lack of security in these games (Figure 1), which creates fertile grounds for cheating. In addition to the inherent security risk, MMPGs are also lacking in terms of legal regulation, security and privacy protection, and other related legislation which can resolve these security issues. As a result, users have taken advantage of this shortcoming and exploited these games through hacks, attacks, and cheats (Debbie, 2011). Virtual Life Network is another system that was designed without security consideration. On the need to secure some basis, security measures are added. That instead led to non-secure solution (Ilja, 2009).

NPSNET-V is a Java-based application with no security beyond the default provided by the Java Virtual Machine (JVM) (Salles et al., 2002). Ernesto et al. (2002) added that an easy assumption upon which to construct networked applications is that any security concern can generally be resolved via existing computer, network and database security mechanisms. Therefore, the desired security level of the application must be ensured by the application itself. Distributed Interactive Simulation (DIS) aims at proposing a common architecture for communication integration and the interconnection of allowing the large scale simulators. After the success of SIMNET (James et al., 1993), DIS (IEEE 1278.1A, 1998) was developed to address the interoperability of heterogeneous simulators. The essence of DIS is the creation of synthetic environment within which humans and simulations interact at multiple networked sites. DIS was not fully distributed; each message must be received

Table 1. State of the art CVE systems.

CVE platform / Evaluation criteria	MMOG	MASSIVE	DIVE	NPSNET	SPLINE	BRICKNET	SIMNET	VLNET	DIS
Scalability	Good	Average	Average	Average	Average	Average	Average	Low	Average
Reliability	Average	Average	Average	Average	Average	Average	Average	Low	Low
Consistency	Good	Average	Good	Average	Average	Good	Low	Average	Good
Responsiveness	Average	Average	Average	Low	Average	Average	Average	Low	Average
Extensibility	Good	Low	Average	Average	Low	Average	Low	Average	Average
Persistency	Good	Low	Average	Low	Low	Average	Low	Low	Low
Security	Low	Low	Low	Average	Low	Low	Low	Low	Average

and treated by each node, which clutter the bandwidth even though not a lot of data is transmitted. DIS does not manage latency and causality that made the reusability of simulations impossible. Latencies were not controlled and no time management service was incorporated which caused data losses due to the rejection of too old packets. That affects the security of the systems because there was systems ware developed without full security considerations.

The current version of MASSIVE is MASSIVE-3. MASSIVE-3 is based on the authors experience on MASSIVE 1,2. According to James et al. (1993), MASSIVE-3 is a multi-user CVE System that supports populated and interactive virtual worlds combining 3D graphics, real time audio and stream video. MASSIVE-3 allows its virtual worlds to be spatially structured as multiple linked locales each of which can be an arbitrary virtual space (e.g. room, building and open region) with its own Cartesian coordinate system. MASSIVE-3 extends the locales by allowing current locales to be linked to recording of other locales.

According to Ta et al. (2010), Distributed Interactive Virtual Environment (DIVE) is one of the most acknowledged Virtual Collaborative System, which is a tool kit for building distributed VR application in a heterogeneous network environment. DIVE allows many users and applications to interact in a real-time through virtual environment. It can also be described as an Internet-based multi-user system that allows remote participants to meet and interact with each other in a virtual 3D space. DIVE was developed at the Swedish Institute of Computer Science. It is one of the early systems that continue to be developed and improved over the years. The DIVE run-time environment consist of a set of communicating processes, running on nodes distributed within a LAN or WAN. The processes which are either a human user or an autonomous application have access to number of databases updating concurrently. The virtual world in DIVE consists of a database containing numbers of description of graphical object. Objects can be added or modified dynamically, and concurrently using a distributed locking mechanism. DIVE uses multicast protocols for the communication simulating a large shared memory for a process group through the network (Chander, 2010; Gupta et al., 2009).

According to Singh et al. (1995), BrickNet enables graphical objects to be maintained, managed, used efficiently, and permits objects to be shared by multiple virtual worlds or clients. A client can connect to a server to request objects of its interest. These objects are deposited by other clients connected to the same server or another server on the network. Depending on the availability and access rights of objects, the server satisfies client requests. BrickNet's object sharing strategy allows users to set-up their own private work-spaces, populated by shared and private objects. BrickNet virtual worlds are not restricted to sharing an identical set of objects. Virtual world manages its own set of objects, some or all of which may be shared with the other virtual worlds on the network. This basic arrangement can be used to implement several types of applications including collaborative, interactive learning systems.

The security of all the above systems lack literature. Table 1 summarizes the state of the art security of CVE systems and other relevant requirements as described earlier. Researchers did less in this area. Now that CVE is applied in many fields to achieve great cost effective group activities even where the participants are far away, the area has gain researchers attention, and it is high time the issues of CVE systems security be researched upon in order to provide workable solutions. In this evaluation, Low implies no security is implemented, Average implies a system with little security consideration and Good represent full implementation of security measures.

Conclusion

Security of CVE systems is becoming a serious topic of

research due to its application in many areas of study. This paper review the general security requirement in CVEs, identify different types of network attacks and security threats related to the different security requirement in CVE systems, and survey the state of the art of some CVE systems security and other requirements. This is because there are some requirements that achieving them without security consideration may lead to unsatisfactory results. However, it is required that a reliable intrusion detection model for CVE systems should be developed and is lacking in literature.

REFERENCES

Ahmed DT, Shirmohammadi S (2008). Performance Enhancement in MMOGs Using Entity Types. pp. 215-229.

Anderson R (2006). Security Engineering: A Guide to Building Dependable Distributed Systems. First Edition. 21:633-678.

Berket K, Essiari A, Thompson MR (2005). "Securing Resources in Collaborative Environments: A Peer-to-peer Approach," in Proc. of the 17th IASTED International Conference on Parallel and Distributed Computing and Systems, 2005.

Brian EM (2007). "Second Life and Other Virtual Worlds: A Road map for Research," International Conference on Information Systems (2007): http://www.bus.iastate.edu/mennecke/CAIS-Vol22-Article20.pdf.

Bu S, Boehm S, Portela M, Jo H (2007). "Collaborative Design Review in Virtual Environment," in Proc. of Korea Computer Congress 2007, C:229-232.

Bullock A, Benford S (1999). "An Access Control Framework for Multi-user Collaborative Environments," In Proc. SIGGROUP Conference on Supporting Group Work, pp.140-149.

Carlini E, Ricci L (2006). Integration of P2P and Clouds to Support Massively Multiuser Virtual Environments. Science. www.pap.vs.uni-due.de/MMVE10/papers/mmve2010_submission_3.pd.

Chander VR (2010). An Improved Object Interaction Framwork for Dynamic Collaborative Virtual Environments. IJCSNS.10(9):91-95

Chen J, Grottke S, Sablatnig J (2010). Scalability of a Distributed Virtual Environment Based on a Structured Peer-To-Peer Architecture. Symposium A Quarterly J. Modern Foreign Literatures. pp. 1-8.

Chen Jfa, Chen T.-han. (2006). VON: A Scalable Peer-to-Peer Network for Virtual Environments. Ieee Network, (August), pp. 22-31.

Corona I, Giachinto G, Mazzariello C, Roli F, Sansone C (2009). Information Fusion for Computer Security: State of the Art and Open Issues. Info. Fusion. 10(4):274-284.

Davie B, Medved J (2009). "A programmable overlay router for service provider innovation," in Workshop on Programmable Routers for Extensible Services of Tomorrow (PRESTO), Aug. 2009.

Debbie J (2011). Security Issues in MMOG. ACC 626 Research Paper June, 2011.

Deng Y, Lau RWH (2012). On delay adjustment for dynamic load balancing in distributed virtual environments. IEEE trans. visualization computer graphics, 18(4):529-37. doi:10.1109/TVCG.2012.52.

Desnoyers PJ, Shenoy P (2007). "Hyperion: High volume stream archival for retrospective querying," in USNIEX, 2007.

Ernesto J, Sallés J, Bret M, Michael C, Don M, Andrzej K (2002). Security of Runtime Extensible Virtual Environments. CVE'02, 2002, Bonn, Germany. ACM 1-58113-489-4/02/0009.

Gul I, Hussaini M (2011)." Distributed Cloud Intrusion Detection Model" Int. J. Adv. Sci. Technol. September, 2011. P.34.

Gupta N, Demers A, Gehrke J, Unterbrunner P, White W (2009). Scalability for Virtual Worlds. Complexity. ICDE '09 Proceedings of the 2009 IEEE International Conference on Data Engineering. pp 1311-1314.

Hiroki O, Yoshitaka S (2008). Assynchronous Collaborative Virtual Environment Support System by Using Revission Tree Presentation Method. IEEE Computer Society, 2008. http://us.blizzard.com/en-us/company/press/pressreleases.html?101007.

Hu X, Liu L, Yu T (2011). A hierarchical architecture for improving scalability and consistency in CVE systems. Int. J. Parallel, Emergent Distributed Syst. 26(3):179-205. doi:10.1080/17445760.2010.49572.

Hu XM, Cai HX, Yu T (2011). A Self-Adaptive Filtering Algorithm Based on Consistency QoS in CVE Systems. Adv. Mater. Res. 225-226:301-306. doi:10.4028/www.scientific.net/AMR.225-226.301.

IEEE 1278.1A-(1998). –"Standard for Distributed Interactive Simulation - Application protocols". E-ISBN 0-7381-0993-2, Print ISBN 07381-0174-5.

Ilja L (2009). VirtualLife Security Infrastructure. Master's Thesis. June, 2009.

James MC, Alan D, Bob G, Paul M, Dale M, Dan O (1993). "The Simnet Virtual world architecture". In VR, pp. 450-455.

John LM, Jon C (2010).. The Near-Term Feasibility of P2P MMOGs. In Proceedings of the 9th Annual Workshop on Network and Systems Support for Games, 2010.

Khoury M, Shen X, Shirmohammadi S (2007). "Peer-to-Peer Collaborative Virtual Environment for E-Commerce," CCECE 2007, pp. 828-831.

Kulkarni S, Douglas S, Churchill D (2007). Badumna : A decentralised network engine for virtual environments. Environments.

Li Y (2011). Determining Optimal Update Period for Minimizing Inconsistency in Multi-server Distributed Virtual Environments. Simulation. doi:10.1109/DS-RT.2011.10.

Lin Q, Zhang L, Ding S, Feng G, Huang G (2008). Intelligent Mobile Agents for Large-Scale Collaborative Virtual Environment. Processing, 7(2):63-72.

Lin Q, Zhang L, Neo N, Kusuma I (2006). Addressing Scalability Issues in Large-Scale Collaborative Virtual Environment. Design, 477-485.

Mecedonia MR, Zyda MJ, Pratt DR, Barham PT (1994). NPSNET: A Network Software Architecture For Large Scale Virtual Environments, 3(4):1-30.

Morillo P, Rueda S, Orduña JM, Duato J (2010). Ensuring the performance and scalability of peer-to-peer distributed virtual environments. Future Generation Computer Systems, 26(7):905-915. Elsevier B.V. doi:10.1016/j.future.2010.03.003.

Nguyen D, Ta B, Zhou S, Cai W, Tang X (2009). Efficient Zone Mapping Algorithms for Distributed Virtual Environments. Technology. doi:10.1109/PADS.2009.10.

Nguyen D, Ta B, Zhou S, Cai W, Tang X, Ayani R (2011). Multi-objective zone mapping in large-scale distributed virtual environments. J. Network Computer Applications, 34(2):551-561. Elsevier. doi:10.1016/j.jnca.2010.12.008.

Pan Y, Francis MT (2004). "A peer-to-peer Collaborative 3D Virtual Environment for Visualization," in Proc. SPIE. 5295:180-188.

Reuda S, Morillo P, Orduna JM, Duato J (2007). A generic approach for adding QoS to distributed virtual environments. Computer Communications. Elsevier. 30:731-739.

Salles EJ, Michael JB, Capps M, McGregor D, Kapolka A (2002). "Security of Runtime Extensible Virtual Environments," in Proc. the 4th International Conference on Collaborative virtual environments, P. 97104.

Sandhu UA, Haider S, Naseer S, Ateeb OU (2011). "A Survey of Intrusion Detection & Prevention Techniques" 2011 Int. Conf. Info. Comm. Manage. IPCSIT vol.16 (2011) IACSIT Press, Singapore.

Scaforne KSA (2007). Guide to Secure Web Services. Retrieved from http://csrc.nist.gov/publications/nistpubs/800-95/SP800-95.pdf.

Shao-Qing W, Ling C, Gen-Cai C (2003). A framework for Java 3D based Collaborative Environment. The 8th international Conf. Computer Supported Cooperative Work Design Proc. 2003. pp. 34-36.

Signh M, Signh S (2010). "A Novel Grid-based resource management framework for collaborative e-learning environments" Int. J. Computer Application. November. 10(4).

Singh G, Serra L, Png W, Wong A, Ng H (1995). BrickNet: Sharing Object Behaviour on the Net. Virtual Reality.

Song J, Kim J, Shin M, Ryu K (2005). "Design and Implementation of Security System for Wargame Simulation System," Korea Information Processing Society, 12-3:369-378.

Ta D, Nguyen T, Zhou S, Tang X, Cai W (2010). A framework for performance evaluation of large-scale interactive distributed virtual environments, (Cit). doi:10.1109/CIT.2010.459.

Tang X, Member S, Zhou S (2010). Update Scheduling for Improving Consistency in Distributed Virtual Environments, 21(6):765-777.

Wang C, Huang C (2009). A Collaborative Network Security Platform in P2P Networks. 2009 International Conference on New Trends in Information and Service Science.

Wang Y (2011). A Fully Distributed P2P Communications Architecture for Network Virtual Environments. Aerospace, pp. 2-5.

Wright TE, Madey G (2010). Discretionary Access Controls for a Collaborative Virtual Environment. 2010. Int. J. Virtual Reality. 9(1):61-71.

www.ayuverda.hubpages.com

Yong S, Moon H, Sohn Y, Fernandes M (2008). A Survey of Security Issues in Collaborative Virtual Environment. IJCSNS. 8(1):14-19.

11

Neuro-fuzzy decision learning on supply chain configuration

J. C. Garcia Infante[1], J. J. Medel Juarez[2] and J. C. Sanchez Garcia[1]

[1]Mechanical and Electrical Engineering School IPN, Col. San Francisco Culhuacan Del. Coyoacán, D. F. ext.73092 Mexico.
[2]Computing Research Centre, Av. 100 m, esq., Venus, Col. Nueva Industrial Vallejo, C. P. 07738 D. F. ext. 56570 Mexico.

This paper describes the computational automatic supply chain configuration (SCC) based on fuzzy logic prediction actualizing automatically the chain stages considering different customer service level petitions. Each level is selected in accordance with the inference and the knowledge base process supplies (KBPS). The SCC model as an intelligent processes selector (IPS), allows dynamical configuration in accordance with the minimum cost supplies configuration (MCSC) described with the SCC functional error. The basic future decisions set as a knowledge base (KB) using fuzzy rules and inferences, transforms the proposed decisions into actions over the elements required by the process supply (PS). The minimal functional error and the SCC best selection, permits excellent client attention. The adaptive model stages operational SCC is described illustratively using Matlab® software.

Key words: Fuzzy digital learning, neural networks, supply chain configuration.

INTRODUCTION

Intelligent supply chain (ISC) structure is built dynamically considering the competitive actions inside the dynamical world, where customers have different services and product requirements. The enterprises should have new approaches that can help to support customer innovation, flexibility, quality, and excellent service maintenance to increase their competence. The multiple scenarios analysis allows developing the strategies for the supply chain configuration (SCC), considering different customer service levels increasing the decision-making support affecting global efficiency (Croom et al., 2000). ISC viewed as an intelligent process requires a mechanism that deduces previously the customer wishes (requirements or specifications) described dynamically as a class of service (CS). In agreement to CS degree the ISC transforms the service into a specific grade, according to the structure conditions with minimum cost

and customer requirements. Therefore, the ISC selected the best attention level answer. An intelligent structure has different supply scenarios with different interpretations, considering the actual customer requirements, having automatic SC changing stage process (Chong and David, 2011). The customer requirements prediction is based on a fuzzy learning system that adjusts dynamically the process stage conditions through the parametre weights with respect to the mean square error criterion or functional error.

The learning system scheme inferences and the knowledge base process is the tool considered as the ISC in accordance with different types of service changes. The learning technique selects the best service for each requirement and obtains the knowledge. Concerning the enterprise processes, the fuzzy intelligent tools viewed as a learning system is an option to obtain

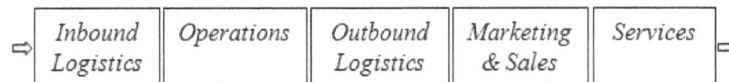

Figure 1. The supply chain stages.

different types of service, dynamically interacting with the customer needs, having an adaptation of the ISC answer in accordance to the possible changes. The customer attention uses a rule set based on the Takagi and Sugeno (TS) mechanism which requires a feedback law, adjusting the process parametres, minimizing the functional error response to updating the process (García et al., 2011).

With this perspective, this paper introduces the Fuzzy learning system (FDF) analyzing and improving the supply chain configuration structure, giving answers levels with respect to the customer requirements, changing the enterprise structure in a natural sense. This selects a specific decision using an ISC configuration in order to give the best service level to each possible scenario. In a novel view, this provides a roadmap improving the enterprise, having a different configuration level support for the supply chain (Korena and Shpitalni, 2010).

SC configuration description

A supply chain integrates a set of links that makes up an economic process from the supplier to the distribution of finished products. Each stage process has to add value as a goal. Each individual stage should be reviewed in order to obtain the best service response. In accordance with the type of service required by the customer and the best response levels at each stage, integrating and optimizing the supply chain for the enterprise structure gives the best service response to the customer. Figure 1 shows the SC in general form (Croom et al., 2000). The SC structure integrates the next stages, as seen in Figure 1:

Inbound logistics: receiving, storing, inventory, control and transportation scheduling.
Operations: machining, packaging, assembly, equipment maintenance, testing.
Outbound logistics: warehousing, order fulfillment, transportation, distribution.
Marketing and sales: channel selection, advertising, promotion, selling, pricing, retailing.
Services: support, repair, installation, training, parts management.

There are different ways to develop a supply chain as an automated intelligent process in order to optimize the enterprise services and make better decisions. Most used is expertise and heuristics. This case used a learning

system integrating all the possible SC configurations to an enterprise structure inferring the type of service needed and dynamically selecting the best SC configuration giving the best customer level answer (Craves and Tomlin, 2003; Korena and Shpitalni, 2010). Each stage is analyzed separately, generating more value to the chain. To decide how to configure the SC needs to consider some characteristics that could be changed in order to obtain the best service level required by the customer, for example, the product type, velocity, category, demand, cost, suppliers and other attributes. The best configuration solution is usually defined as minimizing or maximizing a specific variable at each SC stage (Giunipero et al., 2008).

If an enterprise needs different service levels with a specific type of service required by the customer an intelligent process has to change to customer specifications. It needs to have different specific SC configurations optimizing the service level. The SC will always keep the same stages, but each of these stages will modify its value to configure the SC to a new service level for the customer needs and the enterprise type of service selected with minimum cost. The following stages allow specific service level in accordance with the SC selected (Chong and David, 2011; Yao, 2010):

(i) The customer invokes a specific type of service required
(ii) The learning process infers the actual customer needs
(iii) The selection stage gets the best supply chain configuration
(iv) The process has an update in order to give the corresponding service level

LEARNING SYSTEM DESCRIPTION

The learning system structure has an intelligent structure that infers a real process change with an actual situation. This improves a dynamical process operation because the learning description will give the best answer condition to update the process configuration and work with the optimal operation level in accordance with the changes. The objective of the learning system is the parametre estimation described as $\hat{a}_k$. When the customer selects a specific type of service described as t_k the enterprise offers the learning system that will obtain the specific parametres sequence of $\hat{a}_k$ to update the learning mechanism. This approximation to the

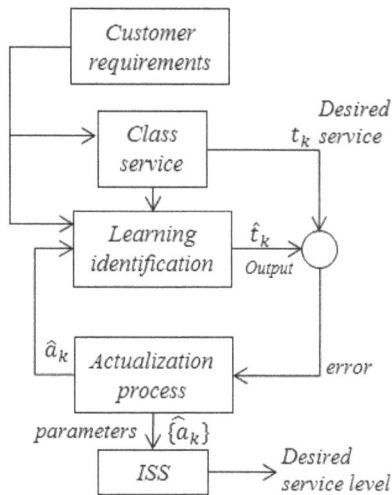

Figure 2. Identification structure services SC model.

desired type of service describes the actual customer requirements automatically (García et al., 2011; Mamdani, 1974).

The learning system has a structure that classifies its answers with different operational levels in order to select the corresponding parametre $\hat{a}_k$ from a knowledge base (KB), using the logic connectors *if-then*, to update the learning system weights describing the actual desired type of service. The next stage called intelligent supply selection (ISS), infers the $\hat{a}_k$ sequence selecting the best SC configuration values with the best service level (Craves and Tomlin, 2003; Zadeh, 1965). Figure 2 shows the learning system operation structure (García et al., 2011; Korena and Shpitalni, 2010; Takagi and Sugeno, 1986).

Figure 2 describes the learning system operation, where t_k represents the actual desired type of service, $\hat{t}_k$ as the signal approximation of the actual service type, e_k is the error between both signals and $\{\hat{a}_k\}$ is the estimated parametres sequence by the learning process representing a specific value sequence identifying the respective type of service. The selection process stage will use the parametres value $\{\hat{a}_k\}$ as a learning process inferring the type of service and selecting the best SC service level (Passino, 1998).

The FLP has a knowledge base, which is limited by the mean square error described in Equation (1). The knowledge base has all the possible values of $\hat{a}_k$ corresponding with the desired input system service. These membership values of $\hat{a}_k$ are selected by a process dynamically update its weights, with the service changes type t_k and the criterion minimizing the

estimation error obtaining the best approximation of the learning output system $\hat{t}_k$ (Mamdani, 1974).

$$k\left\langle J_k, J_k^T \right\rangle = \left[\left\langle \Delta_k, \Delta_k^T \right\rangle + (k-1)\left\langle J_{k-1}, J_{k-1}^T \right\rangle \right] \in \Re_{[0,1]}^+ \tag{1}$$

Theorem

Let the learning system description in Equation (2)

$$t_k = at_{k-1} + w_k \tag{2}$$

Where $t_k \in \Re^+$ as the service changes type, $a \in \Re_{[-1,1]}$ as a specific service parametre and, w_k as the actual situation; has an optimal estimation Equation (3).

$$\hat{a}_k \to a + \varepsilon_k \tag{3}$$

Proof

In agreement to $\Delta_k = t_k - \hat{t}_k$, the quadratic form Δ_k^2 into functional error (1) with Equation (2), has Equation (4).

$$k\left\langle J_k, J_k^T \right\rangle = a^2 \left\langle t_{k-1}, t_{k-1}^T \right\rangle + \left\langle w_k, w_k^T \right\rangle - \left\langle \hat{t}_k, \hat{t}_k^T \right\rangle + 2a\left\langle t_{k-1}, w_k^T \right\rangle$$
$$- 2a\left\langle t_{k-1}, \hat{t}_k^T \right\rangle - 2\left\langle w_k, \hat{t}_k^T \right\rangle + (k-1)\left\langle J_{k-1}, J_k^T \right\rangle \tag{4}$$

In Equation (4) the stochastic gradient with respect to "a" obtaining Equation (5).

$$a = \frac{\left\langle t_{k-1}, \hat{t}_k^T \right\rangle - \left\langle t_{k-1}, w_k^T \right\rangle}{\left\langle t_{k-1}, t_{k-1}^T \right\rangle} \tag{5}$$

Where, the paramtre and perturbation are described respectively in Equation (5).

$$\hat{a}_k \cong \frac{\left\langle t_{k-1}, \hat{t}_k \right\rangle}{\left\langle t_{k-1}, t_{k-1}^T \right\rangle}, \qquad \varepsilon_k \cong -\frac{\left\langle t_{k-1}, w_k^T \right\rangle}{\left\langle t_{k-1}, t_{k-1}^T \right\rangle} \tag{6}$$

Finally, the estimator is optimal and has the form (3). Then the convergence exists when $\left\langle t_{k-1}, w_k \right\rangle \to 0$ ∎.

Where "a", represents the reference variable obtaining a specific service parametre; "m" is the number of

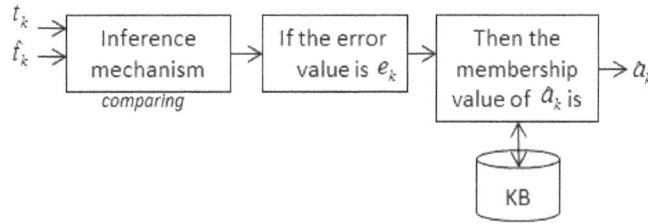

Figure 3. Fuzzy Inference Mechanism based on error values.

elements of each type of service, representing the SC stages number and; "ε_k" represents the region where the results are considered acceptable.

The functional error when $k \to m$, converge to δ_m, where $k\langle \delta_m, \delta_m \rangle = \sum_{i=1}^{m} (\varepsilon_i, \varepsilon_i^T)$ and in recursive form $= \langle \varepsilon_m \varepsilon_m^T \rangle + (m-1)\langle \delta_{m-1}, \delta_{m-1} \rangle$.

For dynamic parametres selection, in accordance with the description of the actual type of service t_k and the FLP $\hat{t}_k$, have the best signal approximation. This is an indicator process with respect to the actual type of service parametres. The different operational levels inside the FLP must accomplish the error criterion described in Equation (1) (García et al., 2011).

To extract the learning identification parametres in fuzzy form uses the mechanism shown in Figure 3, and uses the logic connectors if-then, based on a TS inference dynamically, selecting the best parametres value to the internal FLP classification levels. The different type of services, use the knowledge base that has the membership values of $\hat{a}_k$.

The fuzzy process with respect to the criterion described as $\lim_{k \to m} J_k \to J_{\min}$, makes a selection of the knowledge base parametres, permitting an approximation of the output signal $\hat{t}_k$, to each actual desired service described as t_k. The learning system selection process is in heuristic form, based on probabilistic properties system. This establishes the operational levels bounded by the error functional as: $J_k \subseteq [\delta_{\min}, \delta_{\max})$. To each level, the process selects a specific value of $\hat{a}_k$, having as a goal the best approximation of $\hat{t}_k \cong t_k$ (Craves and Tomlin, 2003; Takagi and Sugeno, 1986).

For parametre selection of the KB into the FLP, it is important for the functional error value to take a parametre value from the knowledge base, when the functional error obtains its minimum value, approximating $\hat{t}_k$ to t_k, and is the smallest distance between both

values. To each operation level the learning approach has a specific parametre configuration value updating the process mechanism. The goal is obtaining the minimum error difference approximation between the type of service required by the customer and the learning identification deducing the SC specific service level configuration (Korena and Shpitalni, 2010).

ISC selection process

The learning system architecture integrates a stage process selection that uses the parametres obtained from the previous learning stage described as $\{\hat{a}_k\}$ to each type of service level. This stage operates as a neural net that deduces the best SC configuration in accordance to the actual type of service. First, training the knowledge base network (KBN), the learning system gets the representative parametre sequences to each level previously at the training stage (implementation). It selects the value sequence to each type of service in $\{a_{kBn}\}$. All the possible $\{a_{KBn}\}$ describes different levels stored in the KBN of the Neural Net. This makes a classification of the different SC service levels with the possible type of service to be selected by the customer (Chong and David, 2011; Marcek, 2004).

This stage selection makes a comparison between the actual learning identification parametres $\{\hat{a}_k\}$ and each sequence $\{a_{kBn}\}$ stored previously in the knowledge base network obtaining the error value. The error rank establishes the correspondence to SC configuration using fuzzy rules which recognize and select the configuration service type. The TS inference has the sequence described as $[\{\hat{a}_k\}, \{a_{kBn}\}]$. Figure 4, shows the network structure to select the SC configuration (Marcek, 2004).

The neural architecture represents the stages, which obtain the parametre information described as $\{\hat{a}_k\}$ in the learning system. This process starts with first layer (input) and continues to the other neurons, in the hidden layer. The neuronal structure process from the previous nodes allows the following stages, having as a goal identifying the corresponding SC configuration (Marcek, 2004):

(i)Inference layer: The error criterion service level described as D_e, into the rank $[0, \varepsilon_k]$, $\varepsilon_k \in \mathfrak{R}^+$, makes a comparison of actual $\{\hat{a}_k\}$ and the $\{a_{kB_n}\}$ stored in the knowledge base. The minimum error distance between values considers the rank error

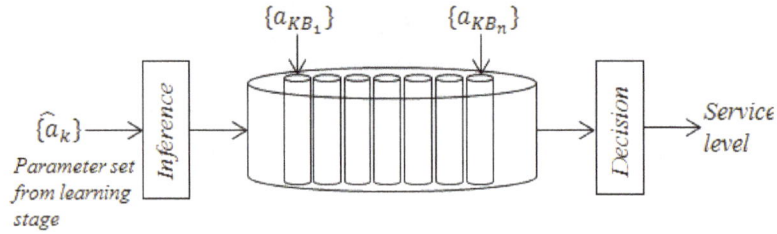

Figure 4. Supply chain configuration selection.

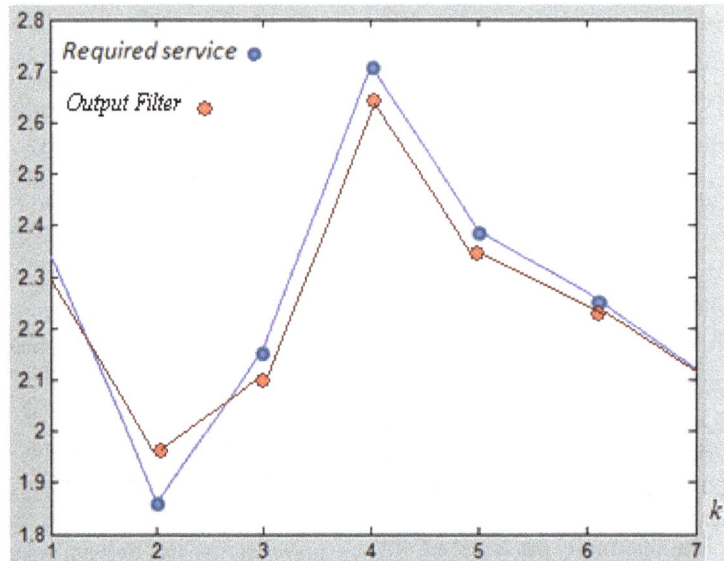

Figure 5. Learning stage applied into identification parametres.

defined previously (García et al., 2011). The error is described by the distance between the $\{\hat{a}_k\}$ as actual parametres value and the stored sample $\{a_{kBn}\}$, given that $d(\{\hat{a}_k\},\{a_{ID}\}) \in \{D_a\}|\{D_a\} \subset R^+_{[0,\varepsilon_k]}$, then $\{\hat{a}_k\}$ has a membership function into the knowledge base neural net corresponding to SC configuration. However, in the case of $d(\{\hat{a}_k\},\{a_{ID}\}) \notin \{D_a\}$, then $\{\hat{a}_k\}$ does not belongs to the network, then there is no type of service required by the customer and it cannot select a corresponding SC configuration.

(ii)Actualization layer: This stage makes the update of the SC configuration value to the service level required for the process in accordance with the actual type of service selected by the customer, using a set of fuzzy rules (if-then) to this inference process (Korena and Shpitalni, 2010). If the $\{a_{kBn}\}$ set value from the Bn is the service level 1, then the SC configuration is selected.

SIMULATION

For simulation there are five possible supply chain options to choose in order to satisfy customer needs with the minimum cost. First, the customer configures their requirements in the database, and the learning identification stage obtains the parametres that corresponds to customer needs. The network gets the parametres and makes a selection of the best corresponding supply chain configuration complying with customer selection with minimum cost for the business, which provides a dynamic decision making different possible service levels (García et al., 2011; Zadeh, 1965). Figure 5, shows the learning approximation requirements in order to obtain the parametres to be used in the selection stage. Figure 6, is the learning stage convergence (1) based on the mean square error in accordance with the approximations. In the selection process, the network has an inference stage that compares the learning stage parametre values with the values stored into the knowledge base. This comparison process is shown in Figure 7.

Figure 8, shows the possible supply chains configurations stored at the knowledge base. It could have more configurations stored, considering the capacity and flexibility of the company. In accordance with customer needs and the error distance of Figure 7,

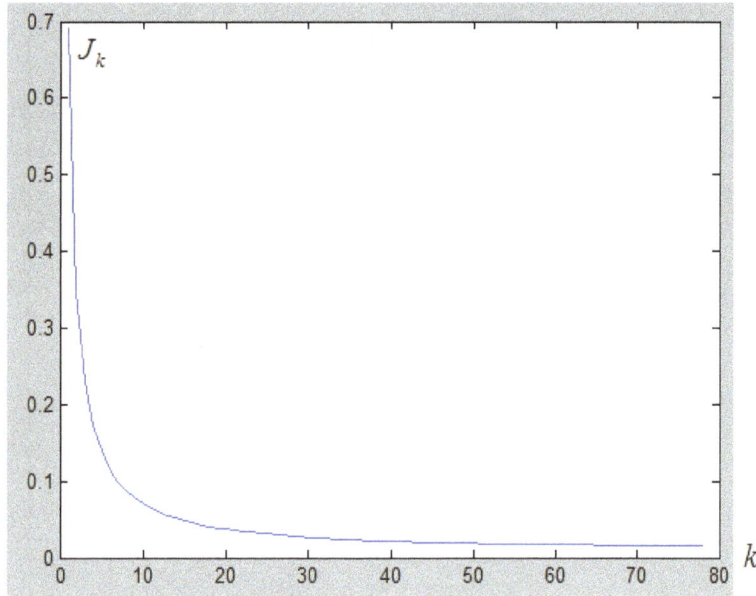

Figure 6. Recursive stage error.

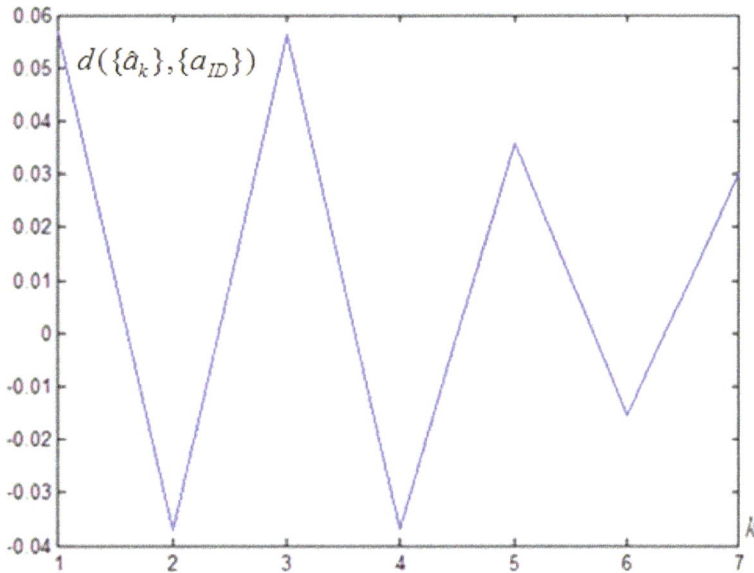

Figure 7. Error inference at the network stage.

one supply configuration will be selected. The service level represents the supply chain selected in the process. Figure 9 shows the configuration with the minimum cost, considering the best service for customer requirements.

CONCLUSIONS

The supply chain in the learning system needs a Neuro-fuzzy model with dynamical decisions offering to the customer different possible service levels. In this paper customer service was viewed with the decisions selected in an intelligent form, estimating the best coefficients required by the supply process. The decisions were defined in the knowledge base using inferences rules and transformed into actions required for the supply process. Automatically, all the supply structure was adjusted in accordance with customer needs and minimum cost. The

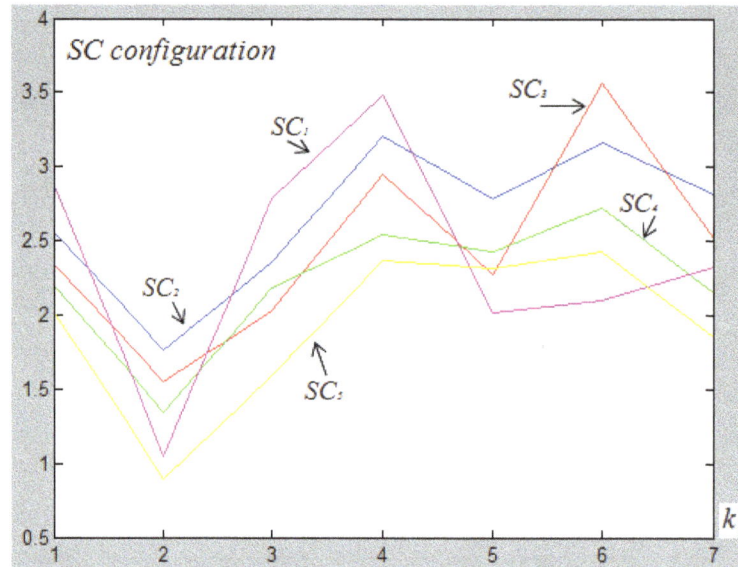

Figure 8. Supply chain set to be selected in the network.

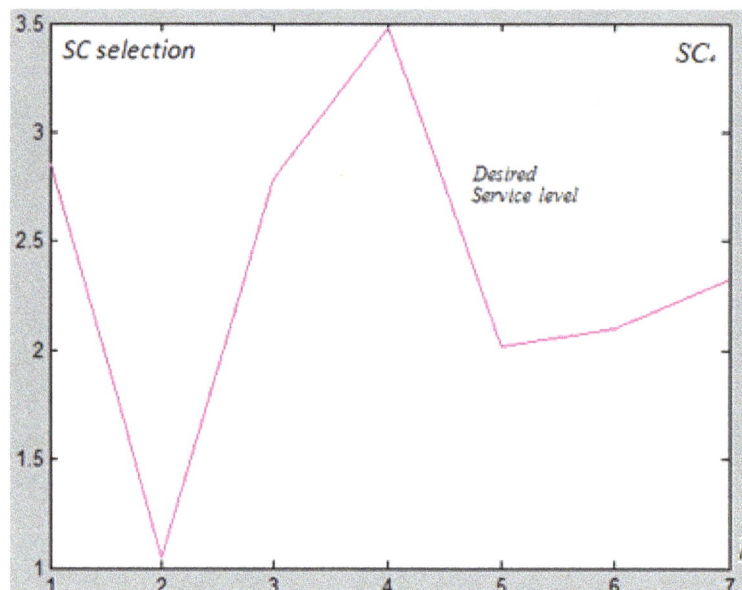

Figure 9. Supply chain selected with the minimum cost as a service level.

simulation of this process shows the operation of the dynamic supply chain using a learning process that describes customer requirements obtaining the parametres, and then the supply chain selection process with a neural network structure.

REFERENCES

Chong W, David B (2011). A literature review of decision-making models and approaches for partner selection in agile supply chains. J. Purchasing Supply Manage. 17(4):256–274.

Craves SC, Tomlin BT (2003). Process Flexibility in Supply Chains, Manage. Sci. 49(7):907–919.

Croom S, Romano P, Giannakis M (2000). Supply chain management: an analytical framework for critical literature review, Eur. J. Purchasing Supply Manage. 6:67-83.

García JC, Medel JJ, Sánchez JC (2011). Filtrado digital neuronal difuso: caso MIMO, Revista Ingeniería e Investigación, 31(1):184-192.

Giunipero LC, Hooker RE, Joseph-Matthews S (2008). A Decade of SCM Literature: Past, Present and Future Implications. J. Supply Chain Manage 44(4):66-86.

Korena Y, Shpitalni M (2010). Design of reconfigurable manufacturing systems, J. Manufacturing Syst. 29:130–141.

Mamdani E (1974). Applications of Fuzzy Algorithms for Control of

Simple Dynamic Plant. Proc. IEEE, 121:1585-1588.

Marcek D (2004). Stock Price Forecasting: Statistical, Classical and Fuzzy Neural Networks., Modeling Decisions for Artificial Intelligence, Springer Verlag, pp. 41-48.

Passino KM (1998). Fuzzy Control, USA, Addison Wesley.

Takagi T, Sugeno M (1986). Fuzzy Identification of Systems and its Applications to Modelling and control. IEEE Trans. Syst. Man. cybernetics, 15:116-132.

Yao J (2010). Decision optimization analysis on supply chain resource integration in fourth party logistics, J. Manufacturing Syst. 29:121–129.

Zadeh L (1965). Fuzzy Sets. Information and control, 8:338-353.

12

Triangular fuzzy based classification of IP request to detect spoofing request in data network

Narayanan Arumugam[1] and Chakrapani Venkatesh[2]

1Anna University, Chennai, India.
[2]Faculty of Engineering, EBET Group of Institutions,Kankayam, Tamilnadu, India, Member IEEE, India.

In data nework data packets are normally forwarded from one router to another through networks until it gets to reach its destination node. According to the internet architecture routers in the internet do not perform any security verification of the source IP address contained in the IP packets. The lack of such a verification opens the door for variety of network security vulnerabilities like Denial-of-Service (DoS) attacks, man-in-the-middle attacks. One of the major threats to the Internet is source IP address spoofing. Different types of IP spoofing detection and prevention approaches are proposed by the research community. In this paper an ant algorithm based traceback approach is proposed to identify the spoofed request origin. In the proposed traceback approach flow level information of each network path is used to identify the origin of the spoofing attack. The significant characteristics of ant algorithm such as quick convergence and heuristic are adopted in the proposed method to find out the origin of the attack.

Key words: IP spoofing, hop count, ant algorithm, pheromone intensity, fuzzification.

INTRODUCTION

Packet forwarding in the Internet is based only on the destination IP address contained in the IP packet. This permits forging of the source IP address, commonly referred to as IP spoofing (Beverly and Bauer, 2005). IP spoofing is a boon for miscreants. Perhaps the most well-known misuse of IP spoofing is in launching Denial-of-Service (DoS) attacks on critical infrastructure such as Web and DNS servers, as evidenced by backscatter analysis (Moore et al., 2001, 2006). Another avenue made possible by spoofing is that of illegal content distribution. UDP-based peer-to-peer (p2p) applications that exploit IP spoofing to mask the identity of the sender already exist. Present approaches to curb IP spoofing researchers have taken two distinct approaches: router-based and victim-based. The router-based approach makes improvements to the routing infrastructure, while the victim based approach enhances the resilience of Internet servers against attacks. The router-based approach performs either off-line analysis of flooding traffic or on-line filtering of DDoS traffic inside routers. But the victim-based prevention methods, which detects and discards spoofed traffic without any router support. Compared to the router-based approach, the victim based approach has the advantage of being immediately deployable. More importantly, a potential victim has a much stronger incentive to deploy defense mechanisms than network service providers. The current victim-based approach protects Internet servers using sophisticated resource management schemes. These schemes provide more accurate resource accounting and fine-grained service isolation and differentiation (Wang, 2007).

Spoofed packets detection methods

A variety of methods are deployed in determining whether a received packet has spoofed source IP address or not. In Internet, when a node receiving a packet can determine whether the packet is spoofed by either an active or passive ways. The term active mean the host must perform some network action but the passive method does not require such action. However, an active method may be used to validate cases where the passive method indicates the packet was spoofed. Among different methods this study considers both IP trace back and hop count based detection method. Since the late 1999 research on IP trace back has been active to detection of DDOS attacks. Several approaches have been proposed to trace IP packets to their origins. IP trace back is usually performed at the network layer, with the help of routers and gateways. The traceback techniques can trace packet paths and help in identifying the perpetrators of the DoS attacks with a high probability. These can be useful forensic tools in law enforcement but do nothing to prevent the occurrence of IP spoofing (Bellovin et al., 2001).

Traceback techniques

IP traceback is a name given to any method for reliably determining the origin of a packet on the Internet by Goodrich (2002). It is a critical ability for identifying sources of attacks and instituting protection measures for the Internet. Probabilistic marking method suggested by Savage et al. (2000) probabilistically marking packets as they traverse routers through the Internet. They propose that the router mark the packet with either the router's IP address or the edges of the path that the packet traversed to reach the router. Deterministic packet marking scheme outlined by Belenky and Ansari (2007) is a more realistic topology for the Internet Snoeren et al. (2001) propose marking details within the router that is to generate a fingerprint is generated with each of the packet. Another method denoted as ant-based traceback approach is proposed to identify the DoS attack origin by Gu Hsin Lai et al. (2008).

Time-To-Live (TTL) methods

When IP packets are routed across the Internet, the Time-To-Live (TTL) field is decremented. This field in the IP packet header is used to prevent packets from being routed endlessly when the destination host cannot be located in a fixed number of hops. It is also used by some networked devices to prevent packets from being sent beyond a host's network subnet. The TTL is a useful value for detecting spoofed packets. Its use is based on several assumptions, which, from our network observations, appear to be true. When a packet is sent between two hosts, as long as the same route is taken, the number of hops will be the same. This means that the initial TTL will be decremented by the same amount. Packets sent near in time to each other will take the same route to the destination. Routes change infrequently. When routes change, they do not result in a significant change in the number of hops (Steven and Templeton, 2003).

The objective of this study is to find out the DOS attack origin (spoofing request) on the network. For the detection process this article uses both the concepts of traceback and hop count of the packet while routing from source to destination on Internet. The IP traceback approach is used to finding out the origin of the spoofing attack using the network data packets traffic flow information on each path. Furthermore, to strengthen the spoofing prevention hop count value of the packet between the source and destination are also validated. An ant-based traceback algorithm is using for finding the traffic flow information as the trace for ants finding the attack path. The hop-count information is indirectly reflected in the TTL field of the IP header, since each intermediate router decrements the TTL value by one before forwarding a packet to the next hop (Stevens and Wright, 1995). The difference between the initial TTL (at the source) and the final TTL value (at the destination) is the hop-count between the source and the destination.

MATERIALS AND METHODS

This study proposes an optimistic method that validates incoming request before it reach the destination without using any cryptographic methodology. The fundamental idea is to utilize inherent network information that each packet carries. The inherent network information this study use here is the flow information and the number of hops of a packet takes to reach its destination. This proposed method uses an ant-based traceback algorithm to find the traffic flow information and hop count value, Since an attacker can forge any field in the IP header, he cannot forged the number of hops an IP packet takes to reach its destination, which is solely determined by the Internet routing infrastructure. The hop-count information is indirectly reflected in the TTL field of the IP header, since each intermediate router decrements the TTL value by one before forwarding a packet to the next hop.

Ant algorithm

Ethnologists states that animals like ants could manage to establish shortest route paths from their colony to feeding sources and back. It was found that the medium used to communicate information among individuals regarding paths and used to decide where to go, consists of pheromone trails. A moving ant lays some pheromone (in varying quantities) on the ground, thus marking

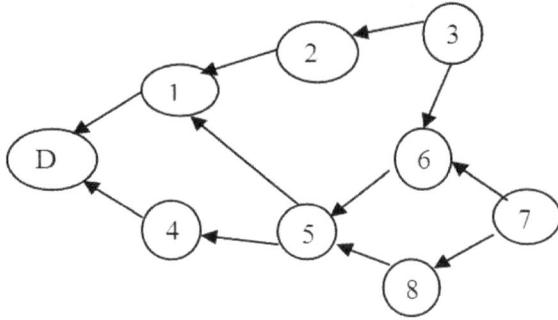

Figure 1. IP trace back of all possible paths.

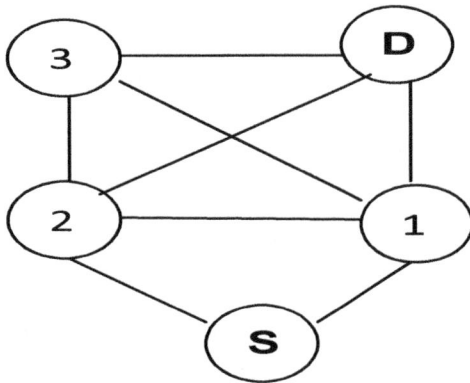

Figure 2. Experimental topology with 5 nodes. S = Source node, D = Destination node.

the path by a trail of this substance. While an isolated ant moves essentially at random, an ant encountering a previously laid trail can detect it and decide with high probability to follow it, thus reinforcing the trail with its own pheromone. The collective behavior that emerges is a form of autocatalytic behavior 1 where the more the ants following a trail, the more attractive that trail becomes for being followed. The process is thus characterized by a positive feedback loop, where the probability with which an ant chooses a path increases with the number of ants that previously chose the same path (Dorigo et al., 1996). The idea is that if at a given point an ant has to choose among different paths, those which were heavily chosen by preceding ants (that is, those with a high trail level) are chosen with higher probability. Furthermore, high trail levels are the same with shortest paths.

Ant based IP traceback

Basically, the attack path reconstruction process involves interrogating the routing packets received at the victim in order to find the immediate upstream node and then systematically repeating the interrogation process at each intermediate upstream node until the attack source is reached. The path reconstruction problem could be solved using the ant-based IP traceback. Figure 1 shows the IP trace back of all possible paths from the source node 3 to the destination node D. Basically the ants lay a pheromone trail along the route they select between the source node (the food source) and the destination (e.g., paths 3-2-1, 3-6-5-4 and 3-6-5-1 in Figure 1) and the relative probability of each path being the actual

path is given by the intensity of the pheromone along the corresponding trail.

As in nature, the isolated ants in the ant algorithm scheme move essentially at random. However, upon encountering a previously laid trail, the ants decide with a high probability to trace it. As a result, the pheromone intensity of this path progressively increases and thus the likelihood of the path representing the actual path also increases. The proposed solution could take the victim host as the starting point and perform IP traceback. It is assumed that the legitimate request might reach the victim node in a shortest path (Lai et al., 2008). The description of the ant-based IP trace back is as follows:

Step 1: Construct network topology,
Step 2: Determine all possible paths between two network nodes (source node to destination node),
Step 3: Find out the shortest path,

The shortest path searching process is done with the exploitation policy as in the Equation (1) chooses the arc with the greatest pheromone intensity and visibility, while the exploration policy as in the Equation (2) is a random decision rule. Thus, an ant located at node i choose the next node j in accordance with the following rule:

$$j = \begin{cases} \arg\max\left\{\lfloor t_{ij}(t)^a \rfloor \lfloor \eta_{ij}(t)^\beta \rfloor\right\} & \text{if } q \le q° \\ S & \text{otherwise} \end{cases} \tag{1}$$

$$S = p_{ij}(t) = \begin{cases} \dfrac{\left[\tau_{ij}(\tau)\right]^\alpha \left[\eta_{ij}(\tau)\right]^\beta}{\sum \left[\tau_{ij}(\tau)\right]^\alpha \left[\eta_{ij}(t)\right]^\beta} \\ 0 \qquad\qquad \text{otherwise} \end{cases} \tag{2}$$

where, $\tau_{ij}(t)$ the pheromone intensity of trail between router i and router j at time $\eta_{ij}(t)$ = the number of routing packets between router i and router j between time (t-1) and time (t) α is the weighting factor of pheromone, β is the weighting factor of visibility.
Ant colony updates the probability density function of feasible attack paths and chooses the right one.

RESULTS

The suggested method have tested using a PC with a specification of Intel Dual core CPU, 3.0G DDR2, 1G of RAM and the MS Windows XP operating system. The experimental topology constructed with 5 nodes as shown in the Figure 2. Where node S considered as a source node and the node D as s destination, the possible path between the node S and D where identified using an algorithm. According to the ant system optimization by a colony of cooperating agent, ants follow a path between the source to destination with all possible paths with equal probability. This process continues until all of the ants will eventually choose the shortest path.

The idea is that at a given point an isolated ant can choose a path among different paths, but according to the ant algorithm, those which were heavily chosen by preceding ants are chosen with higher probability based on with a high trail level. Furthermore, high trail levels are synonymous with shortest paths. It is

Table 1. Experimental value for 5 nodes.

S/N	Possible path	Hop count	Pheromone intensity
1.	s-> 1->d	1	2.129463
2.	s->1->d	1	2.129463
3.	s->1->2->d	2	2.063101
4.	s->1->3->d	2	1.631010
5.	s->2->1->d	2	2.211162
6.	s->1->3->d	2	1.911162
7.	s->1->2->3->d	3	1.327916
8.	s->1->3->2->d	3	1.309615
9.	s->2->1->3->d	3	1.309615
10.	s->2->3->1->d	3	1.512709

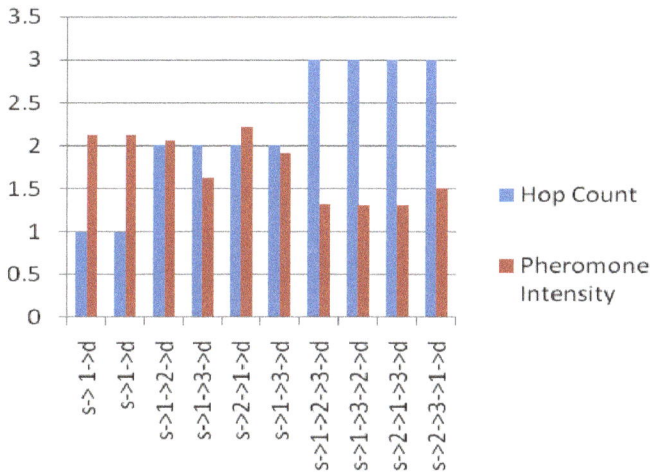

Figure 3. Possible path between sources to destination.

Intuition is used to fuzzify this scalar quantity into the fuzzy or linguistic variables as spoofed request, partially spoofed and legitimate request. The membership function associated with each scalar quantity as defined by intuition is as follows:

$$\mu_{leg} = \begin{cases} 1 & if & p \leq 1.4 \\ \dfrac{1.6-p}{0.2} & if & 1.4 < p < 1.6 \\ 0 & if & p \geq 1.6 \end{cases} \quad (3)$$

$$\mu_{par} = \begin{cases} 1 & if & p \leq 1.9\,(or)\,p \geq 1.6 \\ \dfrac{p-1.9}{0.2} & if & 1.9 < p < 2.1 \\ \dfrac{1.6-p}{0.2} & if & 1.4 < p < 1.6 \end{cases} \quad (4)$$

$$\mu_{s} = \begin{cases} 0 & if & p \leq 1.9 \\ \dfrac{p-1.9}{0.2} & if & 1.9 < p < 2.0 \\ 1 & if & p \geq 2.1 \end{cases} \quad (5)$$

where p is the pheromone intensity, and subscript μ_{leg} denotes legitimate request, μ_{par} denotes partially spoofed request and μ_{s} denotes spoofed request. From the fuzzification condition stated as in Equations (3), (4) and (5) it is understood that the pheromone intensity p below 1.4 is assumed as legitimate request, between 1.6 to 1.9 as partially spoofed request and above 1.9 as spoofed request. A graphical representation of the membership function of IP request is shown in Figure 4.

Table 2 gives the pheromone intensity of all possible shortest paths among the source to destination with the membership function associated with each fuzzy variable, that is, spoofed request, partially spoofed request and legitimate request for each path. For example consider a specific path s->2->1->d and the membership value of each fuzzy set for this path is calculated from Equations (3), (4), and (5) as: μ_{leg} =0, μ_{par} =0, μ_{s} =1. It can be bring

understood that the isolated ant would reach the destination in a shortest way. The shortest path is identified by the isolated ant based on the maximum pheromone intensity. Hence, it is clear that the shortest path may not have fake request. Experimental values are tabulated as in the Table 1 with possible path, hop count and pheromone intensity. From the tabulation it is understood that the legitimate request has minimum hop value and maximum pheromone intensity value. Figure 3 shows the possible path among source and destination with pheromone intensity.

DISCUSSION

From the experimental result this paper classify the IP request either spoofed or legitimate. Fuzzification techniques is used to classify the spoofing request among all possible IP request. Fuzzification is the process of changing a real scalar value into a fuzzy value. This is achieved with the Trapezoidal fuzzifiers.

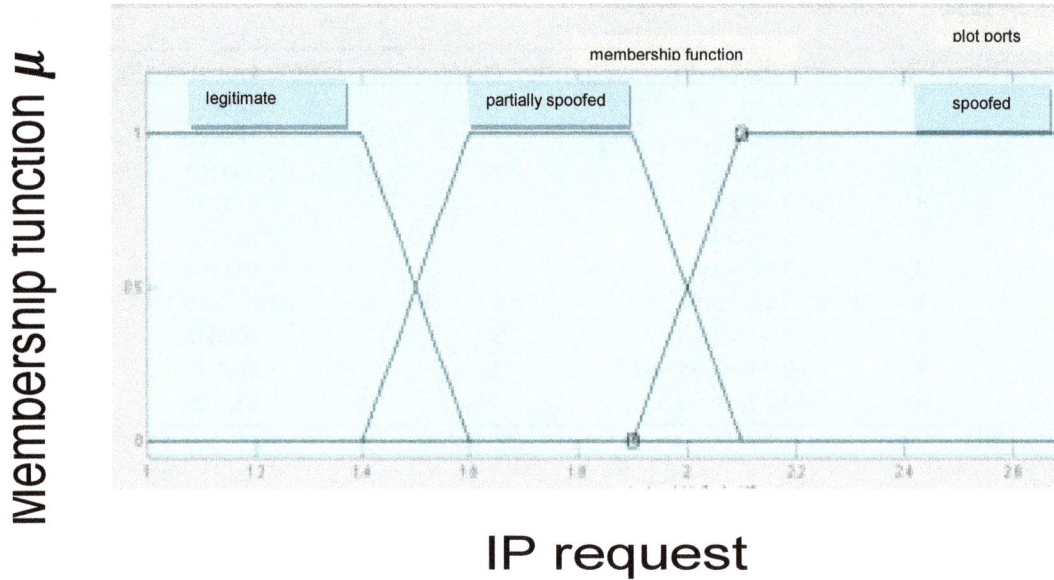

Figure 4. Membership function of IP request.

Table 2. Experimental value for 5 nodes.

S/N	Possible Path	Hop	Pheromone Intensity	μ_l	μ_p	μ_s
1	s-> 1->d	1	2.129463			1
2	s->1->2->d	2	2.063101		0.5	
3	s->2->1->d	2	2.211162			1
4	s->1->2->3->d	3	1.327916	1		
5	s->1->3->2->d	3	1.309615	1		
6	s->2->1->3->d	3	1.309615	1		
7	s->2->3->1->d	3	1.512709		0.5	

to a close from above result that the path request(s->2->1->d) is spoofed by 100%, partially spoofed by 0%, and legitimate request by 0%. In this article the fuzzification techniques is used to classified each request as legitimate, partially spoofed and spoofed. These types of request classification may easy for prevention of spoofed request.

Conclusion

Internet security is a fashionable and fast-moving field at the same time network-based attacks are inevitable. Identifying the source of attack origin is mandatory to protect the network resources. Among different IP spoofing detection method classification of each request is mandatory. The proposed ant algorithm based IP trace back method is identified the attack source effectively and also the spoofing request is classified using triangular fuzzification. This method examined all

possible way to reach destination node and classified the most spoofed request from partially spoofed request.

REFERENCES

Belenky A, Nirwan A (2007). "On deterministic packet marking," Computer Networks: The International Journal of Computer and Telecommunications Networking.

Bellovin S, Leech M, Taylor T (2001). On design and evaluation of intention-driven ICMP traceback. Proceedings of the Tenth International Conference on Computer Communications and Networks, IEEE Xplore Press, Scottsdale, AZ, pp. 159-165. DOI: 10.1109/ICCCN.2001.956234.

Beverly R, Bauer S (2005). The spoofer project: Inferring the extent of Internet source address filtering on the Internet. Proceedings of the Steps to Reducing Unwanted Traffic on the Internet on Steps to Reducing Unwanted Traffic on the Internet Workshop, (TIW '05), USENIX Association Berkeley, CA, USA. pp. 8-8.

Dorigo M, Maniezzo V, Colorni A (1996). The ant system: Optimization by a colony of cooperating agents. IEEE/ACM Trans. Syst. pp. 1-13.

Goodrich MT (2002). Efficient packet marking for large-scale IP traceback. Proceedings of the 9th ACM Conference on Computer and Communications Security, (CCS' 02) ACM New York, NY, USA.

Lai GH, Chen CM, Jeng BC, Chao W (2008). Department of Information Management, National Sun Yat-Sen University, Taiwan," Ant-based IP traceback", Elsevier, pp. 3071-3080.

Moore D, Voelker G, Savage S (2001). Inferring internet denial-of service activity. USENIX Secur. Symp. 24(2):115-139.

Moore D, Shannon C, Brown D, Voelker GM, Savage S (2006). Inferring internet denial-of-service activity. ACM Tran. Comp. Sys. 24:115-139. DOI: 10.1145/1132026.1132027.

Savage S, Wetherall D, Karlin A, Anderson T (2000). Practical network support for IP traceback. Proceedings of the Conference on Applications, Technologies, Architectures and Protocols for Computer Communication, (PCC' 00) ACM New York, NY, USA. pp. 295-306. DOI: 10.1145/347059.347560.

Snoeren AC, Partridge C, Sanchez LA, Jones CE, Tchakountio F (2001). Hash-based IP trace back. Proceedings of the 2001 Conference on Applications, Technologies, Architectures and Protocols for Computer Communications, (PCC' 01) ACM New York, NY, USA., pp. 3-14. DOI: 10.1145/383059.383060.

Steven and Templeton (2003). Detecting Spoofed Packets, http://seclab.cs.ucdavis.edu/papers/DetectingSpoofed-DISCEX.pdf.

Stevens WR, Wright GR (1995). TCP/IP Illustrated: The Implementation. 1st Edn., Addison-Wesley Professional, Reading, Mass, ISBN: 020163354X, P. 1174.

Wang H (2007). Defense against spoofed ip traffic using hop-count filtering. IEEE/ACM, DOI: 10.1109/TNET.2006.890133 Tran. Netw. 15:40-53.

13

Performance evaluation of broadband access network based on subcarrier multiplexing (SCM): Spectral amplitude coding optical code division multiple access

Junita Mohd Nordin[1] , Syed Alwee Aljunid[1], Anuar Mat Safar[1], Amir Razif Arief[1], Rosemizi Abd Rahim[1], R. Badlishah Ahmad[1] and Naufal Saad[2]

[1]School of Computer and Communication Engineering, University Malaysia Perlis, 01000 Perlis, Malaysia.
[2]School of Electric and Electronic Engineering, Universiti Teknologi Petronas, Tronoh, Perak, Malaysia.

A hybrid spectrally amplitude coded (SAC) optical code division multiple access (CDMA) with subcarrier multiplexing (SCM) scheme is a potential candidate for future broadband access solutions in the communication network. In this work, the performance of hybrid technique is investigated. Theoretical derivations are developed based on the zero cross correlation (ZCC) code. The performances are evaluated taking into account the relevant noises such as thermal noise, shot noise and inter modulation noise. Design parameters such as number of optical channels, number of subcarriers, and input power are varied to see their effect on the system performances. Simulation analysis using Optisys simulator is also done and compared to the numerical theoretical results. We measured the performance based on the merits of received power, input power and bit error rates of the received signals. Results revealed that power penalty due to increased number of channels in the hybrid system is lower compared to the conventional OCDMA system, and hence improving the system performance in term of power efficiency. This exhibits the ability of the hybrid system to be one powerful technique as a candidate for future access network.

Key words: Spectral amplitude coding-optical code division multiple access (SAC-OCDMA), subcarrier multiplexing (SCM), zero cross correlation (ZCC), broadband access network.

INTRODUCTION

Recent years have seen the rapid growth of bandwidth hungry broadband access network which keeps the network and service providers busy finding ways of connecting the network up to the last mile solutions. Although fiber optics may seem to be the best option in the near future because of its huge amount of available bandwidth, the main issue of cost-constraint and the fact that it cannot eventually go everywhere makes the standalone access solution unreliable. On the other hand, wireless access solutions, despite of having a congested bandwidth, can eventually go everywhere with cheaper

cost preferable by most end users. Moreover, the overwhelming advance of high speed electronics components makes the RF domain architecture still an ongoing technology. Thus, an ultimate solution that can satisfy the limitations in the broadband access network is the convergence of the optical access and wireless access to exploit the advantages of both architectures rather than their stand-alone architecture. This hybrid technology gives rise to the fiber-wireless (Fi-Wi) access networks, or also known as the Radio over Fiber (RoF) technology. Until date, there are quite a number of published works done on the Fi Wi networks architecture, mostly proposing on the Wavelength Division Multiplexing (WDM) architecture on the optical side (Ghazisaidi and Maier, 2009; Opatic, 2009). In this article, on the other hand, we proposed the application of optical code division multiple access (OCDMA) scheme instead of the WDM to be integrated with the RF domain in subcarrier multiplexing (SCM) manner.

Optical CDMA is getting more attention because of its ability to support asynchronous burst communications with higher level of security. It was employed initially for local-area (Smith et al., 1998), for access network applications (Pearce and Aazhang, 1994; Zhang et al., 1999) and also for emerging networks such as the generalized multiprotocol label switching (Maric et al., 1993). However, OCDMA system networks are limited by the multiple access interference (MAI), which occurs especially when large numbers of users are involved. Thus, in multi-service environments, the transmission capacity and the number of users can be increased by using the SCM technology that enables multiple RF signals to be transmitted simultaneously over the fiber-optic links (Kim et al., 2010). The SCM is an attractive technique where it provides the independency of different channels. This allows for great flexibility in the choice of modulation schemes. With this technique, the information signals are modulated onto different electrical subcarriers in the radio or microwave domain and combined together to modulate the intensity of an optical carrier. At the receiving part, a photo detector will convert the optical signal into an electric current. To retrieve the original signal, the electric current will then be demultiplexed and demodulated using a conventional method. One significant benefit of SCM is that the electrical components and equipments are far less expensive than its optical counterparts. Considering this advantage, we integrated the SCM into the OCDMA access network to investigate the effect on the system's performance. Related work on the area of hybrid SCM OCDMA has been done by Sahbudin et al. (2009, 2010) where the authors evaluated the performance of this hybrid technique based on different detection techniques. From their results, it is shown that the best performance of the hybrid system is by using the direct detection technique. On another related work by Abd et al. (2012), they concluded that the hybrid system can be improved in term of cardinality by selecting an appropriate code in OCDMA. In their study, the researchers utilized the multi diagonal (MD) code which has a zero cross correlation property compared to other codes evaluated in the research. Hence, in our work, we did not compare the results based on number of users or subcarriers since the zero cross correlation (ZCC) code used in this study also have the same property as in MD code. The originality in this paper lies in the evaluation on how increasing the channels in optical domain and subcarrier domain affect the power loss of the hybrid system.

MATERIALS AND METHODS

Hybrid SCM-OCDMA system description

The proposed hybrid OCDMA SCM system configuration is shown in Figure 1. The main difference between the hybrid architecture compared to the conventional OCDMA architecture lies in the added microwave domain in the transmitter and receiver part to represent the SCM architecture. At the transmitter, data with independent unipolar digital signal is mixed with a different microwave carrier (f_i), called the subcarriers. The subcarriers are then combined and optically modulated onto the code words (c_i) using an optical electrical modulator (OEM). In this paper, we utilized the ZCC code (Anuar et al., 2006) in the SAC-OCDMA code family because of its special feature where the cross correlation is always zero. The code structure does not have an overlapping of bit '1' and it will not cause the ZCC code to interfere between users. Eventually, this will definitely suppress the phase intensity induced noise (PIIN). Thus, only shot noise and thermal noise have been considered, ignoring PIIN due to zero cross-correlation between users.

The code words which are based on the ZCC code are generated with different wavelengths provided by the light source. Table 1 shows an example of ZCC code with weight two and three code words, where $λ1$-$λ6$ represents the assigned wavelengths for each code. From Table 1, the assigned code words will be:

$$\text{Codeword} \begin{cases} optical\ channel\ 1 => λ_1\ λ_3 \\ optical\ channel\ 2 => λ_2\ λ_5 \\ optical\ channel\ 3 => λ_4\ λ_6 \end{cases}$$

The encoders consist of a multiplexer which will combine the different wavelengths to generate different code words. The n modulated code words are multiplexed together via a combiner and transmitted through the single mode fiber. Upon this stage, we can observe that each channel is actually assigned a subcarrier frequency, f_i, and a particular code word, c_i, where each pair (f_i, c_i) is unique with respect to every other channel. Each channel of (f_i, c_i) is representing one user in the system. At the receiver, the different modulated code words are separated by an optical decoder consisting of a splitter and multiplexers based on their assigned wavelengths. Then, the decoded signal is detected by the photo detector. A splitter and an electrical band pass filters are used to split the subcarrier multiplexed signals and reject unwanted signals, respectively. In order to recover the original transmitted data, the incoming signal is electrically mixed with a microwave frequency f_i and filtered using a low pass filter (LPF). In the transceiver scheme, it is important for the receiver to decode the correct code sequence and to be tuned to the correct RF frequency to ensure the receiver recovers the desired data signal while rejecting other unwanted signals. Therefore, this enables the hybrid scheme to support high transmission rate with a high level of security. The hybrid system design is based on the direct detection technique. The advantage of direct detection technique is that only one decoder and one detector are needed for each code sequence. This is achievable due to the zero cross correlation properties of the ZCC code where the information can be adequately recovered from any of the chips that do not overlap with any other chips from other code sequences. Accordingly, only the clean chips are filtered out by the decoder and detected by the photo detector.

System performance analysis

In the analysis of this hybrid SCM/OCDMA system, we have considered the effect of shot noise, thermal noise and also the inter-modulation distortion of subcarrier channels on the photo detector. The PIIN is ignored due to the zero cross correlation

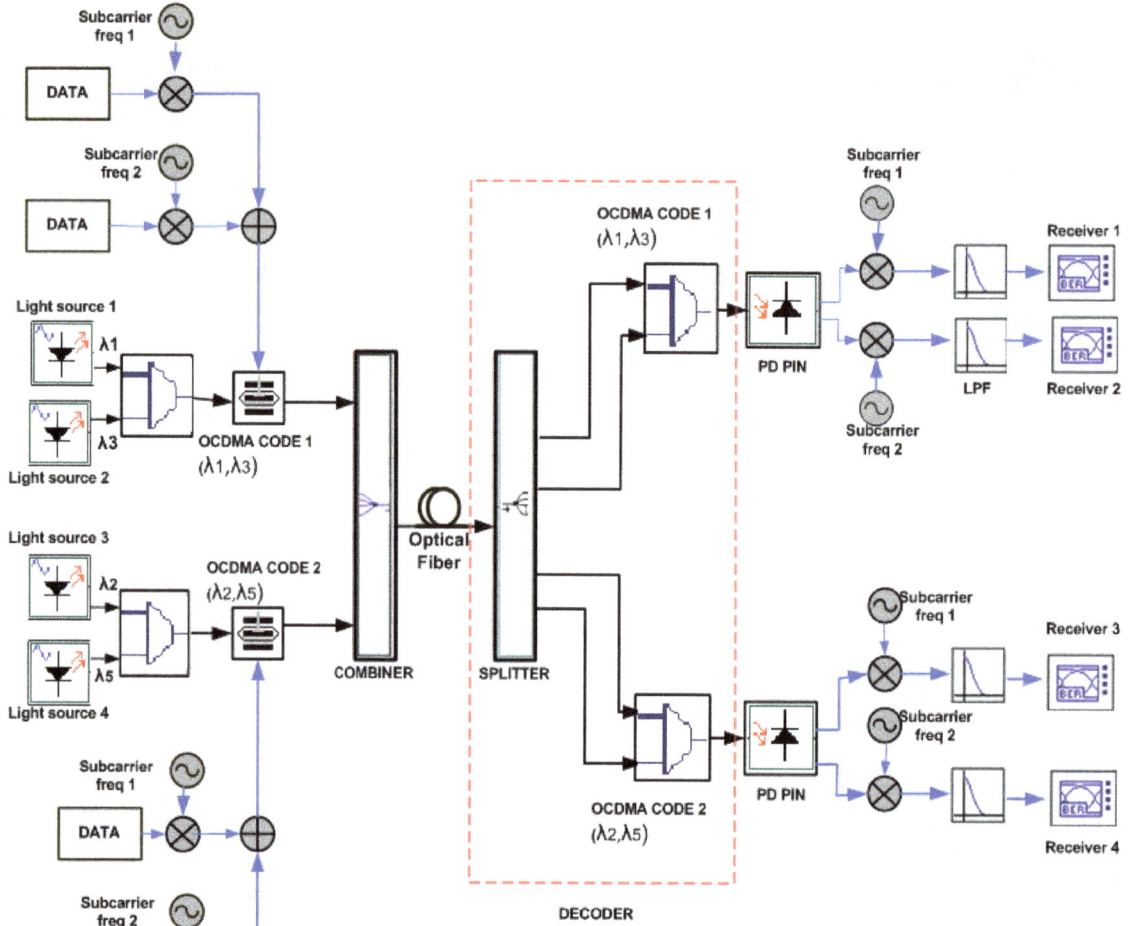

Figure 1. The hybrid SCM/OCDMA system configuration.

condition and no overlapping of spectra from different users. In our analysis, the following assumptions are made (Wei et al., 2001):

(i) Each light source spectra is ideally unpolarized and its spectrum is flat over the bandwidth $[v - \Delta v/2, v_0 + \Delta v/2]$ where v_0 is the optical center frequency and Δv is the optical source bandwidth in Hertz,
(ii) Each power spectral component has identical spectral width,
(iii) Each user has equal power at the receiver,
(iv) Each bit stream from each user is synchronized.

These assumptions are important for mathematical simplicity. Without these assumptions, it is difficult to analyze the system. To simplify our analysis, Gaussian approximation is used for all. The Gaussian's approximation is used in the calculation of bit error rate (BER) because the noises and disturbances that are always present in the communication systems are modeled quite accurately (Kazovsky et al., 1996; Lin et al., 2005). PIN photo detectors are used, and the dark current is assumed to be negligible. The spacing of optical carriers is assumed to be sufficiently wide so that the effect of crosstalk from adjacent channels is negligible (Chao, 1993). The subcarrier channels are equally spaced. Thus, the noise variances at the photo detector due to detection can be denoted as:

$$\left\langle i^2 \right\rangle = \left\langle I_{shot}^2 \right\rangle + \left\langle I_{thermal}^2 \right\rangle + \left\langle I_{IMD}^2 \right\rangle \tag{1}$$

where I_{shot} denotes the shot noise, $I_{thermal}$ is the thermal noise and I_{IMD} is the inter-modulation distortion noise due to the subcarrier channels. When incoherent light fields are mixed and incident upon a photo detector, the phase noise of the fields causes an intensity noise in term of photo detector output, where the source coherence time is expressed as (Smith et al., 1998):

$$\tau_c = \frac{\int_0^\infty G^2(v)\,dv}{\left[\int_0^\infty G(v)\,dv\right]^2} \tag{2}$$

where $G(v)$ is the power spectral density (PSD) of the thermal source. Let $C_{k(i)}$ denotes the ith element of the kth ZCC code sequence, and the code properties for direct detection technique can be written as:

Table 1. ZCC code with weight two and three code words.

Code words	λ1	λ2	λ3	λ4	λ5	λ6
1st code word	1	0	1	0	0	0
2nd code word	0	1	0	0	1	0
3rd code word	0	0	0	1	0	1

$$\sum_{i=1}^{N} C_k(i)C_l(i) = \begin{cases} W \ for \ k = l \\ 0 \ for \ else \end{cases} \tag{3}$$

The power spectral density (PSD) of the received signal is Yang (2004):

$$r(v) = \frac{P_{sr}}{\Delta v} \sum_{k=1}^{K} d_k(t) \sum_{i=1}^{L} C_k(i)C_l(i) rect(i) \tag{4}$$

K represents the number of ZCC code sequences, where each carrying their own subcarrier channels; L is the ZCC code length; P_{sr} is the effective power of broadband source at receiver, with bandwidth $=\Delta v$; $d_{k(t)}$ is the modulated data of the nth subcarrier on the kth optical code; which can be denoted as:

$$d_k(t) = \sum_{n=1}^{Nc} u_{n,k}(t) m_{n,k} \cos(\omega_n t) \tag{5}$$

$u_{n,k(t)}$ denotes the normalized digital signal of "0" and "1" at the nth subcarrier channel of the kth code, while $m_{n,k}$ is the modulation index of the nth subcarrier of the kth code. N_c is the number of subcarrier channels in each code sequence and ω_n is the angular subcarrier frequency. Here, we assume that the modulation index is identical for all subcarrier channels (Hui et al., 2002), thus,

$$0 < m_{n,k} \leq \frac{1}{N_c}; \tag{6}$$

The *rect (i)* function in Equation (4) is denoted as:

$$rect(i) = u\left[v - v_0 - \frac{\Delta V}{2L}(-L + 2i - 2)\right] - u\left[v - v_0 - \frac{\Delta V}{2L}(-L + 2i)\right] = \left\{u\left[\frac{\Delta V}{L}\right]\right\} \tag{7}$$

where $u(v)$ is the unit step function. The PSD at the input of the photo detector at the lth receiver during one data bit period can be expressed as $G_{dd}(v)$, which equals to $r(v)$ as stated in Equation (4). Hence, the total power incident at the input of the photo detector is written as:

$$\int_0^\infty G_{dd}(v)dv = \int_0^\infty \left[\frac{P_{sr}}{\Delta v} \sum_{k=1}^{K} d_k(t) \sum_{i=1}^{L} C_k(i)C_l(i)\left\{u\left[\frac{\Delta v}{L}\right]\right\}\right]dv \tag{8}$$

$$= \frac{P_{sr}}{\Delta v} \cdot \frac{\Delta v}{L} \sum_{k=1}^{K} d_k(t) \sum_{i=1}^{L} C_k(i)C_l(i) \tag{9}$$

Thus,

$$\int_0^\infty G_{dd}(v)dv = \frac{P_{sr}}{L} W d_l \tag{10}$$

The desired signal for a particular user is obtained by expressing the photocurrent, I as:

$$I = I_{dd} = \Re \int_0^\infty G_{dd}(v)dv \tag{11}$$

where $\Re$ is the responsivity of the photodetector denoted as:

$$\Re = \frac{\eta e}{h v_c} \tag{12}$$

η is the quantum efficiency, e is the electron's charge, h is the Planck's constant, and v_c is the central frequency of the original broadband pulse. Substituting Equation (10) into the Equation (11) we obtain

$$I = \Re\left[\frac{P_{sr}}{L} W d_l\right] \tag{13}$$

Substituting Equation (5) into Equation (13), the desired photocurrent signal for the kth channel can be written as:

$$I = \Re \frac{P_{sr}}{L} W \sum_{n=1}^{Nc} u_{n,k}(t) m_{n,k} \cos(\omega_n t) \tag{14}$$

The desired signal is then driven to the RF demodulator and mix coherently with a local oscillator given by 2cos (ω_nt). The signal becomes

$$I = \Re \frac{P_{sr}}{L} W \sum_{n=1}^{Nc} u_{n,k}(t) m_{n,k} \cos(\omega_n t)[2\cos(\omega_n t)] \tag{15}$$

$$= \Re \frac{P_{sr}}{L} W \sum_{n=1}^{Nc} u_{n,k}(t) m_{n,k} \left[2\cos^2(\omega_n t)\right]$$

Simplified by using trigonometric identities, I is expressed as

$$I = \Re \frac{P_{sr}}{L} W \sum_{n=1}^{Nc} u_{n,k}(t) m_{n,k}[1 + \cos(2\omega_n t)] \tag{16}$$

At the receiver's end, the doubled frequency component is filtered out using low pass filter (LPF). Thus the output of the RF demodulator can be expressed as:

$$I = \frac{\Re P_{sr} W}{L} u_{n,k}(t) m_{n,k} \tag{17}$$

$u_{n,k}(t)$ is the normalized digital ≈ 1. For the noise variances, we only consider shot noise, thermal noise and inter-modulation distortion

Table 2. The parameters used for BER calculation.

Symbol	Parameter	Value
η	Photodetector quantum efficiency	0.6
Δv	Linewidth broadband source	3.75THz
λ_0	Operating wavelength	1550nm
B	Electrical bandwidth	80MHz/311MHz
R_b	Data bit rate	155Mbps/622Mbps
T_n	Receiver noise temperature	300K
R_L	Receiver load resistor	1030Ω
E	Electron charge	1.6×10^{-19}C
H	Planck's constant	6.66×10^{-34} Js
K_b	Boltzmann's constant	1.38 10^{-23}J/K

(IMD) noise. The noise power for shot noise is given as:

$$\left\langle I_{shot}^2 \right\rangle = 2eBI_{dd} \tag{18}$$

Substituting I_{dd} in Equation (18), assuming all users transmitting bit "1", we get

$$\left\langle I_{shot}^2 \right\rangle = 2eB\frac{\Re P_{sr}W}{L} \tag{19}$$

The thermal noise is given by Papannareddy (1997);

$$\left\langle I_{thermal}^2 \right\rangle = \frac{4K_B T_N B}{R_L} \tag{20}$$

where $K_b, T_n, B,$ and R_L is the Boltzmann constant, absolute receiver noise temperature, noise equivalent electrical bandwidth of the receiver and receiver load resistor respectively. The inter-modulation distortion noise as expressed in Koshy and Shankar (1997, 1999) is:

$$\left\langle I_{IMD}^2 \right\rangle = P_{sr}^2 \Re^2 m_{n,k}^6 \left[\frac{D_{111}}{32} + \frac{D_{21}}{64} \right] \tag{21}$$

where D_{111} is the three tone third order inter-modulation at $f_i + f_k - f_l$; given by

$$D_{111} = \frac{r}{2}(N - r + 1) + \frac{1}{4}\left[(N-3)^2 - 5\right] - \frac{1}{8}\left[1 - (-1)^N\right](-1)^{N+r} \tag{22}$$

where N is the number of subcarrier channels which in this case is equal to N_c, r is the rth subcarrier. D_{21} is the two tone third order modulation at $2f_i - f_k$; given by

$$D_{21} = \frac{1}{2}\left[N - 2 - \frac{1}{2}\left\{1 - (-1)^N\right\}(-1)^r\right] \tag{23}$$

The total noise here can then be expressed as

$$\left\langle i^2 \right\rangle = \left\langle I_{shot}^2 \right\rangle + \left\langle I_{thermal}^2 \right\rangle + \left\langle I_{IMD}^2 \right\rangle$$

$$\left\langle i^2 \right\rangle = 2eB\frac{\Re P_{sr}W}{L} + \frac{4K_B T_N B}{R_L} + P_{sr}^2 \Re^2 m_{n,k}^6 \left[\frac{D_{111}}{32} + \frac{D_{21}}{64} \right]$$

The SNR of the hybrid SCM/OCDMA using ZCC code can be written as

$$SNR = \frac{(I)^2}{\left\langle i^2 \right\rangle}$$

$$SNR = \frac{(I)^2}{\left\langle i^2 \right\rangle} = \frac{\dfrac{\Re^2 P_{SR}^2 W^2}{L^2} m_{n,k}^2}{\dfrac{2eBP_{SR}W}{L} + \dfrac{4K_B T_n B}{R_L} + P_{SR}^2 \Re^2 m_{n,k}^6 \left[\dfrac{D_{111}}{32} + \dfrac{D_{21}}{64} \right]} \tag{24}$$

The BER can be obtained from the SNR by employing Gaussian approximation:

$$BER = 0.5 * erfc\left(\sqrt{\frac{SNR}{8}}\right) \tag{25}$$

Where the error function erfc (x) can be defined as Larry (1998);

$$erfc(x) = \frac{2}{\sqrt{\pi}} \int_0^x e^{-t^2} dt \tag{26}$$

Typical error rates for optical fiber telecommunication systems is 10^{-9}. This error rate depends on signal to noise ratio at the receiver. The typical parameters used for BER calculation in this study are summarized in Table 2.

Network simulation setup

The hybrid system has been simulated using the software "*Optisys*" Version 6.0. We vary the design parameters such as the number of OCDMA channels and number of subcarrier channels to observe their effect in the simulated environment. The optical channel chip width is equal to 0.8 nm while the subcarrier frequency spacing is

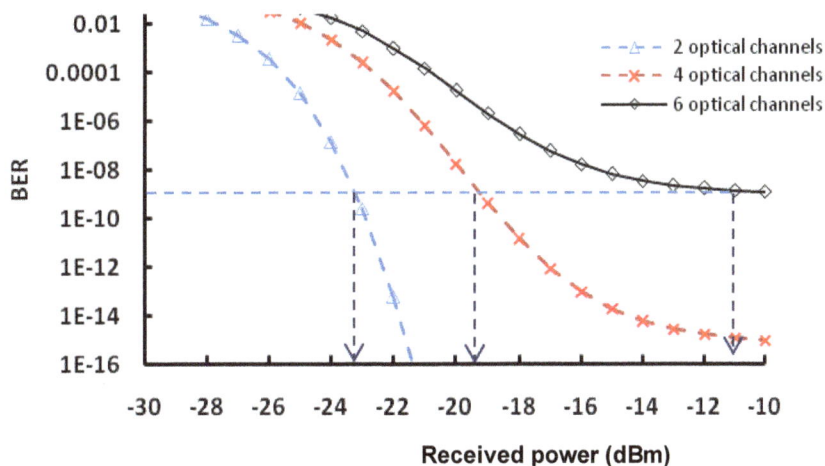

Figure 2. Bit error rates (BER) versus received power for the conventional OCDMA network when optical channels are increased from 2, 4 and 6.

RESULTS AND DISCUSSION

To ensure that the fiber system has sufficient power for correct operation, network designer needs to calculate the span's power budget, which is the maximum amount of power it can transmit. From a design perspective, worst-case analysis calls for assuming minimum transmitter power and minimum receiver sensitivity. This provides for a margin that compensates for variations of transmitter power and receiver sensitivity level. Our objective is to investigate how an increasing number of channels affect the power penalty loss of the system. We then compare the conventional all optical OCDMA with a system with hybrid SCM OCDMA. Figure 2 shows the variation of BER over received power in the conventional OCDMA system as the number of channels is increased. Here, each code word represents one channel (that is, one user or subscriber).

To increase the number of channels, we need to increase the number of code words. We can see that the BER degrades as the received power gets smaller. It must be noted that even a very small fall in optical signal power can deteriorate the BER by some order

set to 600 MHz. In addition, to avoid clipping in the data spectrum, the subcarrier frequencies are set to be equal or larger than twice the bit rate assigned. The transmission medium used in the simulation is the standard single mode fiber (ITU-T G.652). To set the simulation as close to the real environment as possible, the nonlinear effects such as the four wave mixing and self-phase modulation are activated according to their typical industrial values. Accordingly, attenuation and dispersion effects are set to be at 0.25 dB/km and 18 ps/nm-km. For the receiver, we set the photo detector thermal noise coefficient at 1.8×10^{-23} W/Hz, while the dark current value is set to be 5nA. BER analyzers are placed at the end of the receiver's output to evaluate the bit error rates and observe the eye pattern of each received signal.

of magnitudes. It is obvious in Figure 2, at the 10^{-9} BER, the received power at photo detector is -23 dBm for 2 optical channels and -19 dbm for 4 optical channels. The error flooring at BER of 10^{-9} occurs at -11 dBm received power for 6 optical channels. Thus the power penalty loss here is about 12dB when we increase the channels from two to six in the all-optical environments. Clearly, even a small increase in the number of optical channels, the required receiver sensitivity power increased significantly. This leads to a huge difference in the amount of power that needs to be transmitted at the transmitter side. It is also evident that the BER degrades with the increase number of channels because, as channels are increased, the code length of the system will also increase. The larger the code lengths (L), means a larger bandwidth is needed for the system. Noise power is proportional to bandwidth. Larger bandwidth will cause a higher noise power level, thus, lowering the signal –to noise ratio of the system. Hence, BER will degrade. This can be described by SNR and BER equation in the article in Equation (24) and (25). In addition, as the number of channels increase, the MAI from other users will be a limiting factor, and hence, results in degrading the BER.

Figure 3, on the other hand shows the relation of BER and received power with a fixed optical channel (fixed code word) and increase number of subcarrier channels. The received power sensitivity is shown to be increased in smaller range values when the subcarriers are increased from three to six subcarriers. For BER of 10^{-9} error floor, a system with three subcarriers shows a -23dBm receiver sensitivity, while the system with four and six subcarriers only needs an increased receiver sensitivity of -22dBm and -20dBm. Here, the power penalty loss with increased channels from three to six is only about 3dB. This difference ranges is much smaller as compared to increasing the number of optical channels.

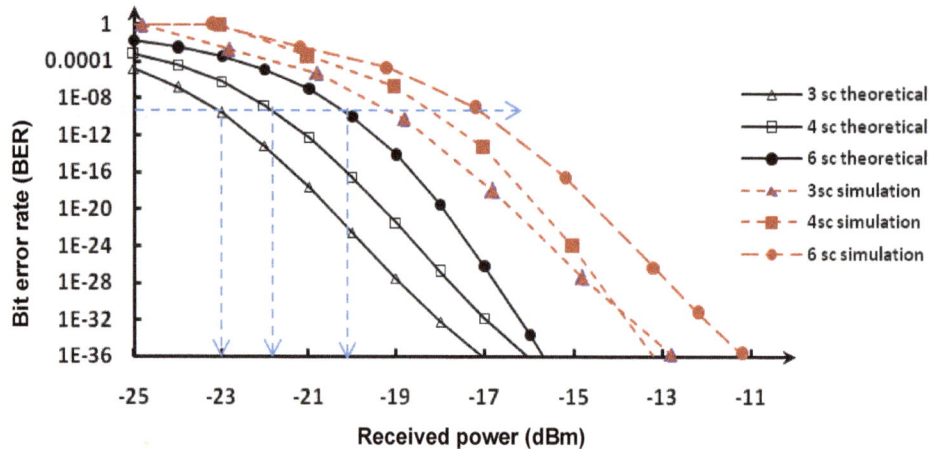

Figure 3. BER against received power with different number of subcarriers for theoretical (black lines) and simulations (red dashes).

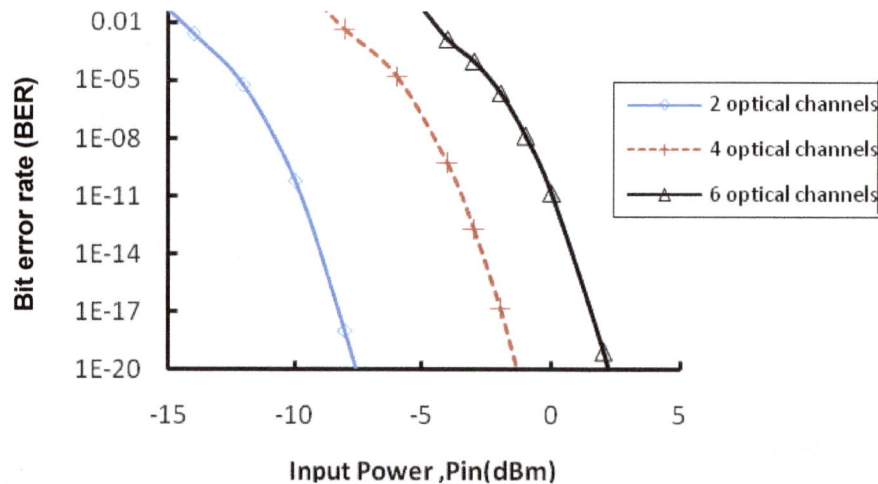

Figure 4. BER against input power simulated with various numbers of optical channels.

Figure 3 also includes the simulation result using Optisys software of a system with fixed optical channel equals to two (two OCDMA codes) and subcarriers channels equal to three, four and six. The BER curves of the simulation parts are plotted in red dashes. As expected, the plot of BER over received power curves of the simulation shows relatively the same patterns as compared to the theoretical curves in black lines. In addition, BER is depicted to be deteriorating when the number of sub channels is increased. In SCM OCDMA system, as the number of sub channels is increased, inter-modulation distortion products of the second and third order harmonics will also increase. Thus, this will degrade the BER as these harmonics can combine or overlap with the original signal and destructively decrease or reduce the signal power of the original signal. Again, the SNR decrease, and finally the BER are deteriorated.

Figures 4 and 5 indicate the simulation result to validate our previous discussion by showing the variations of BER dependences over input power. In Figure 4, to achieve a BER of 10^{-9}, the required input power for two optical channels is -10dBm. As the optical channels are increased to four and six, input power required also increased to -4dBm and 0dBm, respectively. A total power loss of 10dB in the transmitted power is observed when we increase the number of channels from two to six. In contrast, when subcarrier channels in Figure 5 are increased from three to four and finally to six sub channels, the input powers required by the system are from -10 dBm to about -8 dBm, which is only about 2 dB power penalty lost. This is a significant difference in the power range from that in Figure 4. Thus, it is clearly demonstrated that to accommodate a larger number of subscribers, power consumptions are

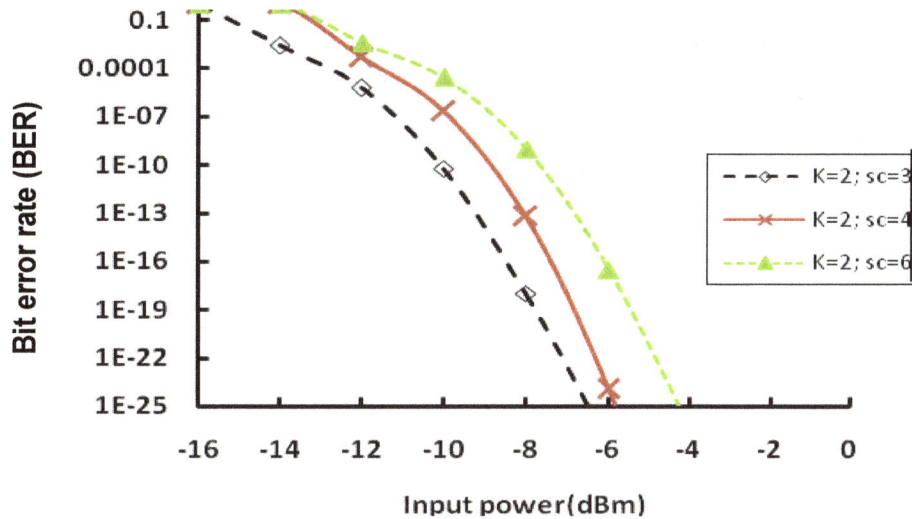

Figure 5. BER against input power simulated with different number of subcarrier channels.

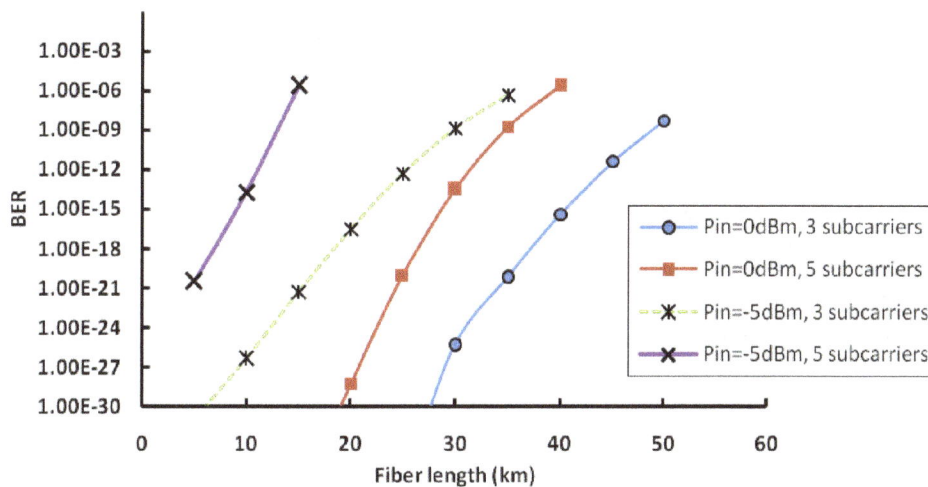

Figure 6. Variation of BER with respect to fiber length for different subcarrier numbers.

improved in a hybrid system compared to the conventional system.

Figure 6 depicts the variation of BER with respect to fiber length for three and five subcarriers when different input powers are fed to the system. It is observed that the BER degrades as the transmission distance increases. The dispersion and attenuation increases as the optical fiber length increases, thus decreasing the BER. The curves show that the system can perform well with an acceptable BER better than 10^{-9} for up to 50 km for the system occupying three subcarriers with input power of 0dBm and 35 km for five subcarriers. It is obvious that larger number of subcarrier channels will degrade the system performance. In addition, lower input power will result in shorter achievable distance for the acceptable

BER of 10^{-9}.The system with three subcarriers can be transmitted up to 30 km with -5dBm input power, while for five subcarriers channel can only be transmitted up to 15 km.

Conclusion

This work presents an analytical and simulated performance of a hybrid subcarrier multiplexing over optical CDMA based on ZCC code. Our work reveals that system performance in term of power penalty loss is improved via the proposed hybrid SCM OCDMA system based on the observation of BER curves over transmitted and received power. In addition, it is apparent that the theoretical and simulation both have almost similar relation patterns and results in the hybrid performances. We have ascertained that to achieve large cardinality broadband access network, this hybrid system is a more preferable architecture compared to conventional all optical OCDMA. This work can be a future guide to more intensive research on hybrid network. The SCM/OCDMA system could be one promising solution to the symmetric high capacity network with high spectral efficiency, cost effective, good flexibility and enhanced security, an attractive candidate for next generation broadband access networks.

REFERENCES

Abd TH, Aljunid SA, Fadhil HA, Radhi IF, Ahmad RB, Rashid MA (2012). Performance improvement of hybrid SCM SAC-OCDMA networks using multi-diagonal code. Sci. Res. Essays 7(11):1262-1272.

Anuar MS, Aljunid SA, Saad NM (2006). Development of a New Code for Spectral-Amplitude Coding Optical Code Division Multiple Access (OCDMA). IJCSNS Int. J. Comput. Sci. Netw. Secur. 6:180-184.

Chao L (1993). Effect of laser diode characteristics on the performance of an SCM OFDM direct detection system. IEE Proc. J. Optoelectron. 140:392-396.

Ghazisaidi N, Maier M (2009). Fiber-Wireless (FiWi) Access Networks:A Survey. IEEE Commun. Mag. pp. 160-167.

Hui R, Zhu B, Huang R, Allen CT, Demarest KR, Richards D (2002). Subcarrier Multiplexing for High Speed Optical Transmission. J. Lightwave Technol. 20:417-427.

Kazovsky L, Benedetto S, Willner A (1996). Optical Fiber Communication Systems. Artech House, Inc.

Kim K, Jeoung J, Lee J (2010). Effect of Fiber Dispersion and Self Phase Modulation in Multi-channel Subcarrier Multiplexed Optical Signal Transmission. J. Opt. Soc. Korea 14:351-356.

Koshy BJ, Shankar PM (1997). Efficient Modeling and Evaluation of Fiber-fed Microcellular Networks in a Land Mobile Channel using a GMSK Modem Scheme. J. IEEE. Selected Areas Commun. 15:694-705.

Koshy BJ, Shankar PM (1999). Spread Spectrum techniques for fiber-fed microcellular networks. J. IEEE. Trans. Vehicul. Technol. 48:847-857.

Larry CA (1998). Special Functions of Mathematics for Engineers 2nd. Ed. SPIE Press and Oxford University Press.

Lin CH, Wu J, Tsao HW, Yang CL (2005). Spectral amplitude-coding optical CDMA system using Mach–Zehnder interferometers. IEEE J. Lightwave Technol. 23:1543-1553.

Maric SV, Kostic ZI, Titlebaum EL (1993). A new family of optical code sequences for use in spread-spectrum fiber-optic local area networks. IEEE Trans. Commun. 41:1217-1221.

Opatic D (2009). Radio over Fiber Technology for Wireless Access. GSDC Croatia.

Papannareddy R (1997). Introduction to Lightwave Communication Systems. Artech House, Inc. P. 246.

Pearce MB, Aazhang B (1994). Multiuser detection for optical code division multiple access system. IEEE Trans. Commun. 42:1801-1810.

Sahbudin RKZ, Abdullah MK, Mokhtar M (2009). Performance improvement of hybrid subcarrier multiplexing optical spectrum code division multiplexing system using spectral direct decoding detection technique. Opt. Fiber Tech. 15:266-273.

Sahbudin RKZ, Abdullah MK, Mokhtar M, Anas SBA, Hitam S (2010). Performance of Subcarrier-OCDMA System with Complementary Subtraction Detection Technique. Int. J. Info. Commun. Eng. 6(4):248-253.

Smith EDJ, Blaikie RJ, Taylor DP (1998). Performance enhancement of spectral-amplitude-coding optical CDMA using pulse-position modulation. J. IEEE. Trans. Commun. 46(9):1176-1185.

Smith GH, Novak D, Lim C (1998). A millimeter-wave full-duplex WDM/SCM fiber-radio access network. OFC '98, paper TuC5.

Wei Z, Shalaby HMH, Ghafouri-Shiraz H (2001). Modified quadratic congruence codes for fiber Bragg-grating-based spectral-amplitude-coding optical CDMA systems. J. Lightwave Technol. 19(9):1274-1281.

Yang CC, Huang JF, Tseng SP (2004). Optical CDMA Network Codecs Structured with M-Sequence Codes over Waveguide-Grating Routers. IEEE Photonics Technol. Let. 16:641-643.

Zhang X, Ji Y, Chen X (1999). Code routing technique in optical network. Beijing Univ. Posts Telecomm. 1:416-419.

14

A learning automata-based algorithm using stochastic minimum spanning tree for improving life time in wireless sensor networks

Chamran Asgari[1] , Rohollah Rahmati Torkashvand[2] and Masoud Barati[3]

[1]Department of Computer Engineering, Payame Noor University, Iran.
[2]Department of Computer Engineering, Islamic Azad University, Borujerd Branch, Lorestan, Iran.
[3]Department of Computer,Islamic Azad University, Kangavar branch, Kangavar, Iran.

Several algorithms have already been provided for problems of data aggregation in wireless sensor networks, which somehow tried to increase networks lifetimes. In this study, we dealt with this problem using a more efficient method by taking parameters such as the distance between two sensors into account. In this paper, we presented a heuristic algorithm based on distributed learning automata with variable actions set for solving data aggregation problems within stochastic graphs where the weights of edges change with time. To aggregate data, the algorithm, in fact, creates a stochastic minimum spanning tree (SMST) in networks where variable distances of links are considered as edges, and sends data in the form of a single packet to central node after data was processed inside networks. To understand this subject better, we modeled the problem for a stochastic graph having edges with changing weights. Although this assumption that edges weights change with time makes our task difficult, the results of simulations indicate relatively optimal performance of this method.

Key words: Data aggregation, stochastic graph, learning automata, minimum spanning tree, life time.

INTRODUCTION

Wireless sensor networks consist of a large number of inexpensive sensor nodes distributed densely in the environment, having limited energy and on the other hand, consuming a great deal of energy in order to send information to central node directly. Thus, in most cases, nodes communicate with central node via their neighbors (Gupta and Kumar, 2000). On one hand, there are different paths to central node from each node, so optimal path must be selected. The frequent use of one path results in energy reduction of sensors located on that path, ultimately resulting in sensor loss. Therefore, we tried to increase networks lifetime by providing an intelligent algorithm and taking such parameters as sensor lifetime, remaining and consumption energies of

sensors and distances between sensors into account, in order to have an almost optimal data aggregation in networks. The proposed algorithm includes some steps at each of which one of possible spanning trees is created randomly.

The proposed algorithm (LA-SMST) is based on distributed learning automata, and each step of algorithm begins with selecting one of graph's nodes randomly in order to discover spanning trees and surveys of distributed learning automata using backtracking technique. Learning automaton related to chosen nodes is activated and selects one action (one edge) based on action probability vector. The edge related to this selection is added to spanning tree just formed. The weight assigned to selected edge is added to total weight of spanning tree. To avoid forming a loop in the tree, each activated learning automaton trims its actions set by disabling actions related to already chosen edges or those edges which may form a cycle. Then, the learning automaton at the other end of selected edge is activated and selects based on of its own actions, activating the automaton located on its end (Asgari and Akbari, 2012). The process of sequential activation of learning automata (or selection of tree edges) is repeated until spanning trees are formed and / or no further action is done by current active learning automaton. Next, it performs data aggregation within middle nodes and sends the result to central node in the form of a single packet.

To create a spanning tree for data aggregating is a promising approach to reduce overhead of broadcast routing where messages are induced among minimum spanning trees. A case wireless network can be modeled as a unit disk graph $G = (V, E)$, in which nodes represent hosts and edges represent relationship between them; hosts must be in each other's transfer ranges (Clark et al., 1990; Marathe et al., 1995). Consider a network of wireless sensors located uniformly in the environment. Assume that nodes have fixed locations and identical transfer ranges. Two sensor nodes communicate directly with each other if they are in each other's transfer ranges; otherwise, they make indirect multistep communications via middle nodes. The aim of algorithm provided is to create minimum spanning trees for data aggregation in wireless sensor networks through finding a relatively optimal solution for problem of minimum spanning trees. In order to implement this approach, at first, a network of distributed learning automata is used to form this network's unit disk graph by equipping each host with a learning automaton. Then, at each step, learning automata select one of their actions randomly, considering their probability vectors until minimum spanning trees are formed. Then the minimum spanning trees formed are evaluated by random environment, and actions probability vectors from learning automata dependent on the response they receive from the environment are updated. In any iterations of this process, finally, learning automata converge to public policy of making minimum spanning trees for network graph.

This paper provides an intelligent algorithm based on distributed learning automata to aggregate data in wireless sensor networks. Each host is equipped with a learning automaton; sink node is considered as root, and then given the action probability vectors, learning automata select next action randomly from variable actions set of learning automata. This process continues until the entire network is covered and minimum spanning trees are formed. Then, the message of data aggregation is sent to all nodes from sink node in minimum spanning tree. Upon receiving the message, all nodes send their data to their parents that must wait until receiving data from all their children. After that, parents aggregate all data and send it to their higher level parents until the aggregated data are being sent to sink node in the form of a single packet. After completing each iteration process, action probability vector of any learning automata is updated. In this study, a proposed algorithm is presented and the experiments results are demonstrated.

LITERATURE REVIEW

Many routing algorithms have been provided for sensor networks. For some of these algorithms, each node may have more than one route to sink node that one of them is selected on the basis of a series of criteria, among which the level of energy consumption along the route can be a proper criterion. Energy saving can be taken into account in two ways: (1) energy consumption is calculated for any separate routes, then the route with minimal energy consumption is chosen (Shah and Rabaey, 2002); and (2) data aggregation is based on provided learning automata, which prevents extra packets from being sent in networks by identifying sensors generating identical data and by activating sensor nodes periodically, thus saving a large amount of energy while increasing network lifetime (Esnaashari and Meybodi, 2010). A solution has been provided in Al-Karaki et al. (2009) for data aggregating and routing with internetwork aggregations in wireless sensor networks in order to maximize network lifetime by using internetwork processing techniques and data aggregation. The relationship between security and data aggregation process within wireless sensor networks has also been investigated in Ozdemir and Xiao (2009). In Soro and Heinzelman (2005), network is first clustered in order to aggregate data, and then head-clusters aggregate data from each cluster separately. A network organized into clusters with the same sizes results in unequal load distribution among head-cluster nodes. Nevertheless, Soro and Heinzelman (2005) provided a model in which clusters are of different sizes, resulting in more uniform energy distribution among head-cluster nodes and increase in network lifetime.

Furthermore, Liao et al. (2008) has offered data aggregation in wireless sensor networks by using the ant colony algorithm, which states the problem of creating data aggregation tree in wireless sensor networks for a group of source nodes to send sensed data to the single sink node. The ant colony system represents a natural method of heuristic search in determining data aggregation. Each ant discovers all possible routes to sink node and a data aggregation tree is created using accumulated pheromone. Lee and Wong (2006) also provided two different tree structures: the lifetime-preserving tree (LPT) and energy-aware spanning tree construction (E-Span) to facilitate aggregation of data of data in wireless sensor networks. In LPT, nodes having more remaining energy are chosen as aggregation parents. The tree is restructured when one node has no long function or when a broken link is identified. E-Span is an aware energy–spanning tree algorithm in which source node with maximal remaining energy is selected as root. Other source nodes select their corresponding parents from their neighbors on the basis of such information as remaining energy and distance to root. In the report of Eskandari et al. (2009), an efficient energy–spanning tree is used to aggregate data in wireless sensor network using two parameters; energy and distance uses route energy average to balance parameters energy an distance while previously provided algorithms have selected only one of these parameters as the main one and gave sound priority to the other.

In the report of Cam et al. (2006), unlike common data aggregation methods, the energy-efficient secure pattern based data aggregation (ESPDA) avoids transmitting redundant data to head–clusters from sensor nodes in order to remove redundancy for improving application of efficient energy and bandwidth in sensor nodes. Li et al. (2010) also presented a scheme of efficient and highly accurate energy to aggregate data securely. The main idea of this is to aggregate data carefully without disclosing or reading secret information of sensors and posing considerable overhead in energy – limited sensors. In Korteweg et al. (2009) aggregation of data in wireless sensor networks is raised to balance latency and communication cost. In Korteweg et al. (2009), spanning tree- based algorithms are provided to create high convergence between data aggregation and efficient energy and low latency in wireless sensor networks. Initially, Upadhyayula and Gupta (2007) provided two algorithms for making Data aggregation enhanced convergecast (DAC) tree. The first algorithm is the kind of minimum spanning tree, and the second of individual source shortest path spanning tree, both of which are used as combined (COM) algorithm stimulator generally based on minimum spanning tree (MST) and shortest path spanning tree (SPT).

PRIMARY DEFINITIONS AND CONCEPTS

Stochastic minimum spanning tree problem

As earlier mentioned, the aim of an absolute MST algorithm is to find minimum spanning trees from graphs, assuming fixed weights of edges (Hutson and Shier, 2006). Although stochastic minimum spanning tree (SMST) algorithm concerns with graphic edges the weights of which are a stochastic variable, most scenarios assume edges weights are fixed (Ishii et al., 1981; Dhamdhere et al., 2005). But this assumption is not always true. Generally, edges of a changing network take various states (several states). Therefore, an absolute graph is not capable of modeling features of such a network really. For this reason, network topology should be modeled by a stochastic graph. As mentioned earlier, several algorithms have been proposed for solving MST problems for which

network parameters are absolute. Anyway, when the graphs are stochastic, MST is considerably difficult to find. Herein, we examine SMST problems and algorithms.

Definition 1: Graph G with stochastic weighed edges is defined by triple <V, E, W>, where V= {V_1,...,V_n} is edges set being a subset of V × V and W= {W_1,...,W_m} is the set of weights assigned to edges set, with positive variable W_i representing the weight of edge $(e_i \in E)$ (Akbari and Meybodi, 2010).

Definition 2: Let G<V, E, W> show a stochastic weighed graph and T= {T_1,T_2,...} show the set of possible spanning trees from graph G, assuming that W'_{Ti} represents the expected weight of spanning tree T_i. A SMST is defined as a stochastic spanning tree with minimum expected weight where $T^* \in T$ if and only if $\min_{\forall T_i \in T}\{W'_{T*} = W'_{Ti}\}$ (Akbari and Meybodi, 2010).

Learning automata

A learning automaton (LA) is an abstract model capable of doing finite actions. Each selected action is evaluated by a probable environment, the result of which is delivered to automata in the form of a positive or negative signal. Learning automata use this response to select their next action. The ultimate goal is for automatas to select the best of their actions. The best action is one maximizing the likelihood of receiving rewards from environment (Narendra and Thathachar, 1989; Thathachar and Sastry, 1997; Thathachar and Harita, 1987).

Probable environment can be expressed mathematically by triple $E \equiv \{\underline{a}, \underline{B}, \underline{c}\}$ where $\underline{\alpha} \equiv \{\alpha_1, \alpha_2,....., \alpha_r\}$ is the set of environment inputs and $\underline{\beta} \equiv \{\beta_1, \beta_2,......, \beta_r\}$ is each actions being penalized. Figure 1 shows the relationship between learning automate and environment. Given the values of $\underline{\beta}$, three different models are defined for probable environments. Whenever $\underline{\beta}$ is a two-members set of [0, 1], the environment is of type P, that is- values of 0 and 1 are selected as environment outputs. In this case, $\beta_1 = 1$ means "being penalized" and $\beta_2 = 0$ means "being rewarded" (Asgari and Akbari, 2012). If $\beta(n)$ is a value bounded to [0,1], the model is of type Q; and if $\beta(n)$ is a stochastic variable within [0,1], the environment is of type S. C_i represents the probability that action α_i receives an undesirable response from environment. The values of C_i do not change in static environments while changing with time in non- static ones (Lakshmivarahan and Thathachar, 1976).

Learning automatas are divided into two groups: (a) those with fixed structures and (b) those with variable structured. In this study, we made use of the variable structured. For learning automata with fixed structures, probabilities of automata actions are fixed, while for learning automata with variable structures, they are updated with each turn of iteration. Learning automatas with variable structures can be denoted by triple $\{\alpha, \beta, P, T\}$ where $\alpha \equiv \{\alpha_1, \alpha_2,....., \alpha_r\}$ is automata's action set; $\beta \equiv \{\beta_1, \beta_2,......, \beta_r\}$ is its inputs; $P \equiv \{P_1, P_2,....., P_r\}$ is probability vector of each automata's action; and $\tau \equiv P(x+1) = T[\alpha(x), \beta(x), \rho(x)]$ is the learning algorithm.

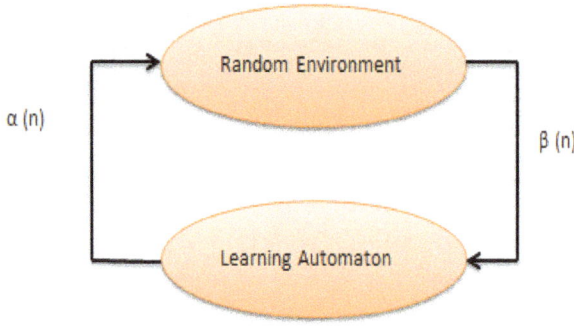

Figure 1. The relationship between learning automata and environment.

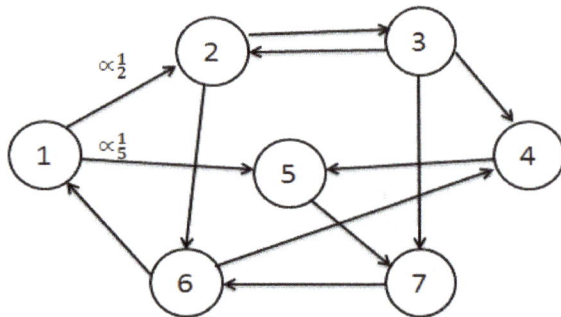

Figure 2. Network of distributed learning automata.

Automatas choose one of their actions randomly on the basis of probability vector P_i and exercise it on the environments from which they get a response. If the action selected by learning automata is action α_i, then, automata updates its action probabilities according to Equation (1) in the case of receiving desirable response from environment, while it does this according to Equation (2) in the case of receiving undesirable one.

$$\begin{cases} P_i(n+1) = P_i(n) + a[1 - P_i(n)] \\ P_j(n+1) = (1-a)P_j(n) \quad \forall j, \quad j \neq i \end{cases} \quad (1)$$

$$\begin{cases} P_i(n+1) = (1-b)P_i(n) \\ P_j(n+1) = (b/(r-1)) + (1-b)P_j(n) \quad \forall j, j \neq i \end{cases} \quad (2)$$

Where r is the number of automata's actions, and b is penalty parameter. The following algorithms can be available on the basis of different values considered for parameters a and b of learning:

1) If a = b, linear reward-penalty (L$_{R-P}$) scheme is obtained.
2) If the value of b is many times smaller than that of a, the resulting learning method is called liner reward epsilon scheme (L$_{R\_\varepsilon P}$).
3) If b = 0, the algorithm is called linear reward inaction scheme (L$_{R-I}$).

Distributed learning automata

A distributed learning automaton (DLA) (Narendra and Thathachar, 1980; Beigy and Meybodi, 2006) is a network of LAs cooperating to solve a particular problem. Within this network of cooperating automata, only one automaton is active at a time. In DLA, the number of actions each automata is able to do is equal to the number of automata connected to that one. When an automata selects an action in the network, other automata connected to it is activated. In other words, choosing an action by an automata in this network corresponds to activation of another automata present. The model considered for DLA network is graphical, each vertex of which is an automata as shown in Figure 2.

In this graph, the presence of edge (LA$_i$, LA$_j$) means that choosing the action α_j^i by LA$_i$ activates LA$_j$. The number of actions LA$_K$ can select is denoted as $P^k = \{P_1^k, P_2^k, \ldots\ldots\ldots, P_{rk}^k\}$. Within this set, P_m^k represents probability related to action α_m^k.

Selecting the action α_m^k by LA$_k$ activates LA$_m$. r_k shows the number of actions LA$_k$ is able perform.

PROPOSED STOCHASTIC MINIMUM SPANNING TREE ALGORITHM

In this paper, we proposed a heuristic algorithm called LA-SMSTA to find an optimal solution from SMST problems where edges' weights are unknown. When the weights of edges change with time, finding optimal solution from MST problem becomes too difficult. Suppose that G(V,E,W) represents entries of stochastic graph, where V={V$_1$,...,V$_2$} is nodes set, E = {e$_1$, e$_2$,...,e$_m$} $\subseteq$ V×V is edges set, and matrix W represents the weights assigned to edges set. In this algorithm, a network of distributed learning automata is formed by equipping each node of the graph with a learning automaton. Network results can be described with triad $<A, \alpha, W>$ where A= {A$_1$,...,A$_n$} represents a set of learning automata; $\alpha = \{\alpha_1,...,\alpha_n\}$ is a set of possible actions where $\alpha_i = \{\alpha_i^1, \alpha_i^2,..., \alpha_i^{ni}\}$ defines a set of actions that can be selected by learning automata A$_i$ (for each $\alpha_i \in \alpha$), and V$_i$ is cardinality of action set αi. Edge e(i,j) relates either to action α_i^j of learning automata A$_i$ or to action α_i^j of learning automata A$_j$. This means that each of learning automata can select each of edges as an action. Selecting action α_i^j by automata A$_i$ adds edge e(i,j) to MST. Weight W$_{i,j}$ is the weight assigned to edge e(i,j) and assumed to be a positive stochastic variable. For the proposed algorithm, all learning automata are in a passive state in the primary set.

The proposed algorithm includes some steps at each of which one of possible spanning trees is identified randomly. The algorithm is based on distributed learning automata, which surveys them by means of backtracking technique in order to discover spanning trees. Any steps of LA-SMSTA algorithm begins randomly with selecting one of graph's nodes as a sink node. Learning automata related to chosen node are activated and one action is selected based on actions probability vector. The edge related to this selection is added to spanning tree already made. The weight assigned to the chosen edge is added to total weight of spanning tree. To avoid forming a loop in a tree, each of active learning automata trims its own actions set. Then, the learning automata located at other end of chosen edge is activated, which also selects one of its own actions and activates the automata located at its end.

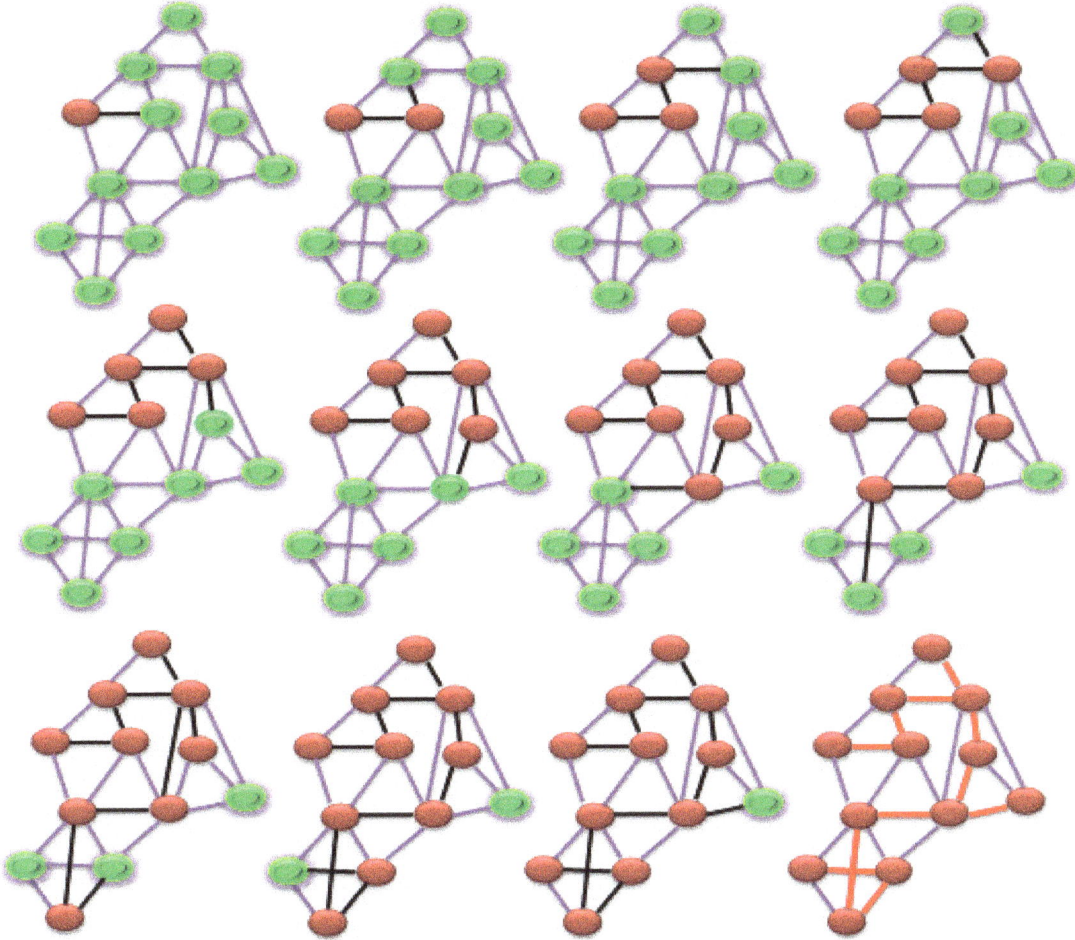

Figure 3. Steps of forming spanning trees.

The process of sequential activation is repeated from learning automata (or from selection of tree edges) until it leads to two following states: in the first state, spanning trees are formed, and in the second, current active learning automata has no action to choose. In the former, the current step is completed successfully by finding a solution for the problem of spanning trees with minimum weights (this happens when the number of selected edges $\geq$ n-1, where n shows cardinality of nodes set), and in the latter, learning automata are found through backtracking process, are activated again, and actions set of automata is updated by disabling the last chosen action. Afterward, the activated automata resume the current step by selecting one of possible actions. The process of activating learning automata continues until spanning trees are formed. Then, data aggregation is performed within middle nodes and the results are sent to central node in the form of a single packet. By means of backtracking technique, each of learning automata may activate more than one of its neighbors at each step. In other words, any learning automata can select more than one action. As stated earlier, respective edge is added to spanning tree, and this task is chosen by learning automata. Also, the weight assigned to selected edge is added to total weight of spanning tree. Figure 3 shows the step of forming spanning trees. Since the weight assigned to graph edge was assumed to be a positive stochastic variable, a particular spanning tree may experience different weights. Therefore, the proposed algorithm is concerned

with the average weight of spanning trees at each step instead of the trees' own weights. To do this, at the end of the step, average weight of selected spanning tree is calculated. We assumed that spanning tree T_i was selected at step X. Average weight from spanning tree T_i to step x is calculated as follows:

$$
\begin{cases}
W_{Ti}^{j} = \sum_{\forall (s,t) \in Ti} W \, e^{j}(s,t) \\
W_{e}^{j}(s,t) = 1/De(s,t) \\
\overline{W}_{Ti}^{x} = \frac{1}{x_i} \sum_{j=1}^{x_i} W_{T_i}^{j}
\end{cases}
$$

(3)

Where $W_{T_i}^{j}$ represents the weight of sample j^{th} from spanning tree T_j, $W_{e(s,t)}^{j}$ shows the weight of edge e(s, t) as a part of sample j^{th} taken from spanning tree T_i, $D_{e(s,t)}$ represents the length

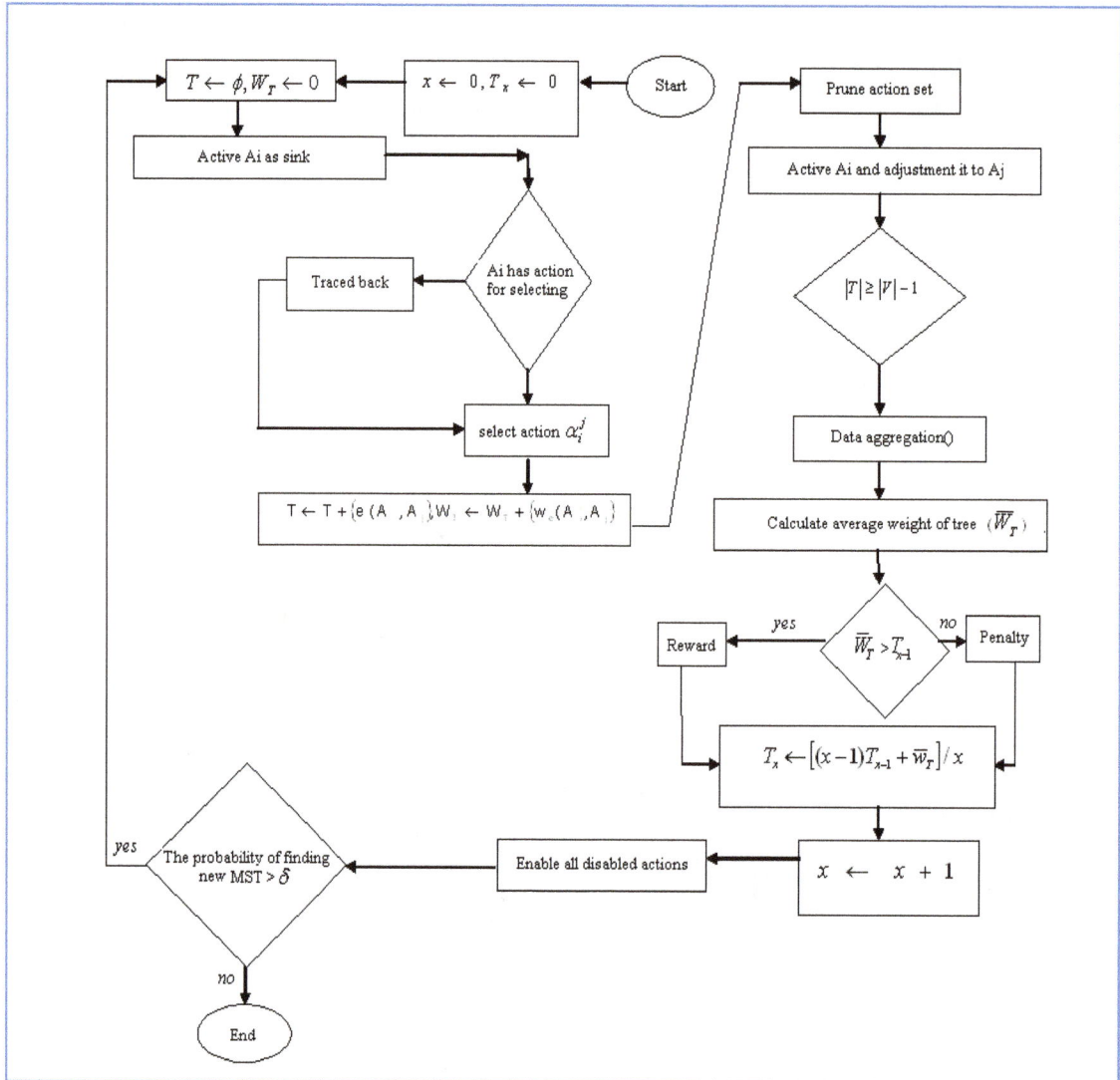

Figure 4. Flowchart of the proposed algorithm.

(distance between s and t) of edge e(s,t), and $\overline{W}_{Ti}^{X}$ is the average weight of spanning tree T_i to step x and X_i shows the times of forming spanning tree T_i until step X.

To estimate convergence of proposed algorithm to optimal solution (minimum spanning tree), average weight of formed spanning tree is compared with dynamic threshold. At each step, T_x is compared with dynamic threshold at step x>1 as follows:

$$T_X = \frac{1}{r} \sum_{i=1}^{r} \overline{W}_{Ti}^{X}$$

(4)

Where r shows the number of spanning trees discovered until step x. Since the weights of edges changes, a given spanning tree may be made several times, having a different weight at each time. At each step, the average weight of selected spanning tree is compared with dynamic threshold. If the average weight of selected spanning tree is bigger than the dynamic thresholds, then all learning automata reward their chosen actions, otherwise, they penalize them. Although each of learning automata updates its

action probability vector by means of learning algorithm, when learning algorithm is penalized, probability vector remains unchanged. At the final step, inactive actions need to be activated again. The process of forming spanning trees and updating action probabilities is repeated until the action probability of formed spanning trees is greater than a specific threshold called stopping threshold. Prior to stopping the algorithm, selected spanning tree is the one with minimum expected weight among all spanning trees of stochastic graph. After rewarding selected action, action probability vector must be updated again by activating all inactivated actions. Since L_{R-I} is the supporting scheme with which learning automata update their own action probability vectors, action probabilities of activated learning automata remain unchanged upon receiving penalty message. In this case, inactivated action of each learning automaton is activated again.

We can comprehend from the aforementioned that with increasing T_x, life time is increases because the relationship between life time and distances between nodes is opposite. On the other hand, when the distance between two nodes is more, the consumption energy for transferring a packet between two nodes will be more. The flowchart that depicts the proposed algorithm is shown in Figure 4.

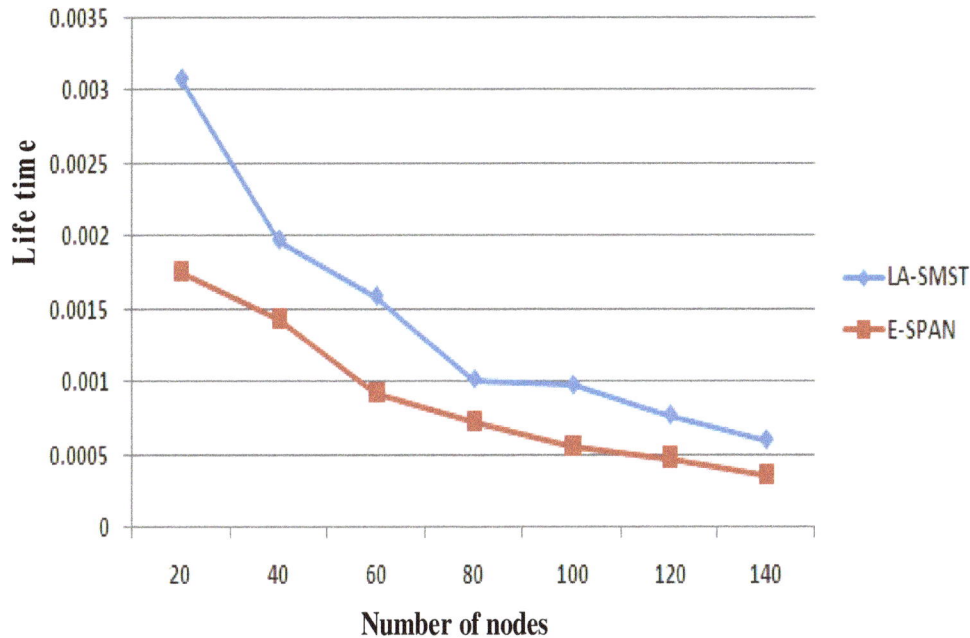

Figure 5. The relationship between life time and number of nodes.

EXPERIMENTAL RESULTS

In this paper, NS2 simulator was used to simulate wireless sensor network. Simulation was performed in a square area of 150 × 150 m^2. We used L_{R-I} model for our learning automata and we assumed that the learning rate is 0.2 and the weight for each edge is allocated randomly. The maximum weights assumed were 3000. For this simulation, the threshold of SMST process and max iteration were set at 0.9 and 100, respectively. For assessing the proposed algorithm, we evaluated our simulation with respect to lifetime by increasing distances between nodes and increasing the number of nodes.

Here, for evaluating our algorithm (LA-SMST), we compared our algorithm with the proposed algorithms of Lee and Wong (2006). Lee and Wong (2006) provided two different tree structures LPT and E-Span to facilitate aggregation of data in wireless sensor networks. E-Span is an aware energy–spanning tree algorithm in which source node with maximal remaining energy is selected as root. Other source nodes select their corresponding parents from their neighbors on the basis of such information as remaining energy and distance to root.

The relationship between life time and different network scales

Here, we evaluated our simulation with respect to SMST lifetime by increasing the number of nodes. We assumed that the maximum distance between two nodes is 20 m and the number of nodes increases from 20 to 140 nodes. As show in Figure 5, the lifetime decreases when the number of nodes increases.

The relationship between life time and distances between nodes

Here, we assumed that the number of nodes is 50 and distances between nodes increase from 10 to 20 m. As show in Figure 6 with increasing distances the life is decreasing. Also, comparing our algorithm (LA-SMST) with the proposed algorithm in Lee and Wong (2006) will determine how much our method performs well.

CONCLUSION AND FUTURE WORK

In this paper, we proposed learning automata based algorithm for improving life time in wireless sensor network. Herein, we used stochastic minimum spanning tree to make a backbone. The process of making created tree was done according to the rate of distance between two nodes in the network and we tried to use route that have higher lifetime to make SMST. We also evaluated the algorithm proposed with increasing the number of nodes and distances between nodes. In addition, we compared our algorithm with other proposed algorithm and found that our algorithm always outperforms in term of the life time. The future studies will focus on increasing security of the proposed method and also fault tolerance while failing each of the sensors after forming a SMST.

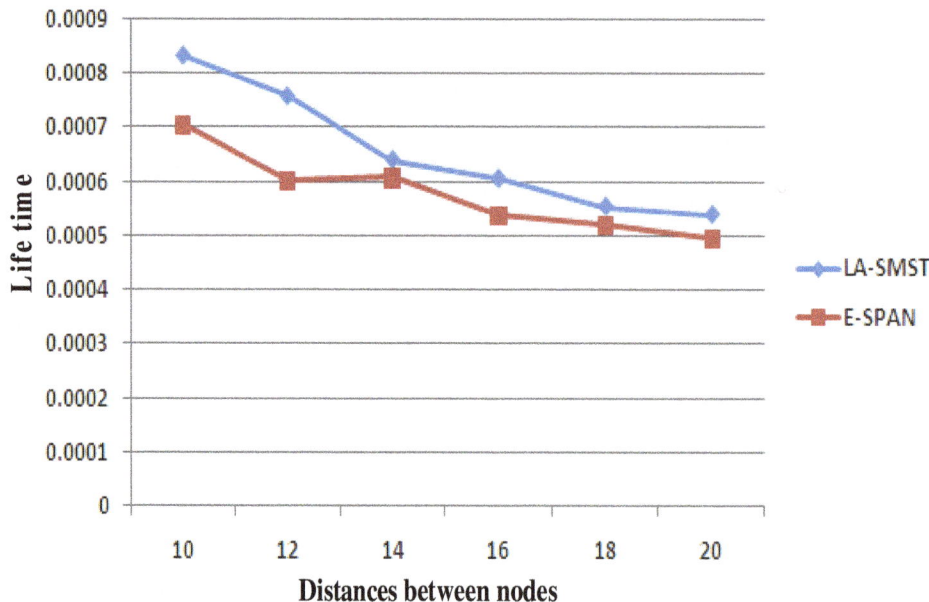

Figure 6. The relationship between life time and distances between nodes.

REERENCES

Akbari TJ, Meybodi MR (2010). A Learning Automatabased Heuristic Algorithm for Solving the Minimum Spanning Tree Problem in Stochastic Graphs. Comput. Netw. 54:826-843.

Al-Karaki JN, Ul-Mustafa R, Kamal AE (2009). Data aggregation and routing in Wireless Sensor Networks: Optimal and heuristic algorithms. Comput. Commun. pp. 945-960.

Asgari CH, Akbari TJ (2012). A New Approach to the Data Aggregation in Wireless Sensor Networks. International J. Comput. Sci. Iss. (IJCSI) 1:328-335.

Asgari CH, Akbari TJ (2012). A new approach for improving lifetime in wireless sensor networks based on distributed learning automata. Int. J. Phys. Sci. (14):2209-2219.

Beigy H, Meybodi MR (2006). Utilizing distributed learning automata to solve stochastic shortest path problems. Int. J. Uncertainty Fuzz. Knowl. Based Syst. 14:591-615.

Cam H, Ozdemir SH, Ozgur S, Prashant N (2006). Energy-efficient secure Pattern based data aggregation for wireless sensor networks. Comput. Commun. pp. 446-455.

Clark BN, Colbourn CJ, Johnson DS (1990). Unit disk graphs. Discrete Math. 86:165-177.

Dhamdhere K, Ravi R, Singh M (2005). On Two-Stage Stochastic Minimum Spanning Trees. Springer-Verlag Berlin. pp.321-334.

Eskandari Z, Yaghmaee M (2009). Energy Efficient Spanning Tree for Data Aggregation in wireless sensor networks. Wireless Sensor Network, 2009. pp. 316-323.

Esnaashari M, Meybodi MR (2010). Data Aggregation in wireless Sensor Networks using Learning Automata. Wireless Netw. 16:687-699.

Gupta P, Kumar PR (2000). The capacity of wireless networks. IEEE Transac. Inform. Theor. 46(2):388-404.

Hutson KR, Shier DR (2006). Minimum Spanning Trees in Networks with varying Edge Weights. Ann. Oper. Res. 146:3-18.

Ishii H, Shiode S, Nishida T, Namasuya Y (1981). Stochastic Spanning Tree Problem. Discr. Appl. Math. 3:263-273.

Korteweg P, Marchetti-Spaccamela A, Leen S, Andrea V (2009). Data aggregation in sensor networks: Balancing communication and delay costs. Theor. Comput. Sci. pp. 1346-1354.

Lakshmivarahan S, Thathachar MAL (1976). Bounds on the convergence probabilities of learning automata. IEEE T. Syst. Man Cy. (SMC) 6:756-763.

Lee WM, Wong VWS (2006). E-Span and LPT for data aggregation in wireless sensor networks. Comput. Commun. pp. 2506-2520.

Li H, Lin K, Li K (2010). Energy-efficient and high-accuracy secure data aggregation in wireless sensor networks. Comput. Commun. 34: 591-597.

Liao WH, Kao Y, Fan CM (2008). Data aggregation in wireless sensor networks using ant colony algorithm. Netw. Comput. Appl. pp. 387-401.

Marathe MV, Breu H, Hunt III H.B, Ravi SS, Rosenkrantz DJ (1995). Simple heuristics for unit disk graphs. Networks 25:59-68.

Narendra KS, Thathachar KS (1989). Learning Automata: An Introduction. Prentice-Hall. New York.

Narendra KS, Thathachar MAL (1980). On the behavior of a learning automaton in a changing environment with application to telephone traffic routing, IEEE T. Syst. Man Cy. SMC-I0 5:262-269.

Ozdemir S, Xiao Y (2009). Secure data aggregation in Wireless Sensor Networks: A comprehensive overview. Comput. Netw. pp. 2022-2037.

Shah R, Rabaey J (2002). "Energy Aware Routing for Low Energy Ad Hoc Sensor Networks," in the Proceedings of the IEEE Wireless Communications and Networking Conference (WCNC), Orlando, FL, March, 2002.

Soro S,Heinzelman WB (2005). Prolonging the Lifetime of Wireless Sensor Networks via Unequal Clustering. Parallel and Distributed Processing Symposium. Proceedings. 19th IEEE International.

Thathachar MAL, Harita BR (1987). Learning automata with changing number of actions. IEEE T. Syst. Man Cy. (SMG) 17:1095-1100.

Thathachar MAL, Sastry PS (1997). A hierarchical system of learning automata that can learn the globally optimal path. Inf. Sci. 42:743-766.

Upadhyayula S, Gupta SKS (2007). Spanning tree based algorithms for low latency and energy efficient data aggregation enhanced converge cast (DAC) in wireless sensor networks. Ad Hoc Netw. pp. 5626-648.

15

A novel bandwidth efficient technique for ICI cancellation in OFDM system

Alka Kalra[1] and Rajesh Khanna[2]

[1]Department of Electronics and Communication Engineering Haryana College of Technology and Management, Kaithal Haryana, India.
[2]Department of Electronics and Communication Engineering, Thapar University, Patiala, Punjab, India.

A well-known problem of orthogonal frequency division multiplexing (OFDM) is its sensitivity to offset between the transmitted and received carrier frequencies. In OFDM communication systems, the frequency offsets in mobile radio channels deform the orthogonality between subcarriers which causes inter carrier interference (ICI). ICI causes power leakage among subcarriers and it further degrades the system performance. In this paper a novel bandwidth efficient method for combating the effects of ICI is presented. In the present scheme, the self cancellation mechanism is used to compress ICI signals without any equalization. At transmitter and receiver parallel Fast Fourier Transform (FFT) factorization is performed using Radix-2 decimation in frequency (DIF) FFT algorithm derived from the well known Cooley-turkey factorization. The proposed scheme cancels the ICI coefficient effectively and further improves system performance as shown through extensive simulations.

Key words: Orthogonal frequency division multiplexing (OFDM), intercarrier interference, carrier frequency offset, Radix2, fast Fourier transform (FFT).

INTRODUCTION

Orthogonal frequency division multiplexing has gained considerable attention due to its high speed data applications. It is emerging as most preferred digital modulation technique due to its high bandwidth efficiency and strong ability to handle multipath interference. In an OFDM system, the whole available bandwidth is divided into N parallel streams, and a block of N data symbols are modulated on N corresponding subcarriers which are orthogonal to each other. The spectra of the subcarriers are overlapping; therefore precise frequency recovering is needed. However, in the mobile radio environment, the relative movement between transmitter and receiver causes Doppler frequency shift, in addition the carriers can never be perfectly synchronized. These random frequency errors in OFDM system distort orthogonality between subcarriers, as a result intercarrier interference (ICI) occurs. Literature shows that in such systems, the

bit error rate (BER) increases rapidly with increasing frequency offsets (Pollet et al., 1993). Therefore, ICI has negative impact on data throughput. Researchers have proposed several methods to reduce ICI. The first approach is frequency-domain equalization in which weighting coefficients of equalizer need be chosen by channel state information (CSI) estimation (Ahn and Lee, 1993). The second one is time-domain windowing (Muschallik, 1996). The third approach is ICI self cancellation, in which redundant data is transmitted onto adjacent subcarriers such that the ICI between adjacent sub carriers cancels out at the receiver (Zhao et al., 1998). This paper is focused on ICI-self-cancellation technique. There are various self-cancellation techniques with different ICI cancelling modulation methods and the corresponding demodulation methods proposed to reduce the ICI caused by carrier frequency offset (CFO).

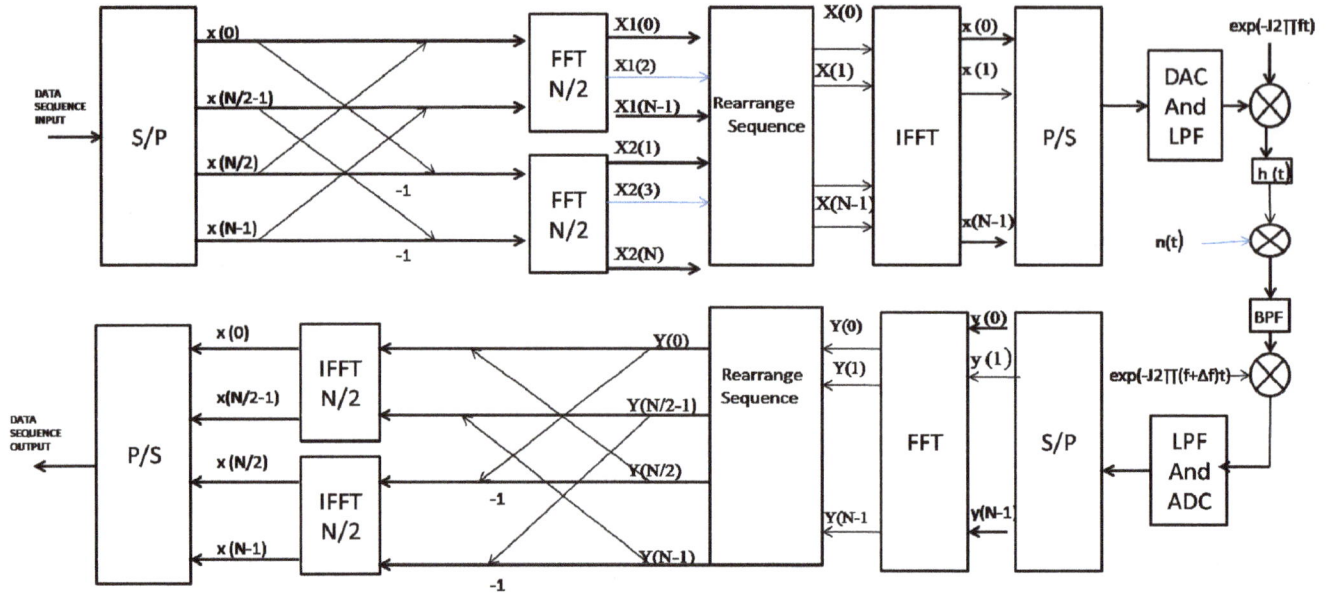

Figure 1. Bandwidth efficient model for ICI Suppression using Radix-2.

These methods include the adjacent data-conversion method (Zhao and Haggman, 2001), the symmetric data-conversion method (Sathananthan and Tellambura, 2001), the adjacent data-conjugate method and the symmetric data-conjugate method (Sathananthan et al., 2004). The aforementioned schemes can suppress the ICI effectively but the spectral efficiency will obviously be reduced as data is replicated on two or more subcarriers.

However, to increase spectral efficiency various schemes has been proposed in literature. DFT based bandwidth efficient scheme for ICI cancellation was proposed by Bing Han et al. (2003). In this scheme at the transmitter side, before modulated onto a group of adjacent subcarriers, the parallel data symbols are divided into two groups, each discrete Fourier trans-formed. At receiver side, after multicarrier demodulation, the received symbols are divided into two groups again and each inverse-discrete-Fourier transformed res-pectively to rebuild the transmitted symbols. However, this scheme outperforms only for small frequency offset values ($\varepsilon<0.2$). In this direction another scheme based on two path algorithm (TPA) was proposed by Kamboj et al. (2009). The first path sends a specially modulated data symbols which results from weighted subtraction of an even numbered modulated symbol and its consecutive symbol. The second path uses the conjugate of a similar type of specially modulated data symbols which results from weighted addition of an even numbered modulated symbol and its consecutive symbol. With additional signal processing, bandwidth efficiency of system can be achieved. A new bandwidth efficient ICI cancellation scheme to cancel interference in OFDM system was proposed by equalization of interference coefficients without requiring any training data (Kumar and Pandey, 2009).

In this paper, a new bandwidth efficient scheme is proposed in which parallel Fast Fourier Transform factorization is performed using Radix -2 DIF FFT algorithms derived from the well known Cooley-turkey factorization using butterfly algorithm. The DIF radix-2

FFT partitions the DFT computations into even indexed and odd indexed outputs which can be computed by shorter length DFT's of different combinations of input samples. The proposed scheme cancels the ICI coefficients and improves system performance of OFDM system. The paper is organized as: The system model of the present scheme is shown. Discussion of the simulation results for the proposed scheme and a comparison with previous bandwidth efficient schemes is further presented. Finally, the overall findings of the study are summarized.

SYSTEM MODEL AND DESCRIPTION

In this section, we propose bandwidth efficient model for ICI cancellation in OFDM system as shown in Figure 1. At transmitter each block of N information symbols are converted from serial to parallel form as sequence of vectors $X = [X_0, X_1 \ldots \ldots X_{N-1}]$, where N is assumed to be equal to IDFT length of OFDM modulator and its integer power of two. DIF FFT algorithm decomposes the DFT by recursively splitting the sequence elements $X(k)$ in frequency domain into sets of smaller subsequences as

$$X_1(k) = \sum_{n=0}^{\frac{N}{2}-1} \left[X(n) + X(n + \tfrac{N}{2}) \right] W_{N/2}^n \qquad (1a)$$

$$X_2(k) = \sum_{n=0}^{\frac{N}{2}-1} \left[X(n) - X(n + \tfrac{N}{2}) \right] W_N^n W_{N/2}^n \qquad (1b)$$

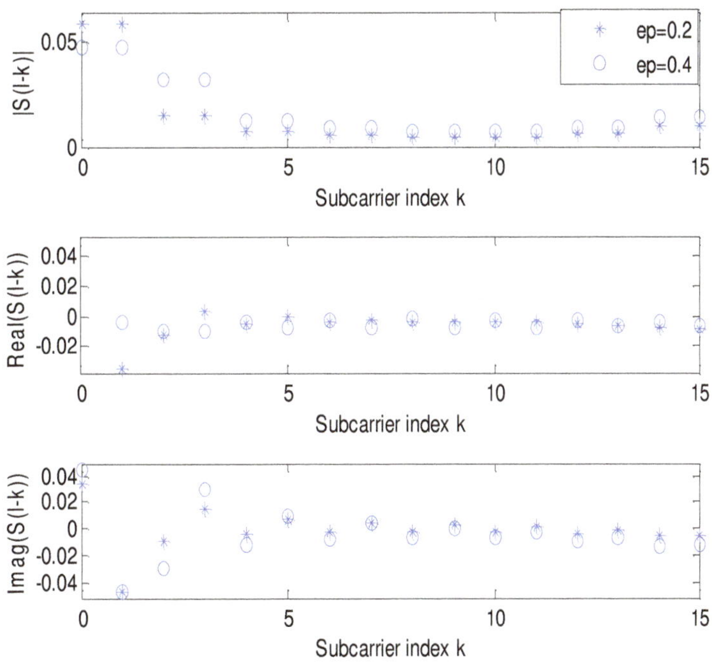

Figure 2. a) Amplitude of $S(l-k)$; b) Real part of $S(l-k)$; c) Imaginary part of $S(l-k)$.

Let $X(k)$ be combined effect of reordered sequence of $X_1(k)$ and $X_2(k)$. Inverse Fourier transform (IFFT) is applied on $X(k)$. The received signal at subcarrier k is represented as

$$x(n) = \frac{1}{N}\sum_{k=0}^{N-1} X(k)W_N^{-nk} exp^{\frac{j2\pi\varepsilon n}{N}} + w(n) \qquad (2)$$

We have assumed that signal is transmitted in additive white Gaussian noise (AWGN) channel and received signal is only affected by carrier frequency offset (CFO) error. Let ε is normalized frequency offset $\Delta f.NT_s$, where Δf is the carrier-frequency offset of the local oscillators between the transmitter and the receiver and Ts denotes the subcarrier symbol period. The effect of this frequency offset on the received symbol stream can be understood by considering the received symbol $Y(k)$ on the kth subcarrier. In the following analysis, we will not consider the guard interval and assume that there is no overlap between different OFDM symbols. The received signal at subcarrier index k can be expressed as:

$$Y(k) = \sum_{l=0}^{N-1} x(n)e^{\frac{-j2\pi kn}{N}}$$

$$= X(k)S(0) + \sum_{\substack{l=0 \\ l \neq k}}^{N-1} X(l)S(l-k) + W(k)$$

$$\quad \text{I} \qquad\qquad \text{II} \qquad\qquad \text{III} \qquad\qquad (3)$$

On right hand side of above equation (I) is desired sequence, (II) is ICI and (III) is AWGN noise introduced in the channel. Let $X(l)$ is transmitted symbol on l^{th} subcarrier and sequence $S(l-k)$ is defined as the ICI co-efficient between l^{th} and k^{th} sub carrier which is expressed as

$$S(l-k) = \frac{sin[\pi(l+\varepsilon-k)]}{Nsin[\frac{\pi}{N}(l+\varepsilon-k)]} exp^{j\pi\left(1-\frac{1}{N}\right)(l+\varepsilon-k)} \qquad (4)$$

Without frequency error $\varepsilon=0$, $|S_0|$ takes its maximum value. It is evident that as ε becomes larger, the desired part decreases and the undesired part increase.

In Figure 2 effect of ICI coefficients at receiver of proposed scheme is presented taking N=16 at different offset values 0.2 and 0.4 respectively. The ICI co-efficients are cancelled out with each other and its effect becomes consistent with increase in frequency offset. The proposed scheme reduces ICI component even at high frequency offset. Applying N/2 IFFT to received signal we have

$$x_1(k) = \frac{1}{N}\sum_{\substack{k=0 \\ l\neq k}}^{\frac{N}{2}-1} \left[Y(k) + Y\left(k+\frac{N}{2}\right)\right] W_{\frac{N}{2}}^{-lk} + W_{k1}^{'}$$

$$= \frac{1}{N}\sum_{k=0}^{\frac{N}{2}-1}\sum_{\substack{=0 \\ l\neq k}}^{N-1} X(K)\left[S(l-k) + S\left(l-k+\frac{N}{2}\right)\right]W_N^{-lk} + W_{k1}^{'} \qquad (5)$$

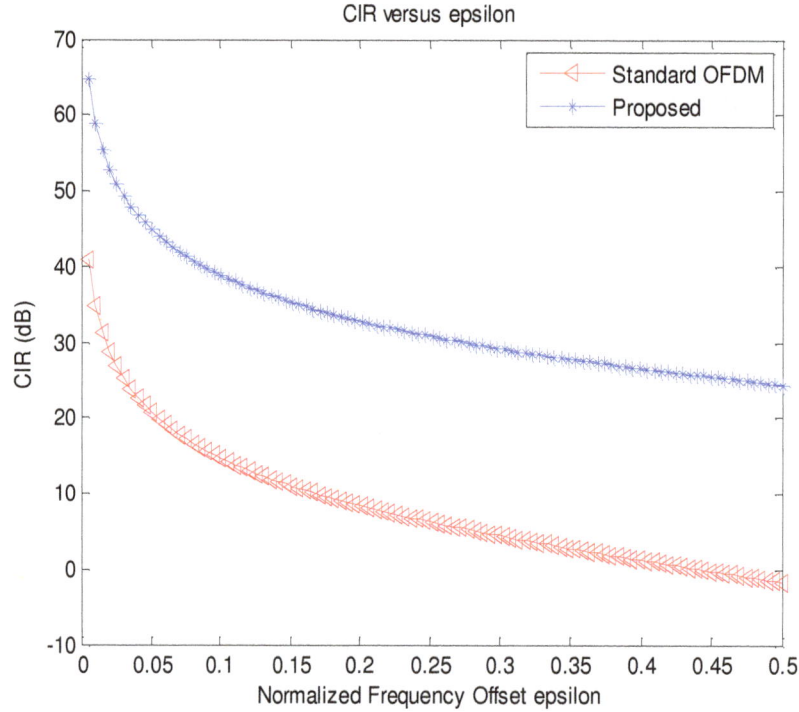

Figure 3. CIR comparison.

$$x_2(k) = \frac{1}{N} \sum_{k=0}^{\frac{N}{2}-1} \left[Y(k) - Y\left(k + \frac{N}{2}\right) \right] W_{\frac{N}{2}}^{-lk} + W_{k2}'$$

$$=$$

$$\sum_{k=0}^{\frac{N}{2}-1} \sum_{l \neq k}^{N-1} X(k) \left[S(l-k) - S\left(l-k+\frac{N}{2}\right) \right] W_N^{-l(k+1)} + W_{k2}' \quad (6)$$

where W_{k1}' and W_{k2}' are N/2 point IFFT of noise samples. Applying symmetry and periodicity property in (5), the ICI coefficients will be zero for even subcarriers and $S(l-k)$ for odd subcarriers. Similarly from (6), ICI coefficients will be zero for odd subcarriers and $S(l-k)$ for even subcarriers. Thus overall effect of ICI is reduced at even symbols and odd symbols respectively.

SIMULATION RESULTS AND DISCUSSION

The simulations have been conducted for performance comparison of various bandwidth efficient techniques is used for ICI cancellation in OFDM system. Figure 3 shows significant improvement in carrier to interference ratio (CIR) of proposed system compared to standard OFDM system. We have performed simulations using MATLAB for BPSK/OFDM with 128 subcarriers. The normalized frequency offset is set to be ε=0.1, 0.3 and 0.5 respectively. The performance of various schemes of ICI cancellation is compared on basis of bit error rate (BER). We have assumed ideal reception of OFDM

symbols for all schemes. BER of standard OFDM, DFT-OFDM, TPA algorithm and proposed (Radix-2) algorithm is plotted in Figure 4. It shows DFT-OFDM, TPA algorithm and present scheme has SNR gain of 8,6,13 dB respectively at BER of 10^{-2} for low frequency offset that is ε=0.1 than standard OFDM system. In Figure 5, it can be observed DFT-OFDM and proposed algorithm has significant improvement in performance than other schemes at ε=0.3. The SNR gain of 10dB, 5dB, 13dB is achieved at 10^{-2} for DFT-OFDM, TPA algorithm and proposed algorithm respectively over standard OFDM performance. From Figure 6 it can be observed clearly the proposed algorithm performs better as compared to another schemes in Rayleigh channel also. The proposed scheme outperforms than other bandwidth efficient algorithms at high frequency offset values. Moreover, spectral efficiency of proposed scheme is maintained without much increase in system complexity.

CONCLUSION

In this paper, ICI self-cancellation in OFDM systems has been investigated. The bandwidth efficiency for normal ICI self cancellation schemes suffers as redundant information is required for such schemes. In proposed Radix 2 based scheme repetition of data symbols and pilot insertion is not required. Thus bandwidth efficiency of system is maintained. At high frequency offset DFT-OFDM, TPA scheme do not perform well. Through

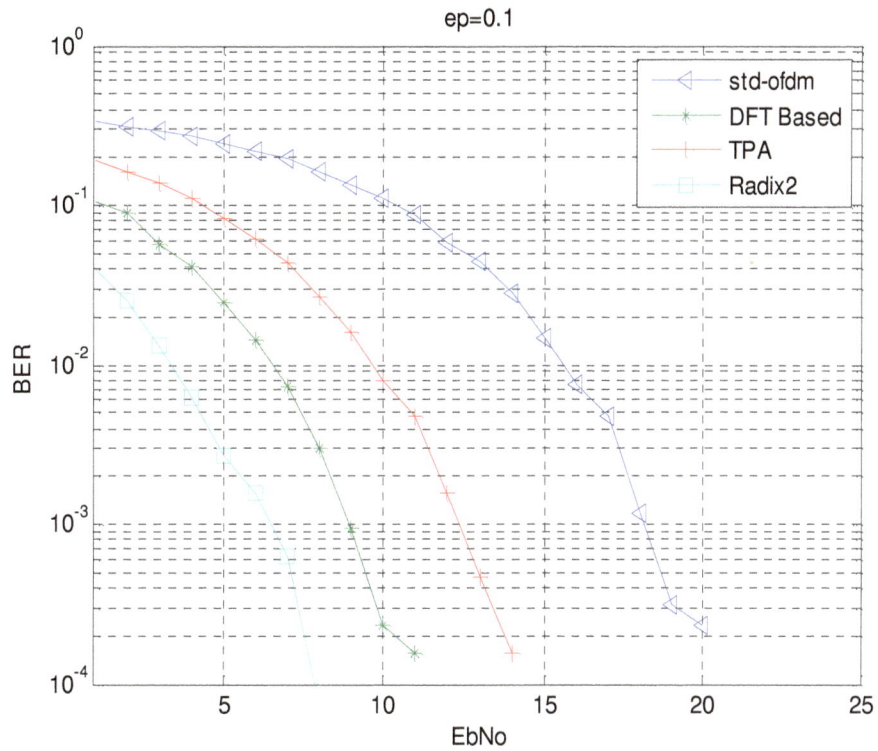

Figure 4. BER comparison with BPSK, AWGN, ε=0.1.

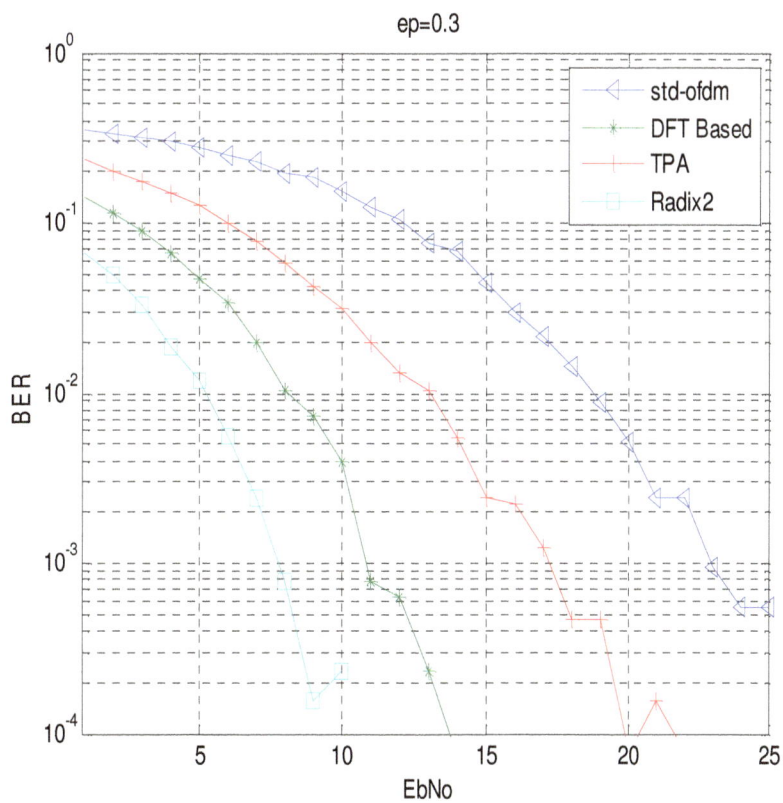

Figure 5. BER ccomparison with BPSK, AWGN Channel, ε=0.3.

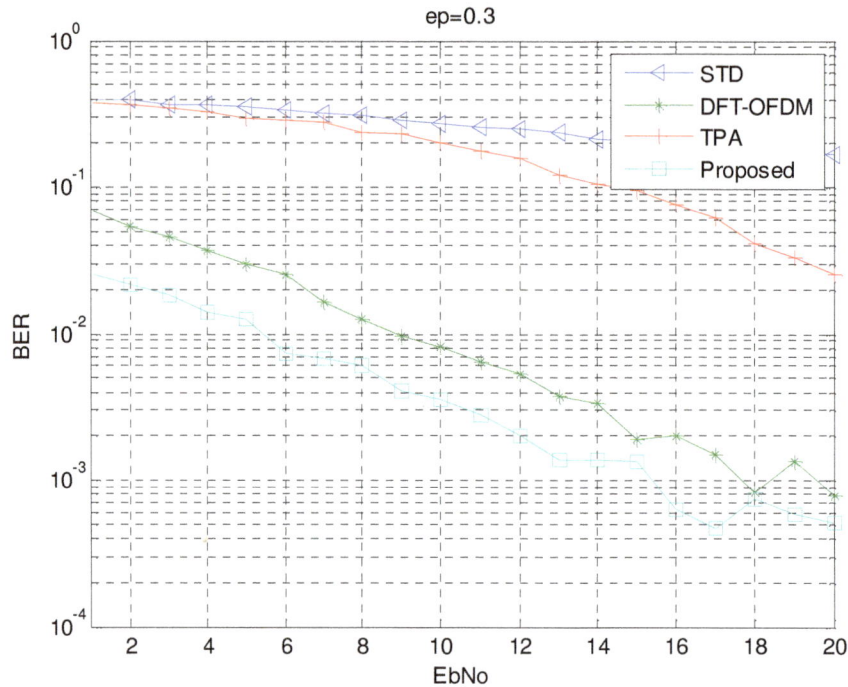

Figure 6. BER comparison with BPSK, Rayleigh Channel, $\varepsilon = 0.3$.

simulations and system model it has been verified that proposed scheme mitigates ICI efficiently at even high frequency offset ($\varepsilon<0.4$) without sacrificing system bandwidth efficiency. Since no equalization procedure is needed, the system can be significantly simplified.

REFERENCES

Ahn J, Lee HS (1993). Frequency domain equalization of OFDM signal over frequency nonselective Rayleigh fading channels. Electron. Lett. 29(16):1476-1477.

Han B, Gao X, You X, Costa E (2003). A DFT-based ICI Self-cancellation Scheme for OFDM Systems. Commun. Technol. Proc. ICCT. 2:1359-1362.

Kamboj A, Keshari A, Dwivedi VK, Singh G (2009). Bandwidth Efficient Intercarrier Interference Cancellation Technique for OFDM Digital Communication Systems. PIERS Proceedings, 5:1244-1248.

Kumar A, Pandey R (2009). A bandwidth-efficient method for cancellation of ICI in OFDM systems. Int. J. Electron. Commun. 63:569-575.

Muschallik C (1996). Improving an OFDM reception using an adaptive Nyquist windowing. IEEE Trans. Consum. Electron., 42:259-269.

Pollet T, Van Bladel M, Moeneciaey M (1993). BER sensitivity of OFDM system to carrier frequency offset and Wiener phase noise. IEEE Trans. Commun. 43:191-193.

Sathananthan K, Athaudage CRN, Qiu B (2004). A novel ICI cancellation scheme to reduce both frequency offset and IQ imbalance effects in OFDM. Proc. IEEE 9th International Symposium on Computers and Communications, pp. 708-713.

Sathananthan K, Tellambura C (2001). Probability of error calculation of OFDM systems with frequency offset. IEEE Trans. Commun. 49(11): 1884-1888.

Zhao Y, Haggman SG (2001). Intercarrier Interference self-cancellation scheme for OFDM mobile communication systems. IEEE Trans. Commun., 49(7):1185-1191.

Zhao Y, Leclercq JD, Häggman SG (1998). Intercarrier interference compression in OFDM communication systems by using correlative coding. IEEE Commun. Lett. 2:214-216.

16

Using social network systems as a tool for political change

Jihan K. Raoof, Halimah Badioze Zaman, Azlina Ahmad and Ammar Al-Qaraghuli

Institute of Visual Informatics, UKM-Malaysia.

Social network sites like Facebook, Twitter and YouTube play a significant role in the political arena nowadays. They are growing engagement tools that assist in improving the political process by helping electoral candidates in communicating their political programs and thoughts to the community, as well as in rallying their campaign supporters. On the other hand, voters can also use social media sites to unconditionally communicate with the candidates. This paper shows the importance of online social networking in modern society by reviewing the literature on social networks usage in politics, and showing how this usage has grown dramatically in different aspects of political life during the past few years. The growth in the use of social network sites was clearly seen after Obama's 2008 US presidential election win, which uncovered the significance of social media in political campaigns and presented new ideas about the utilization of different web 2.0 technologies in politics. Clarifying the relationship between social networking and political life will also assist researchers to study the political behaviors of society and the motivations behind political participation.

Key words: Political elections, social network systems, political campaigns, politics, web 2.0 technologies, social networking and politics.

INTRODUCTION

Before the origination of Web 2.0 technologies, the web was used mainly to search for information and to acquire knowledge (Lewis, 2006; O'Reilly, 2006). Yet, with the internet revolution which began in the mid-2000s and the development of social networking sites (Agre, 2002), users' participation and interaction took different forms such as commenting, reviewing and ranking content, sharing photos and videos, voting and surveying, building special interest groups, making new friends, etc. (Kim et al., 2010; Lilleker and Jackson, 2012).

The term Web 2.0 describes the second generation of the World Wide Web (WWW), which focuses more on the ability of people to collaborate and share information online in contrast to the first web version, where people were mainly obtaining information (O'Reilly, 2006).

One of the most popular Web 2.0 technologies are the Social Network Sites (SNSs) like Facebook, Twitter, Google+, MySpace, LinkedIn, etc. (Click and Petit, 2010).

A social network is a connection network between a set of actors (organization, users, etc.), represented as graph nodes, and the relationships that tie these actors (friendship, common interests, trading partnerships, etc.) represented as graph edges. Social networking sites such as Facebook and Twitter were launched in 2004 and 2006; they allow users to register and create profiles, upload media, contribute to message threads, and keep in touch with friends, family and colleagues. Each Facebook profile has a "wall" where other users can add their posts. Since the wall is viewable by the user's friends, wall postings are basically a public conversation centered on an individual user or group.

Candidates and parties nowadays use social network sites for political purposes by communicating directly with the voters, and thereby benefiting from the lower costs of this communication mode when compared to traditional media. Therefore, they make their campaign information easily accessible to everyone especially young people (Smith, 2011), and they can mobilize supporters, gain more votes, get some attention in the traditional media like television and newspaper, and get the attention of other politicians and political journalists (Karlsen, 2012), as well as facilitate fund raising (Carpini, 2000; Vergeer, 2012). Although television remains the leading source of campaign information and election news for many candidates and parties and even for the people (Smith, 2009), some parties utilize the web 2.0 features and SNSs instead of the traditional media because they are marginalized by mainstream media (Lilleker et al., 2011).

This paper is part of a research project by the institute of visual informatics IVI in UKM-Malaysia to demonstrate the importance of the internet and social networks in politics generally and in elections specifically, to open the doors for scholars to do further studies around this subject, and to investigate the ways SNSs help in identifying potential voters for an electoral candidate and predicting the winning candidates in political elections.

A review of various resources indicates that Obama's victories in 2008 and 2012, and his extensive use of social network sites before and during the elections had attracted the attention of researchers. This paper is organized as follows: the next section explores the use of SNSs in political elections during the past decade and how it is expected to be used in the future. The following section covers the findings and conclusions of the research.

Social network sites and their political use

There is a large volume of published studies describing the role of social network systems in politics generally and elections specifically. In recent years, much more politically relevant information has become available on social network sites. The usage rate of social media, especially Facebook, by politicians increased dramatically after 2008 (Leuschner, 2012; Williams and Gulati, 2009; Williams and Gulati, 2007; 2008). Recently, Facebook ranked first in the most visited sites in the world according to Top (2011).

The increase in politically relevant information on Facebook comes from the fact that Facebook supplies electoral candidates with an enormous opportunity to have contact with the public in a very effective and inexpensive manner, and without any limitations. In addition to that, Facebook allows the public to share their opinions, and participate and engage in the political process freely (Westling, 2007; Williams and Gulati, 2007).

From a political point of view, it is very easy to employ SNSs as a tool to communicate directly with the voters. The increase in the number of supporters and voters using Facebook to get political information has led to the increasing number of politicians using SNSs (Smith, 2009; Smith et al., 2008).

Fast, easy, cheaper and with no control, the internet is spreading information make rumors easily created and reach vase audience which can effect on the number of support the candidate gains from the internet generally and from social media specially but still the significant use of internet in the election and campaigns dramatically increased in the last 10 years according to some statistics submitted by Pew Internet and American Life Project (Garrett, 2010). Statistics in the Associated Press (Press, 2012) showed that, the number of active Facebook users has increased significantly over the past few years: At the end of 2004, Facebook had 1 million registered users only, but this went up to 12 million users by the end of 2006 and 100 million by 2008, and this number jumped to 1 billion in September 2012. Other resources show that, Twitter had a similar surge in the number of users, which increased from around one million users in 2008 (Arrington, 2008) to 500 million users in 2012 (O'Carroll, 2012). These statistics on the use of these free media resources can be considered as evidence for the increasing awareness and knowledge of people.

Politicians utilize social networks and smartphones to reach out to as many voters as possible. By observing political candidates' engagements on social network sites, especially on Twitter, Hong and Nadler found indicators that, political events affect the number of people responding positively to candidates (Hong and Nadler, 2012).

A study done by Lilleker et al. (2011) showed that political parties within Great Britain, France, Germany and Poland that stood in the 2009 European parliamentary elections followed a strategy adopting all the features of web 2.0 and the internet to give their supporters the chance to talk to each other or to talk with party leaders directly.

In countries with a restricted media environment, people try to find alternative forms of media to state their opinions and to engage with political issues. In Malaysia, the number of Facebook users is increasing rapidly. Over 50% of Malaysia's population uses Facebook; this rate constitutes 78% of the Malaysian online community (Bakers, 2012). Facebook has become the most popular website in the country since 2010, ahead of the popular search engines Google and Yahoo and the online video sharing site YouTube (Alexa, 2012; Nardi, 1996).

The increase in the usage of the social media can be considered as an indicator of the increased political awareness in Malaysia. Besides, opposition supporters are using this technology as an alternative way to engage in politics and express their opinions freely (Smeltzer and

Keddy, 2010). Using Malaysia as a case study, Smeltzer and Keddy (2010) looked into the potential of Facebook being used as a tool for political change. They examined if, how, and to what extent Facebook usage can support critical political activities inside and outside formal electoral politics in a restricted media environment. Since the start of Malaysia's revolution in IT in the mid-1990s, Smeltzer and Keddy (2010) found that, there is an increasing number of opposition parties and candidates who use SNSs, especially after Obama's victory in the US 2008 election. Many parties and candidates have leveraged on Obama's effective utilization of web 2.0 technologies in politics, especially in the elections, by employing these technologies as tools for increasing the awareness of people, especially the youth, to spread their political views and to attract voters and gain their support. SNSs have helped these politicians to break the barriers imposed on the domestic media by the governing leading parties.

Facebook as a political support tool

Social networks can impact not only the share of the votes, but also the awareness of people and provide an appropriate venue for them to participate in the political process (Mascaro and Goggins, 2010; Vitak et al., 2011), to hold rallies to demand for their rights and to let their voices reach the specialists and politicians in order for change to occur (Gil de Zúñiga, 2012).

Social media plays an expanding role in increasing people's awareness and knowledge about their rights and the necessity of their engagement in the political scene through their participation in SNSs. This has prompted some governments to reconsider polices and to introduce restrictions on the internet and social media. A good example of this was witnessed in Singapore through the changes in policies that started in 2008 and continued until the 2011 parliamentary elections, due to the wide use of social networks (Skoric et al., 2012). They found in their study of 2000 respondents from Singapore that, the ruling party's increasing control of traditional media and newspapers had led to a growing number of people using social media to express their political opinions, and that in turn led to an increasing number of political activists participating in political rallies during the election.

A study that was conducted by Robertson et al. (2009), a group of researchers in Hawaii University aimed at better understanding social networking in the context of politics, showed the linkage patterns of posters which added comments to the Facebook walls of three major candidates: Barack Obama, Hillary Clinton and John McCain in the period before the 2008 U.S. presidential election. This study showed how the posters used links to different information sources to communicate their points or thoughts to others (Robertson et al., 2009; 2010).

Social network tools, especially Facebook and blogs,

play a significant role in changing society's viewpoints, which in turn helps in promoting people's revolution against non-democratic governments, an example of which was clearly seen in Pakistan in November 2007. The government controlled all the television channels and newspapers so, the youth overcame these barriers by using Facebook, blogs and mobile text messages to communicate and pass political information to others (Shaheen, 2008).

Besides that, Facebook recently played a major role in the Arab Spring. This event has attracted many researchers to find out the impact of the political messages posted on Facebook walls on the users' political trends, their political interest, and on the political arena in general (Yousif and ALsamydai, 2012).

To show how the people in Arab countries (Egypt, Libya, Tunisia, and Syria) employed SNSs in politics and how they used Facebook to participate in political discussion, Khashman (2011) analyzed the Facebook pages of users in the countries that witnessed a political unrest during the still-continuing Arab Spring. Focusing on Egyptian pages on Facebook, Khashman (2011) found that there were more negative pages than positive ones within Facebook about the ousted Egyptian President (Khashman, 2011).

Mobile systems like smartphones and tablets help people to access the internet easily from public places. This in turn facilitates the use of social media. These systems played a big role in the 2011 revolutions in Tunisia and Egypt, by making it easier to pass information among the protesters, and to deliver messages faster to vast audiences, especially the youth. The increasing number of Egyptian people using the internet and social media, the people's anger caused by the accumulated likelihoods of unemployment for youth, and the restrictions imposed by the government on local media, made Facebook the most appropriate space to express and discuss opinions freely. This led to calls for participation in a revolt aimed at political change which succeeded in altering the country's 40 years long regime, even though the Egyptian government tried to control the traditional media, and mobile access to the internet (Attia et al., 2011; Skoric et al., 2012; Tufekci and Wilson, 2012).

As a result of Egypt's revolution and the increasing use of Facebook and its effects, many governments, like the Chinese government, have taken precautions such as blocking access to social media sites to limit their impact on their people (Ho, 2011). During the Iranian presidential election in 2009, the government also blocked access to Facebook because the opposition candidates were using the site in their political campaigns (Bazzi, 2009).

Social network sites in recent elections

The United States was the first country to use the Internet on a large scale in the mid-1990s (Leuschner, 2012;

Vergeer, 2012). After Obama's victory in the 2008 US election, researchers paid more attention to social media tools. Numerous studies were done, examining how Obama influenced the electoral politics by using social media tools, and how these technologies secured his victory and helped in raising enough funds to win the elections.

In 2008 Obama used various web-based tools such as the Obama'08 Web site (barackobama.com), Twitter, Facebook, MySpace, E-mails, iPhone application, and the Obama-Biden transition project (change.gov) site for his campaign. Foreman listed some of the reasons behind Obama's winning of the 2008 presidential election: Obama raised more money than his competitors and utilized social media effectively. That facilitated the dissemination of information especially with the availability of smart phones among enthused youths, winning the support of the youths and leveraging on their ability to influence their families and friends to vote for Obama (Foreman, 2012).

Obama used these tools in an ingenious way that converted online participants to fans and supporters who showed their support by voting for him (Cogburn and Espinoza-Vasquez, 2011; Gil de Zúñiga, 2012; Lilleker and Jackson, 2012). He established a dramatic increase in the use of ICT in politics (Borins, 2009; Kuusk, 2012; Milliken, 2011). Again, similar patterns led to Obama's more recent victory in the 2012 presidential election.

The 2008 presidential elections witnessed several contrasts between how Obama and other candidates utilized social network sites in their campaigns. Obama's strategy led him to reach electorates and win the 2008 election, but other candidates regarded applying social network sites in their strategy as a less important issue since there were no previous facts to prove the relationship between these technologies and success in a presidential election, nor evidence that these technologies contribute in spreading information about their campaigns (O'Brien, 2012).

Despite the fact that traditional media like television and newspapers still play an important role in political campaigns, many candidates and parties show an increased interest in social media, especially after President Obama benefited from the utilizing of SNSs to mobilize and gain access to a greatest number of voters in the 2008 US presidential election (Milliken, 2011).

Milliken also examined how four of the candidates running in the 2012 presidential election: Michele Bachman, Ron Paul, Mitt Romney, and President Obama benefited from employing social media and the Internet on a large scale in their campaigns. These media provided a platform for candidates' words and messages without any kind of filtering or modification which usually happens in traditional media, although each candidate used them in different ways.

European parties used social media as they were influenced by the American success story (Karlsen,

2012). Lilleker et al. (2011) investigated how UK political parties utilized and employed the internet and Web 2.0 in their campaigns during the 2010 UK general elections. He noted that some UK parties were, to some degree, mirroring the strategy of Obama in their campaigns (Lilleker and Jackson, 2012).

Research has continued to ask whether the social networks, especially Facebook, will continue boosting user participation in politics and encourage a strong relationship between the participants, and whether the social networks will make the relationship superficial and isolated from real life. Researchers have found that Facebook is considered by politicians as a chance to attract people to use social networking as a public venue to meet others and to form groups to exchange political information, as well as to communicate with other users who share the same viewpoints. Forming such groups on Facebook helps many users who are seeking political information as noted by Gil de Zúñiga, (2012); Mascaro and Goggins, (2010).

This paper has revealed how the use of social networking sites in politics showed different trends, especially after Obama's victory in the 2008 presidential elections. Previous research showed how users, politicians and candidates used different social media networks for different purposes such as communication, dissemination of information, arrangements for engagement and requests for donations, etc. Several studies showed statistical results that demonstrated a rise in the use of these technologies by politicians and users.

Williams and Gulati (2008) found that 32% of candidates for the U.S. senate and 13% of candidates for the parliament house updated their Facebook profiles at least once for the 2006 elections, and this helped them in gaining extra votes. Furthermore, all presidential candidates in 2008 had social network profiles.

Obama's campaigns were distinguishable in that, he utilized SNSs in a very professional way, and in the 2012 elections, he tried to use each facility available even more extensively than he had in the 2008 campaign. Obama often updated his Facebook page and his web site in the 2008 election, trying to communicate and reach young voters as he believed that the young voters were important, whereas his competitor McCain only updated his Facebook page three times over the election period, and these three updates were made nearer to the election date (Payne, 2009). This lowly administration of social media left a gap between McCain and his followers and giving Obama great opportunity in attracting voters and boosting their attendance(Choy, 2012) (Figure 1).

Subsequent to the 2008 elections, Obama used social media tools extensively again in the 2012 presidential election, gaining more online supporters than his competitor, as illustrated in Figure 2.

Obama's competitor, Mitt Romney, paid more attention to social media and consequently gained a bigger share

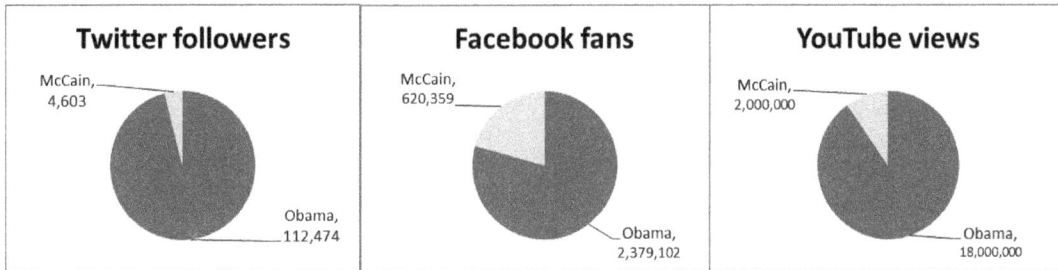

Figure 1. How each candidate used social media in the 2008 US presidential election (Metzgar and Maruggi, 2009).

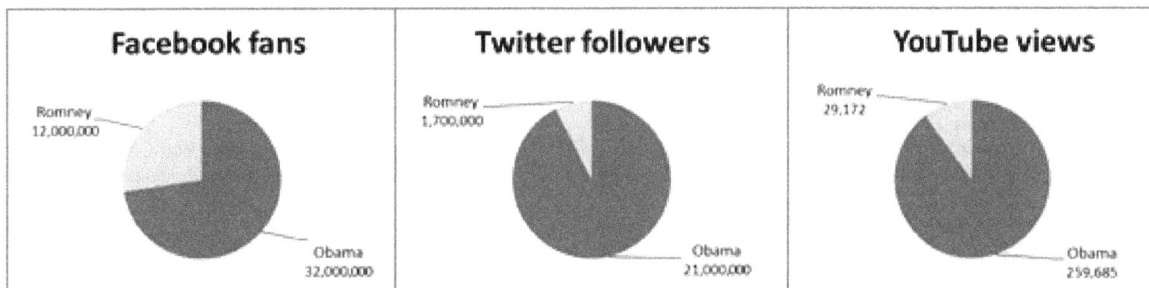

Figure 2. How each competitor in the 2012 presidential election used social media (Burrus, 2012).

of the supporters than Obama's former competitor, John McCain; however, Romney's campaign did not utilize social media as a core and base strategy during the elections as Obama did (Burrus, 2012), and this still left a relatively big gap between Obama and his competitor.

A study conducted in Norway provided statistical evidence that, the younger candidates in the 2009 Norwegian election campaign used Facebook for their campaigns more than the older ones. Over 80% of the candidates aged 25 years and below used Facebook as a part of their campaigns, and 73% of the candidates between 26 to 35 years old used it for the same purpose. There was a lower usage rate of Facebook by candidates between 36 to 50 years old, whilst only a quarter of the candidates aged 51 and above used the site.

In contrast, there were differences in the usage of Twitter, with candidates between 36 to 50 years old using Twitter more extensively than candidates below 35 years old (Karlsen, 2012), and this indicates that micro blogging is more attractive to older people while the content-rich Facebook attracts younger people.

Conclusion

This paper concludes that, social media is used as a continuously expanding communication channel between candidates and voters. The research has discussed the effectiveness of Facebook as a social network site in political elections to rally candidates' supporters, as well

as to make political information available to users of social network systems.

Traditional media like television and newspapers is still significant and important in political campaigns for fast circulation of political information through the internet; this in turn makes rumors easily spreadable. Notwithstanding this, the research predicts that, the number of politicians, especially election candidates employing SNSs in their campaigns will rise after Obama's successful re-election as the US president in 2012 since his campaign focused significantly on employing web 2.0 features effectively to influence potential voters and gain their support.

This paper will give a better insight to politicians, journalists, political analysts, and electoral candidates about how others used SNSs as well as how far they may benefit from these technologies. This will, in turn, make them use SNSs as another source of information for further analysis and understanding of the political process.

This review illuminate the need for further study on the possibility of politicians and specialists' using web 2.0 technologies to predict potential voters for electoral candidates. Consequently, this may lead to predicting the winning candidate in a political election.

REFERENCES

Agre PE (2002). Real-time politics: The Internet and the political process. Inform. Soc. 18(5):311-331.

Alexa (2012). Top Sites in Malaysia Retrieved 13 Nov 2012, from http://www.alexa.com/topsites/countries/MY

Arrington M (2008). End Of Speculation: The Real Twitter Usage Numbers Retrieved 23 January 2013, from http://techcrunch.com/2008/04/29.

Attia AM, Aziz N, Friedman B, Elhusseiny MF (2011). Commentary: The impact of social networking tools on political change in Egypt's. Elect. Com. Res. Appl. 10(4):369-374.

Bakers S (2012). Malaysia Facebook Statistics. URL: http://www.socialbakers.com/facebook-statistics/malaysia Retrieved 18-Nov, 2012.

Bazzi M (2009). Iran elections: latest news.

Borins S (2009). From online candidate to online president. Int. J. Public Admin. 32(9):753-758, 32(9):753-758.

Burrus D (2012). Did Social Media Play a Role in Obama's Victory?

Carpini MXD (2000). Gen. com: Youth, civic engagement, and the new information environment. Pol. Commun. 17(4):341-349.

Choy M (2012). US Presidential Election 2012 Prediction using Census Corrected Twitter Model. arXiv preprint arXiv:1211.0938.

Click A, Petit J (2010). Social networking and Web 2.0 in information literacy. Int. Inform. Lib. Rev. 42(2):137-142.

Cogburn DL, Espinoza-Vasquez FK (2011). From networked nominee to networked nation: Examining the impact of Web 2.0 and social media on political participation and civic engagement in the 2008 Obama campaign. J. Pol. Market. 10(1-2):189-213.

Foreman SD (2012). Top 10 reasons why barack obama won the us presidency in 2008 and what it means in the 2012 election. Fla. Political Chronicle 20(1549-1323).

Garrett RK (2010). Rumors and the Internet in the 2008 US Presidential election. Commun. Res. 37(2):255-274.

Gil de Zúñiga H (2012). Social media use for news and individuals' social capital, civic engagement and political participation. J. Computer-Mediated Commun. 17(3):319-336.

Ho S (2011). China Blocks Some Internet Reports on Egypt Protests. Vooice of America News.

Hong S, Nadler D (2012). Which candidates do the public discuss online in an election campaign?: The use of social media by 2012 presidential candidates and its impact on candidate salience. Government Information Quarterly.

Karlsen R (2012). A Platform for Individualized Campaigning? Social Media and Parliamentary Candidates in the 2009 Norwegian Election Campaign. Policy Internet 3(4):1-25.

Khashman N (2011). The Facebook Revolution: An Exploratory Analysis of Public Pages during the Arab Political Unrest.

Kim HN, Jung JG, El Saddik A (2010). Associative face co-occurrence networks for recommending friends in social networks.

Kuusk L (2012). Social media we can believe in: How social media helped Barack Obama to become the president.

Leuschner K (2012). The Use of the Internet and Social Media in US Presidential Campaigns: 1992-2012 more.

Lewis D (2006). What is web 2.0? Crossroads 13(1):3-3.

Lilleker DG, Jackson NA (2012). Towards a more participatory style of election campaigning: The impact of web 2.0 on the UK 2010 general election. Policy Internet 2(3):69-98.

Lilleker DG, Koc-Michalska K, Schweitzer EJ, Jacunski M, Jackson N, Vedel T (2011). Informing, engaging, mobilizing or interacting: Searching for a European model of web campaigning. Eur. J. Commun. 26(3):195-213.

Mascaro CM, Goggins SP (2010). Collaborative Information Seeking in an Online Political Group Environment.

Metzgar E, Maruggi A (2009). Social media and the 2008 US presidential election. J. New Commun. Res. 4(1):141-165.

Milliken K (2011). Media use in the 2012 presidential campaign. Ball State University.

Nardi BA (1996). Studying context: A comparison of activity theory, situated action models, and distributed cognition. Context and consciousness: Activity theory and human-computer interaction. pp. 69-102.

O'Brien (2012). Social media in 2012 elections will make 2008 look like digital dark ages Retrieved 23 January 2013, from http://www.siliconvalley.com.

O'Carroll L (2012). twitter-users-pass-200-million Retrieved 23 January 2013 from http://www.guardian.co.uk/technology.

O'Reilly T (2006). Web 2.0 compact definition: Trying again.

Payne A (2009). The New Campaign: Social Networking Sites in the 2008 Presidential Election.

Press TA (2012). Number of active users at Facebook over the years Oct 23, 2012. Retrieved Jan 22, 2013, from http://bigstory.ap.org/article/number-active-users-facebook-over-years-3.

Robertson SP, Vatrapu RK, Medina R (2009). The social life of social networks: Facebook linkage patterns in the 2008 US presidential election.

Robertson SP, Vatrapu RK, Medina R (2010). Online video "friends" social networking: Overlapping online public spheres in the 2008 US presidential election. J. Inform. Technol. Polit. 7(2-3):182-201.

Shaheen MA (2008). Use of social networks and information seeking behavior of students during political crises in Pakistan: A case study. Int. Inform. Lib. Rev. 40(3):142-147.

Skoric MM, Pan J, Poor ND (2012). Social Media and Citizen Engagement in a City-State: A Study of Singapore. Paper presented at the Sixth International AAAI Conference on Weblogs and Social Media.

Smeltzer S, Keddy D (2010). Won't You Be My (Political) Friend? The Changing Face (book) of Socio-Political Contestation in Malaysia. Can. J. Dev. Stud./Revue canadienne d'études du développement. 30(3-4):421-440.

Smith A (2009). The Internet's role in campaign 2008. Pew Internet & American Life Project, 15.

Smith AW, Rainie H, Internet P, Project AL (2008). The internet and the 2008 election: Pew Internet and American Life Project.

Smith KN (2011). Social Media and Political Campaigns. [Thesis Projects].

Top A (2011). 500 Global Sites, 2010 Retrieved 15 Nov, 2012, from URL: http://www.alexa.com/topsites.

Tufekci Z, Wilson C (2012). Social media and the decision to participate in political protest: Observations from tahrir square. J. Commun. pp. 363–379.

Vergeer M (2012). Politics, elections and online campaigning: Past, present... and a peek into the future. New Media and Society.

Vitak J, Zube P. Smock A, Carr CT, Ellison N, Lampe C (2011). It's complicated: Facebook users' political participation in the 2008 election. Cyber Psychol. Behav. Soc. Netw. 14(3):107-114.

Westling M (2007). Expanding the public sphere: The impact of Facebook on political communication. Dissertations and Theses.

Williams C, Gulati G (2009). Social networks in political campaigns: Facebook and Congressional elections 2006, 2008.

Williams CB, Gulati GJ (2007). Social networks in political campaigns: Facebook and the 2006 midterm elections.

Williams CB, Gulati GJ (2008). The political impact of Facebook: Evidence from the 2006 midterm elections and 2008 nomination contest. New York 1(1):272-291.

Yousif RO, ALsamydai MJ (2012). The Impact of the Political Promotion via Facebook on Individuals' Political Orientations. Int. J. Bus. Manage. 7(10):85.

17

A Google map-based traffic accident reconstruction system

Chun-Chia Hsu[1], Chih-Yung Lin[2] and Chin-Ping Fung[3]

[1]Department of Cultural Creativity and Digital Media Design, Lunghwa University of Science and Technology, Gueishan, Taoyuan County, Taiwan, Republic of China.
[2]Department of Multi-media and Game Science, Lunghwa University of Science and Technology, Gueishan, Taoyuan County, Taiwan, Republic of China.
[3]Department of Mechanical Engineering, Oriental Institute of Technology, Panchiao, New Taipei City, Taiwan, Republic of China.

Both traffic accident reconstruction and responsibility confirmation of traffic accident depend on crash scene diagramming and crash simulation. However, mistakenly recorded road geometry of the crash scene usually causes misunderstandings in the crash scene diagramming and thus the result of accident reconstruction is untrustworthy. Research has thus been continuously done to solve the problem. This study proposed a Google Map-based accident reconstruction system integrating functions of crash scene diagramming and crash simulation. The modulus of crash scene diagramming includes positioning of accident location, Google satellite map and Google Sketch Up to accurately present road geometry and accident vehicle position information. The modulus of crash simulation refers to the vehicle dynamic differential equations generated using the Newton-Euler formulation, and enhances with the calculation of momentum conservation in the collision to predict the vehicle dynamics after collision. The system developed in this study was validated using a real case. The results showed that a crash scene diagramming correctly drawn on the Google satellite map was an ideal platform to present the crush simulation, and help people understand or make a judgment for accident authentication.

Key words: Accident reconstruction, Google map, crash scene diagramming, crash simulation.

INTRODUCTION

The crash scene diagramming is important for making a judgment on accident authentication as it is used to interpret the crash scene and to be frequently contrasted with the police accident record to reconstruct the accident scene. In order to clean up the obstacles on the road after a traffic accident and restore the passage as soon as possible, the police used to make a sketch of crash scene first at the accident site, using paper and pen. Then based on the sketch the police would further draw a crash scene diagramming using computer software for a formal police report. Nowadays the software of crash scene diagramming has developed well to support police and it has been on the market for a period of time. Crash Zone (The CAD Zone, Inc., 2012) is a full-functioned software of crash scene diagramming and crash simulation. It is used to draw a 2-D crash scene diagramming according to the data measured by police from the accident site. The 2-D diagramming can be further transformed to a 3-D scene. With the built-in tools, Crash Zone calculates the sliding speed and angle of each vehicle after crash. Then the traffic accident is reconstructed and showed with an animation after setting specific before-crash driving paths to separate vehicles. Easy Street Draw (A-T Solutions Inc., 2012) can be used to draw a 2-D crash scene diagramming by selecting a pre-drawn street diagram and editing the diagram according to the real road geometry from the accident scene. With the items in its built-in traffic symbol database, the process of crash scene diagramming is speeded. The office of Traffic Accident Management in China developed the traffic accident scene scale map software (2012) that could quickly draw a scale map of crash scene diagramming according to measurements recorded in a crash scene sketch and using built-in objects in the software such as people, vehicle, road, tyre skid mark, and debris. An Hui Keli Information Industry Co., LTD (2012) also developed a crash scene diagramming system with a laser range finder for police officers to quickly manage traffic accident. The system offers functions of zoom-in and zoom-out, symbols of automobile, non-automobile, human body, path, road, and accident type, and geometry database of line, curve, rectangle, circle, and ellipse. The commercially available software has its own characteristics, however, the accuracy of the crash scene diagramming in road geometry done by those software is based on a crash scene sketch and measurements recorded by police officers. In addition, not all package software has the function of CAD to set relations between different dimensions and maintain the same scale when plotting multiple lines and curves, and therefore, the defect usually causes misunderstanding in the relative positions of the human body, vehicle, tyre skid mark, and debris. If the road geometry of the crash scene is not completely or mistakenly recorded, it usually causes scale distortion in the crash scene diagramming. Most of the problems are not easily detected in the process of drawing crash scene diagramming, and the correction of drawing always takes much more time to finish.

The crash scene diagramming can also be made from photographs. Fenton and Kerr (1997) presented a technique that enables the user to create an accurate accident scene diagram from one photograph of the accident scene, by using a combination of processes called Inverse Camera Projection and Photographic Rectification. Chen and Chao (2006) also developed a fast mapping system for road accident by using digital close range photogrammetric technology. The software includes some functions such as reading the accident image, picking up the calibration points, calibrating, picking up the measurement points, computing the measurement points and mapping. In combination with iWitness and Crash Zone software, Hamzah et al. (2010) reconstructed accident scene using close-range photogrammetric technique to accurately map the crash scenes.

It is difficult to come to a decision for traffic accident judgment just from an accident scene and crash scene diagramming. The statements of the parties in inquiry

record are therefore always used to help understand the driving path and direction of the party before traffic accident, the reaction behavior of the party handling vehicle to respond to emergency. The accident is then reconstructed with all the relative materials from crash scene and party statements to infer a conclusion. However, the party statements may not be completely trusted and lead to a distorted accident analysis as the party cannot correctly describe his driving path and reaction behavior because he was terrified during an accident, or the party describes a false statement because he intends to flinch his legal responsibility for an accident. To overcome the problem, a crash simulation constructed on crash scene diagramming can offer the requirements on accident reconstruction and responsibility confirmation of traffic accident. The crash simulation programming in commercial software is well developed. PC-Crash (MEA Forensic Engineers and Scientists, 2012) is a collision and trajectory simulation tool that presents 2D or 3D motor vehicle collisions with its own modulus of crash scene diagramming. Kinematics and dynamics moduli are adopted in the software to simulate vehicle dynamic behaviors. The kinematics modulus is used to calculate the average acceleration of the vehicle from the accelerating or braking force acting on tyres. Then the acceleration is integrated to obtain the velocity and displacement of the vehicle. The dynamics modulus is used to calculate the longitudinal and lateral forces acting on tyres according to the tyre slip angle, accelerating or braking force, and reaction force from suspension system. The forces acted on tyres are then transformed to vehicle body to obtain vehicle's linear and angular acceleration. PC-Crash was first validated by Cliff and Montgomery (1996). The staged collisions were reconstructed using PC-Crash and the trajectories were compared to actual measurements of the skid marks and rest positions. HVE (Engineering Dynamics Corporation, 2012) is a platform for 3D simulation. The EDC's accident reconstruction packages, EDCRASH and EDSMAC, are operated in HVE. EDCRASH calculates impact velocity and impact gravity based on accident side and vehicle damage measurements. EDCRASH is suited for two-vehicle accidents and collisions with immovable barriers. EDSMAC analyzes vehicle response before, during and after impact. Accident investigators can also determinate impact velocity and impact gravity using EDSMAC. M-smac and M-crash (McHenry Software, 2012) are accident reconstruction packages developed by McHenry software, Inc. using SMAC and CRASH3 model, respectively. The packages can present their simulation results on crash scene diagramming plotted using general CAD software. Johnson et al. (2009) reconstructed delta-V, the vehicle change in velocity, for a series of side impact crash tests using reconstruction code CRASH3, and then compared the reconstructed delta-V with the delta-V recorded by the crash test instrumentation to determine the accuracy of the reconstructed value. WinSMAC and WinCRASH (Trantech Corporation, 2012) in ARSoftware are accident reconstruction programs developed by Trantech Corporation improving SMAC and CRASH3 algorithms for Windows version. The programs can predict velocity change before and after collision and display results in a 2D diagram. WinSMAC and WinCRASH are priced at $769 and $469, respectively. Niehoff and Gabler (2006) investigated the accuracy of WinSmash delta-V estimates as a function of crash mode, vehicle body type, and vehicle stiffness. WinSmash, a direct descendant of crash reconstruction software CRASH3, was found to underestimate delta-V by 23% on average. Johnson and Gabler (2012) further used vehicle damage to estimate absorbed energy and applies momentum conservation to estimate ΔV.

The function and feature of the software for crash scene diagramming and accident reconstruction mentioned above are listed in Table 1. The table indicates that only a few packages own both functions of crash scene diagramming and crash simulation, however, these packages are closed systems. Most packages own one function only, and cannot be extended or integrated with other packages. Therefore, this study developed a Google Map-based accident reconstruction system that consisted of crash scene diagramming and crash simulation. With the system, the crash diagramming can be easily and precisely drawn on a Google map by on-duty police officers using built-in objects from database, and the crash simulation is then clearly animated on the Google map for making a judgment of accident authentication.

METHODS

The system structure of a Google Map-based accident reconstruction system is shown in Figure 1. The system consists of crash scene diagramming and crash simulation. The modulus of crash scene diagramming integrates road network digital maps, Google satellite view, Google SketchUP, and the modulus of crash

simulation includes vehicle dynamic simulation and crush simulation programs.

The road network digital maps belonging to Ministry of Transportation and Communication, Taiwan, is a database of map layers with latitude and longitude coordinates. The latitude and longitude coordinates of a location in urban area with an address, an intersection, and a point in rural area without address can be found using different methodologies. By entering address of a location, the latitude and longitude coordinates of the location with address is easily found. By disassembling the names of the streets meeting at an intersection and comparing the words in the street name with database, the latitude and longitude coordinates of the intersection is also easily found. For a point in rural area without address or street name, the coordinate information on Taiwan Power Company's grid numbers, available on every electric pole and switching box throughout Taiwan, can be transformed into latitude and longitude coordinates by grid conversion computation.

With the latitude and longitude coordinates of a location wherever in urban or rural area, the satellite map of an accident site can be obtained from Google's mapping service. The satellite map is then imported into Google SketchUp as a base map, and the modulus of crash scene diagramming is done in the environment of Google SketchUp. Google SketchUp is a 3D model builder software, and it offers functions of line, curve, color etc. in graphic tools to draw traffic lane lines or zebra stripes on a crash scene diagramming easily step by step. In addition, traffic symbols such as cars and trucks, traffic signs and road markings, roadway objects etc. can be built in advance in database. Then a symbol needed in accident reconstruction is selected, stamped onto a diagram, and modified by adjusting its magnitude and direction to simplify the process of drawing traffic symbol. Locating the vehicle after accident and size marking are the last steps to finish a crash scene diagramming. A fixed object such as a traffic light pole in the accident scene is usually used as a datum point to locate the position of the vehicle after accident. After selecting a datum point and importing horizontal and vertical distances between the vehicle's front and rear tyres and the datum point, the accident vehicle can be automatically located to exact position on the diagram in Google SketchUp. Finally, size marking can be done using size marking tool in Google SketchUp.

The vehicle dynamic simulation program calculates driving path before the vehicle crushed. The formulas of vehicle dynamics in the program are derived referring to Light Vehicle Driving System (Andrzej, 1992). Based on Newton-Euler Formulation, the vehicle dynamic differential equations are generated to obtain the vehicle's longitudinal acceleration $\dot{U}$, lateral acceleration $\dot{V}$, and yaw angular acceleration $\dot{r}$:

$$\dot{U}(t_0) = \frac{F_{x1}(t_0) + F_{x2}(t_0) + F_{x3}(t_0) + F_{x4}(t_0)}{m} + V(t_0)r(t_0)$$

$$\dot{V}(t_0) = \frac{F_{y1}(t_0) + F_{y2}(t_0) + F_{y3}(t_0) + F_{y4}(t_0)}{m} - U(t_0)r(t_0) \tag{1}$$

$$\dot{r}(t_0) = \frac{a\left(F_{y1}(t_0) + F_{y2}(t_0)\right) - b\left(F_{y3}(t_0) + F_{y4}(t_0)\right) + \frac{T_r}{2}\left(F_{x2}(t_0) - F_{x1}(t_0)\right) + \frac{T_r}{2}\left(F_{x4}(t_0) - F_{x3}(t_0)\right)}{I_{zz}}$$

where

$$F_{xi}(t_0) = -\frac{1}{4}mga'_x(t_0)\cos\delta_i(t_0), \quad i=1, 2, 3, 4$$

$$F_{yi}(t_0) = -\frac{1}{4}mga'_x(t_0)\sin\delta_i(t_0), \quad i=1, 2, 3, 4$$

F_{xi} and F_{yi} are the force acting on the vehicle's wheel. a'_x is the driving or braking acceleration from accelerator or brake pedal. δ_i is the steer angle of each wheel.

Then the velocity U, V, r of the vehicle are obtained by integrating the vehicle's acceleration:

Table 1. The function and feature of the software for crash scene diagramming and accident reconstruction.

Software	Crash scene diagramming	Crash simulation	Database	Price
Crash Zone	√	√	√	$699
Easy Street Draw	√	—	√	$199
The office of traffic accident management in China	√	—	√	
AnHui Keli Information Industry Co., LTD	√	—	—	
PC-Crash	√	√	√	$4,995
HVE	√	√	√	
m-crash/ m-smac	—	√	—	
WinCRASH/ WinSMAC	—	√	—	$469/$769

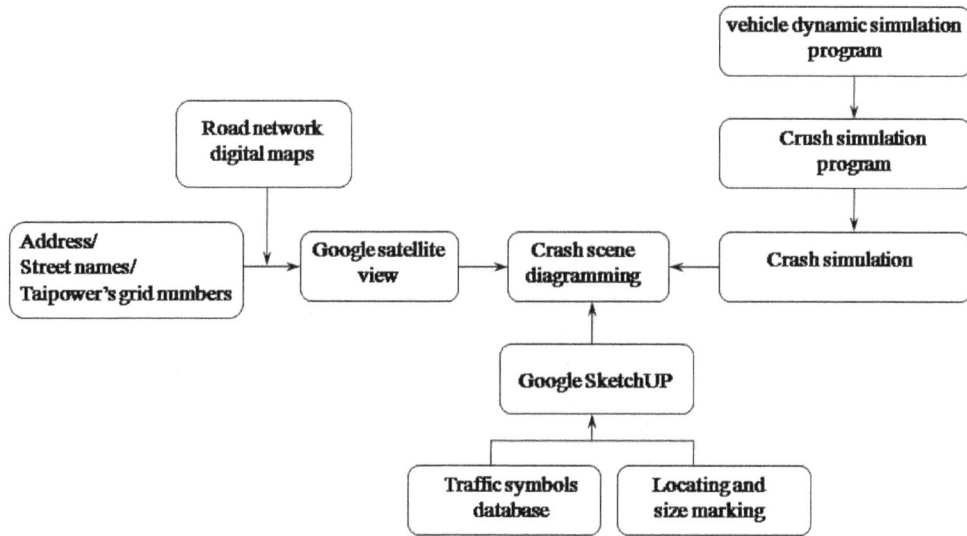

Figure 1. The system structure of a Google Map-based accident reconstruction system.

$$U(t_1) = \dot{U}(t_0)dt + U(t_0)$$
$$V(t_1) = \dot{V}(t_0)dt + V(t_0) \qquad (2)$$
$$r(t_1) = \dot{r}(t_0)dt + r(t_0)$$

and the position X, Y, Ψ of the vehicle are obtained by integrating the vehicle's velocity:

$$X(t_1) = [U(t_0)Cos\,\Psi(t_0) - V(t_0)Sin\,\Psi(t_0)]dt + X(t_0)$$
$$Y(t_1) = [U(t_0)Sin\,\Psi(t_0) + V(t_0)Cos\,\Psi(t_0)]dt + Y(t_0) \qquad (3)$$
$$\Psi(t_1) = r(t_0)dt + \Psi(t_0)$$

The crush simulation program, based on the law of momentum of conservation, calculates velocity vectors $(\dot{X}'_a, \dot{Y}'_a)$ and yaw angular velocity r'_a of vehicle A after vehicle collision:

$$(\dot{X}'_a(t_n), \dot{Y}'_a(t_n)) = (V'_{Na}x_{ab} + V'_{Ta}x_{ab}, \; V'_{Na}y_{ab} - V'_{Ta}y_{ab})$$

$$r'_a(t_n) = r_a(t_n) = r_a \qquad (4)$$

Where

$$V'_{Na} = \frac{m_a - em_b}{m_a + m_b}V_{Na} + \frac{(1+e)m_b}{m_a + m_b}V_{Nb}$$

V'_{Na} and V_{Na} are the velocity components of vehicle A in the normal direction of collision plane after and before collision, respectively. m_a and m_b are the mass of vehicle A and B, and e is the coefficient of restitution. $(x_{ab}, -y_{ab})$ is the tangent directional unit vector of collision plane of vehicle A.

The position X_a, Y_a, Ψ_a of the vehicle A after collision are

$$X_a(t_{n+1}) = \dot{X}'_a(t_n)dt + X_a(t_n)$$
$$Y_a(t_{n+1}) = \dot{Y}'_a(t_n)dt + Y_a(t_n) \qquad (5)$$
$$\Psi_a(t_{n+1}) = r'_a(t_n)dt + \Psi_a(t_n)$$

The velocity and position of vehicle B after collision are derived in

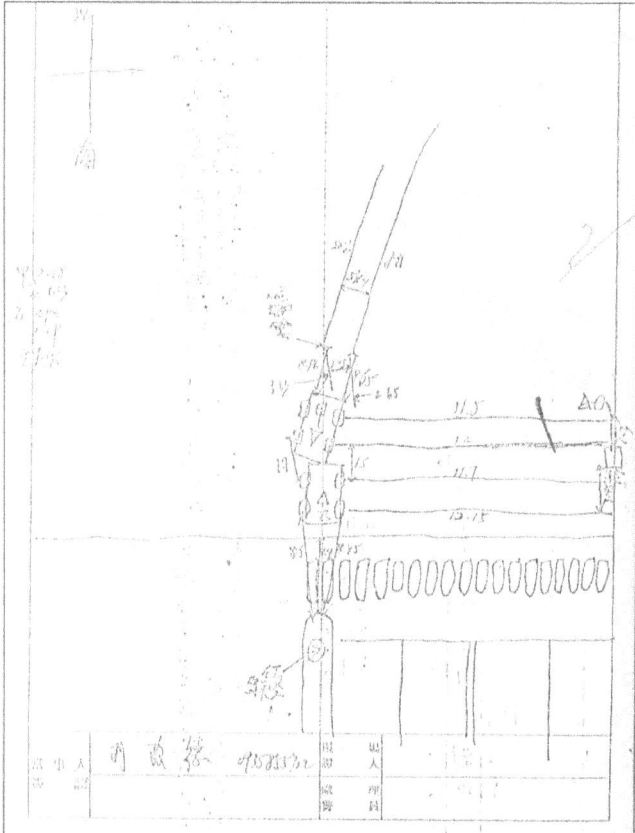

Figure 2. The original hand-drawn crash scene diagramming for a real traffic accident case.

the same process.

RESULTS

Owing to the complexity of traffic accident, the accident reconstruction and analysis are quite difficult. To exclude unreasonable or unexplainable phenomena caused by unpredictable conditions, the validation of the newly developed system in this study first focuses on a relatively simple real case analyzed by a specialist previously (Chen, 2005).

The real traffic accident case for system validation is selected from a case study of crush between two passenger cars (Chen, 2005). The original hand-drawn crash scene diagramming is shown in Figure 2.

Crash scene diagramming

The first step of drawing a crash scene diagramming is downloading a Google Map as a base map. To save time and avoid typing error, a pull-down menu is developed in the system for importing street names. By clicking on the pull-down menu, the names of two streets meeting at the

intersection where an accident was happened are selected one by one, as shown in Figure 3a and b. In addition, an application programming interface is programmed to be used as an interface by road network digital maps and Google satellite maps to communicate with each other. Without entering the website of Google Map API Service, Google satellite view and GPS coordinates are directly shown in the system, as shown in Figure 3c and d, using the application programming interface.

The second step is importing the Google satellite view of the intersection where the real accident case as a base map, shown in Figure 4a. Then road markings such as compulsory ahead only, compulsory turn right ahead, yellow box junction etc. beforehand built in the traffic symbol database of Google SketchUp are selected and stamped onto the base map. A crash scene diagramming is finished by adjusting the road marking's magnitude and direction to fit the base map, shown in Figure 4b.

Crush simulation

Follow the driving conditions derived from the analysis in the reference (Chen, 2005); the driving routes and driving behavior sequence of the two passenger cars were set in the system. The ranges of speed entering the intersection are 80~84 and 32~34 km/h, respectively for car A and car B, respectively. By trying out various speeds for car A and car B until the simulation results, including the location of vehicle impact, and the locations of the vehicles coming to rest after impact, coincide with the locations recorded by police at accident scene. It was found that setting the initial velocity of car A and car B entering the intersection to 83 and 34 km/h, respectively results in a simulation relating well with the known facts. Figure 5a~c show the results of simulation presented on the Google map for vehicles before, at, and after crush, respectively.

DISCUSSION

The diagram shown in Figure 2 clearly shows the positions of the two accident vehicles, however, only part of the intersection is drawn. In addition, road dimensions are not recorded completely and road geometry is thus not correct. The drawbacks and mistakes of the hand-drawn crash scene diagramming are clearly revealed by the diagram. As contrasted with the hand-drawn sketch of crash scene from police, it is evident that the proposed system in this study offers more information about the road geometry for further making a judgment on accident authentication. The system is also an accurate, speedy, free and opened software for police officers to manipulate traffic accident.

For the real accident case of crush between two passenger cars, the analysis in the reference (Chen,

(a)

(b)

Figure 3. To select the (a) first and (b) second street name at the intersection where an accident happened by clicking on the pull-down menu.

2005) derived the drivers' behaviors before the crush from reports made by police, and evidences left behind at the accident scene. Car A entered the interaction at the speed of 80~84 km/h, then applied brakes to reduce speed with a deceleration of around 10 m/s^2, and a skid mark was left around 14 m. On the other hand, car B entered the interaction at the speed of 32~34 km/h, then applied brakes to reduce speed with a deceleration of around 10 m/s^2, and a skid mark was left around 4 m. When the two cars collided with each other, the speeds of car A and car B were around 53~58 and 10~15 km/h, respectively. The location of vehicles impact, and the location of vehicles coming to rest after impact, analyzed by the reference are shown in Figure 6. As the initial velocities of car A and car B entering the intersection are set to 83 and 34 km/h in a simulation using the system

Figure 3c. GPS coordinates are directly shown in the system without entering the website of Google Map API Service.

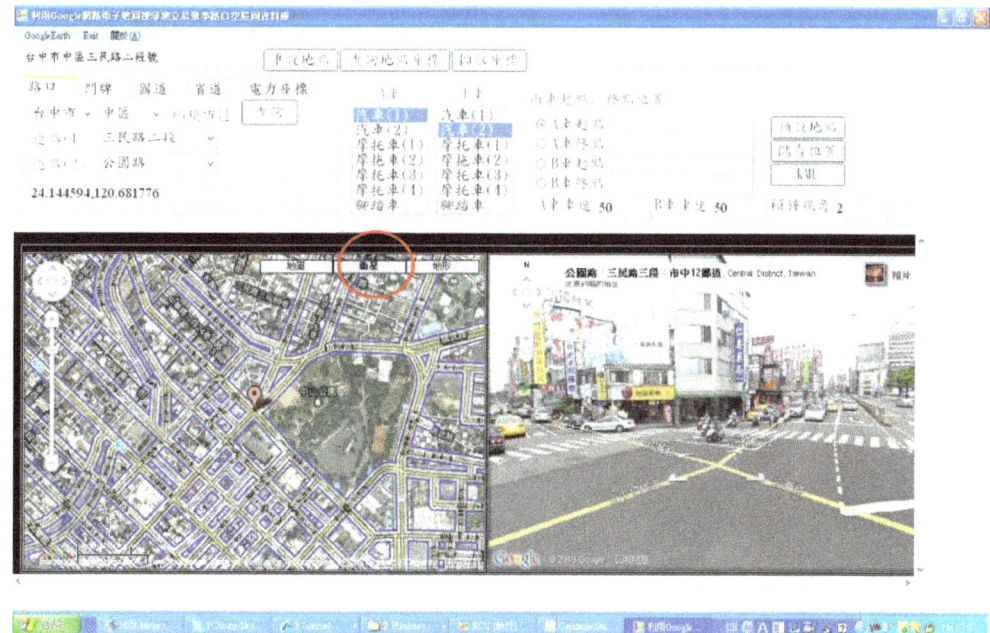

Figure 3d. Google satellite view is directly shown in the system without entering the website of Google Map API Service

proposed in this study, the location of vehicles impact can be presented on the crash scene diagramming. The results are further compared with that from the reference. Figure 7a shows the diagram of vehicle impact location given in this study and that from the reference separately, and Figure 7b is the result of superimposing the two diagrams. The figure shows that the location obtained from this study coincides quite well with that from the reference.

Conclusions

This study developed a Google Map-based accident

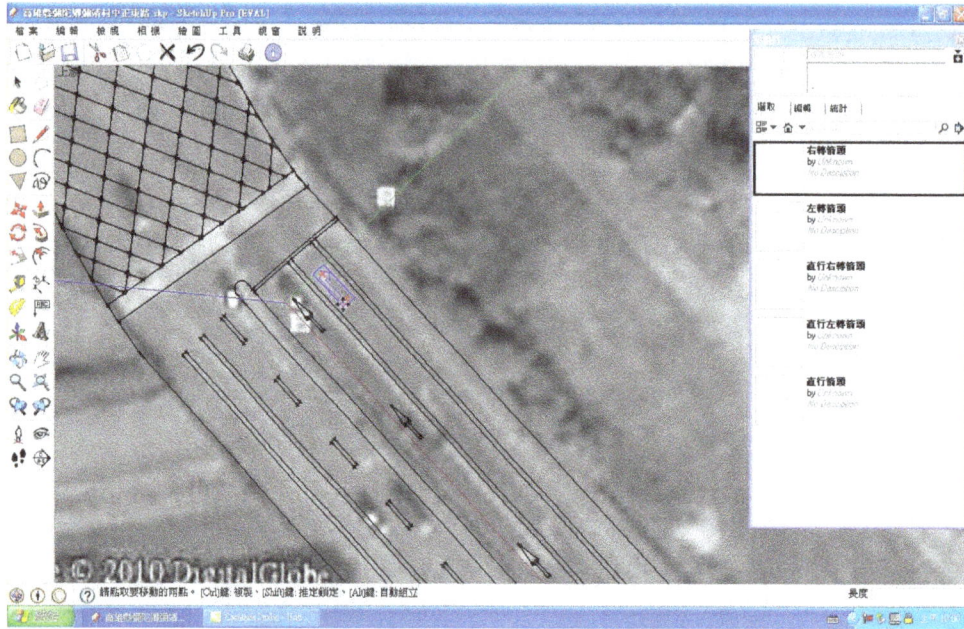

Figure 4a. Importing the Google satellite view of the intersection as a base map and selecting road markings.

Figure 4b. A finished crash scene diagramming based on Google satellite map.

reconstruction system. The system is a free, opened, and extendable software that consisted of crash scene diagramming and crash simulation. The results are summarized as follows:

1. The modulus of crash scene diagramming integrates road network digital maps, Google satellite view, and Google SketchUP. Without entering the website of Google Map API Service, the Google satellite view and GPS coordinates of the accident location are directly shown in the system by operating a pull-down menu. Then a crash scene diagramming is done in the environment of Google SketchUp by selecting pre-drawn traffic symbols and stamping onto the Google satellite map.

2. The modulus of crash simulation includes vehicle

(a)

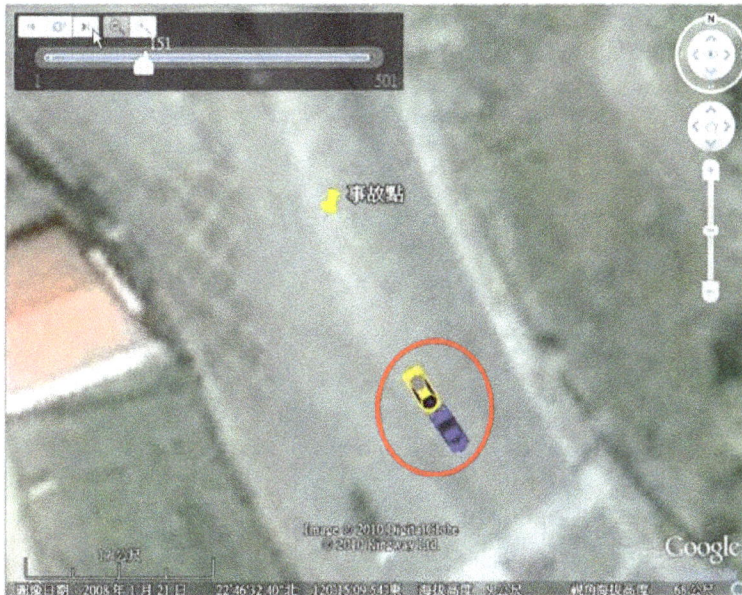

(b)

(c)

Figure 5. Crush simulation presented on the Google map for vehicles (a) before (b) at, and (c) after crush.

dynamic simulation and crush simulation programs. The vehicle dynamic simulation program calculates driving path before the vehicle crushed based on Newton-Euler Formulation, and the crush simulation program calculates velocity vectors and displacements after vehicle collision using the law of momentum of conservation to predict the vehicle motion.

3. The results of crush simulation could be presented on

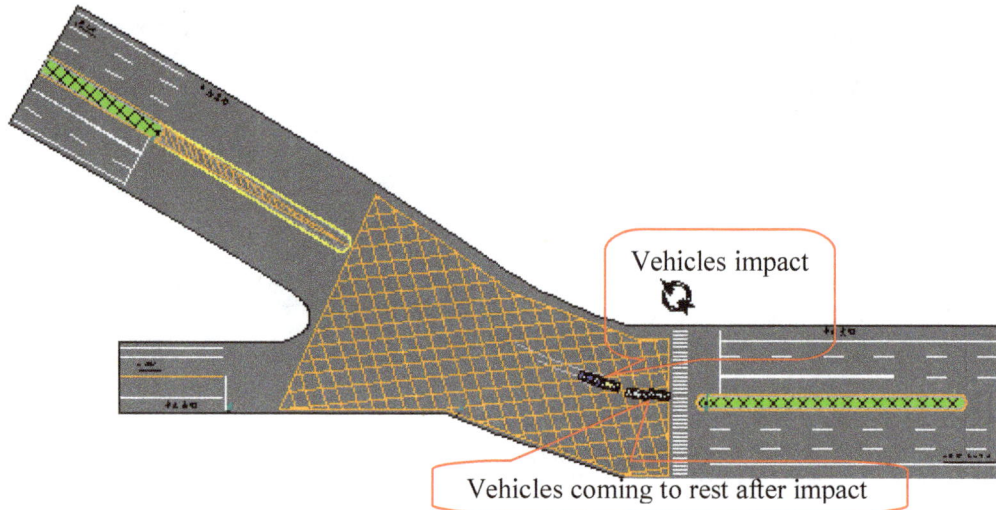

Figure 6. The location of vehicles impact, and the location of vehicles coming to rest after impact, analyzed by the reference.

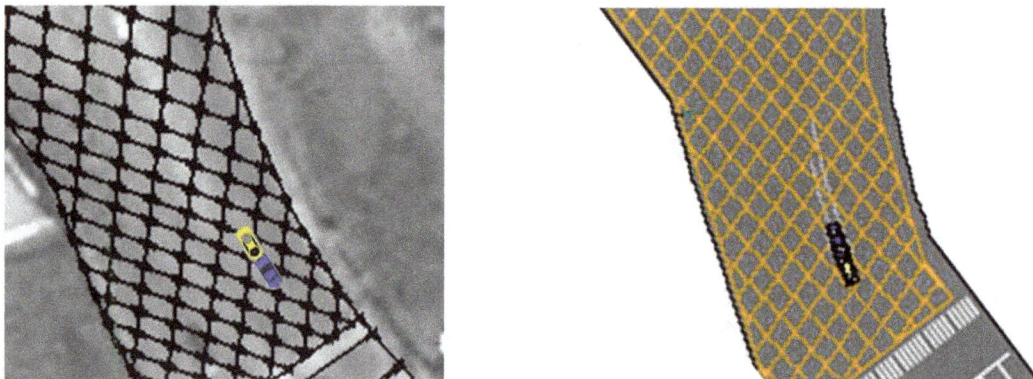

Figure 7a. The diagram of vehicle impact location given in this study (left) and that from the reference (right) separately.

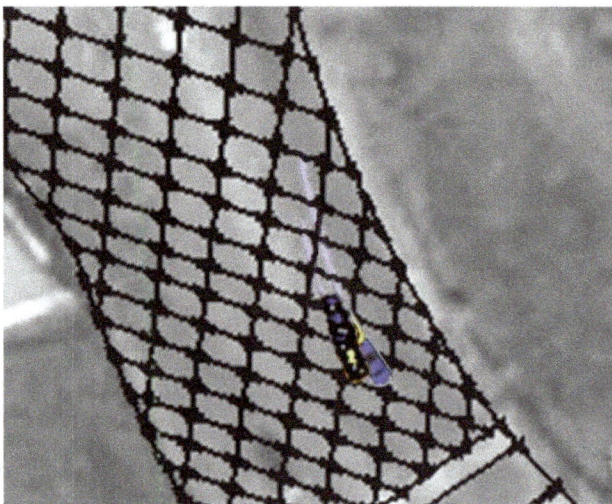

Figure 7b. The result of superimposing the diagram of vehicle impact location given in this study on that from the reference.

the Google map in the form of animation to help people understand or make a judgment for accident authentication.

REFERENCES

Andrzej NG (1992). Develop and Validation of Light Vehicle Dynamics Simulation (LVDS). SAE Paper No. 920056.

AnHui Keli Information Industry Co., LTD (2012). Crash scene diagramming system [online]. Avail from: http://www.ahkeli.com/productcontent.aspx?id=59 [Accessed 12 June 2013].

A-T Solutions Inc. (2012). Easy Street Draw [online]. Avail from: http://www.trancite.com/pro_esd.php [Accessed 12 June, 2013].

Chen KT (2005). Management and authentication of traffic accident (in Chinese). Sun-Wei Press Inc., New Taipei City, Taiwan, pp. 306-329.

Chen Q, Chao KN (2006). Research on Fast Surveying and Mapping System for Road Traffic Accident. J. Shandong Jiaotong Univ. 14:17-21.

Cliff WE, Montgomery DT (1996). Validation of PC-Crash - A Momentum-Based Accident Reconstruction Program. SAE Paper No. 960885.

Engineering Dynamics Corporation (2012). HVE [online]. Available from: http://www.edccorp.com/products/hve.html [Accessed 12 June 2013].

Fenton S, Kerr R (1997). Accident scene diagramming using new photogrammetric technique. SAE Paper No. 970944.

Hamzah NB, Setan H, Majid Z (2010). Reconstruction of traffic accident scene using close-range photogrammetry technique. Geoinf. Sci. J. 10:17-37.

Johnson N, Gabler HC (2012). Accuracy of a damage-based reconstruction method in NHTSA side crash tests. Traffic Inj. Prev. 13:72-80.

Johnson N, Hampton C, Gabler HC (2009). Evaluation of the accuracy of side impact crash test reconstructions. Biomed. Sci. Instrum. 45:250-255.

McHenry Software (2012). M-smac and M-crash [online]. Avail from: http://www.mchenrysoftware.com/ [Accessed 12 June 2013].

MEA Forensic Engineers and Scientists (2012). PC-Crash [online]. Available from: http://www.pc-crash.com/product_pccrash.php [Accessed 12 June 2013].

Niehoff P, Gabler HC (2006). The accuracy of WinSmash delta-V estimates: the influence of vehicle type, stiffness, and impact mode. Annu. Proc. Assoc. Adv. Automot. Med. 50:73-89.

Office of Traffic Accident Management in China (2012). The traffic accident scene scale map software [online]. Available from: http://www.e122.com/drawing.htm [Accessed 12 June 2013].

The CAD Zone, Inc. (2012). Crash Zone [online]. Available from: http://www.cadzone.com/products/the-crash-zone [Accessed 12 June 2013].

Trantech Corporation (2012). WinSMAC and WinCRASH [online]. Avail from: http://www.arsoftware.com/products.htm [Accessed 12 June 2013].

18

Electrical study of plasticized carboxy methylcellulose based solid polymer electrolyte

M. N. Chai[1] and M. I. N. Isa[1,2]

[1]School of Fundamental Sciences, University Malaysia Terengganu, 21030 Kuala Terengganu, Terengganu, Malaysia.
[2]Center of Corporate Communication and Image Development, Chancellery, Universiti Malaysia Terengganu, 21030 Kuala Terengganu, Terengganu, Malaysia.

The electrical conductivity and thermal conductivity of carboxyl methylcellulose doped with oleic acid and plasticized with glycerol have been measured by the electrical impedance spectroscopy method in the temperature range of 303 – 393 K. The composition of glycerol was varied between 0 and 50 wt. % and the samples were prepared via solution casting technique. The highest ionic conductivity at room temperature, σ_{rt} (303K) is 1.64 x 10^{-4} S cm^{-1} for sample containing 40 wt. % of glycerol. The system was found to obey Arrhenius rule where $R^2 \approx 1$. The dielectric study (ε^*, M^*) shows a non-Debye behavior.

Key words: Solid polymer electrolyte, carboxyl methylcellulose, oleic acid, glycerol.

INTRODUCTION

Solid Polymer Electrolytes (SPEs) are the great interests for researchers nowadays, due to their wider range of tremendous applications in electrochemical devices. SPEs also offer numerous of advantages, for example, they can eliminate corrosive solvent and harmful gas formation, have wider electrochemical and thermal stability range as well as low volatility with easy handling. Recently, biodegradable materials attract enormous attention worldwide as a result of white pollution, one of the environmental crises.

According to Guo et al. (2011), plasticizers would make the polymer softer and more flexible, and enhance the chemical and mechanical stability of membranes since they could penetrate and increase the distance of molecules and decrease the polar groups of polymer. In addition, it could overcome the main shortcoming of synthetic polymer, which is mostly insoluble in the solvents. In SPEs, the polymer acts as solvent for a salt which will be partially dissociated in the matrix, leading to ionic conductivity. The electrochemical properties of such polymers are limited by the solvent and the conductivity occurs via interconnected structures of solvent and ions.

To fulfill the stipulations on biodegradable and environmental friendly materials which are significant toward the development of a green nation, carboxy methylcellulose (CMC) has been chosen in this research due to its superior properties as the polymer host. To enhance the conductivity of SPEs, the contribution of mixing polymer and ionic dopant based on the modified double lattice (MDL) is necessary (Pai et al., 2005) thus oleic acid (OA) was chosen to be used as the ionic dopant. In addition, glycerol as plasticizer is introduced into the polymer-dopant system in order to enhance the conductivity and the mechanical properties of the SPEs.

EXPERIMENTAL METHODS

Sample preparation

The CMC based SPEs were prepared by using the solution casting technique. 1 g of CMC (weighted by a digital mass balance) was mixed with 33 ml distilled water and stirred continuously until the CMC dissolved. 20 wt. % (0.25 g) of OA was dissolved in the CMC solution. The CMC-OA composition is following the research of Chai and Isa (2013) of the highest ionic conductivity SPE. In addition to the previous work, the CMC-OA solution was added with different composition of glycerol (0 – 50 wt. %) as plasticizer for each sample respectively. The mixture of CMC-OA-glycerol was poured into Petri dishes and dried in an oven at 60°C. The samples were then kept in desiccators for further drying process.

Electrical impedance spectroscopy (EIS)

The EIS (HIOKI 3532-50 LCR Hi-Tester) are interfaced to a computer with frequency of 50 Hz to 1 MHz. The films was cut into a fitting size and placed between the stainless steel blocking electrodes of the sample holder which connected with the LCR (Inductance, Capacitance, and Resistance) tester. The software controlling the measurement also calculates the real and imaginary impedance. The bulk impedance, R_b value was obtained from the plot of negative imaginary impedance; $-Z_i$ versus real part, Z_r of impedance and the conductivity of the sample was calculated as follow:

$$\sigma = \frac{t}{R_b A} \tag{1}$$

Where A = area of film–electrode contact and t =thickness of the film. Complex dielectric constant, and complex electrical modulus, are evaluated from the recorded complex impedance data, for each temperature. The complex permittivity and complex electrical modulus is given by:

$$\varepsilon_r(\omega) = \frac{Z_i}{\omega C_o (Z_r{}^2 + Z_i{}^2)} \tag{2}$$

$$\varepsilon_i(\omega) = \frac{Z_r}{\omega C_o (Z_r{}^2 + Z_i{}^2)} \tag{3}$$

$$M_r(\omega) = \frac{\varepsilon_r}{(\varepsilon_r{}^2 + \varepsilon_i{}^2)} \tag{4}$$

$$M_i(\omega) = \frac{\varepsilon_i}{(\varepsilon_r{}^2 + \varepsilon_i{}^2)} \tag{5}$$

Here, $C_o = \varepsilon_o A / t$ (ε_o is permittivity of free space, A is electrode–electrolyte contact area, and t is thickness of the electrolyte) and $\omega = 2\pi f$

RESULTS AND DISCUSSION

Conductivity study

The ionic conductivity is depending of various factors,

such as cation and anion types, salt concentration and temperature. The ionic conductivity, σ, of CMC-OA-glycerol was depicted in Figure 1. From Figure 1, it is noted that by the addition of plasticizer affected the conductivity of the CMC-OA-glycerol SPEs. It can be observed that the ionic conductivity in this system increases until 40 wt. % of glycerol and decreases with the addition of higher than 40 wt. % of glycerol. The dependence of ionic conductivity on the plasticizers concentration provides more information on the specific interaction among ionic dopant, polymer matrix and plasticizer. The initial increase of ionic conductivity can be explained by association of ions at higher plasticizer concentration, which leads to the formation of ion clusters and the number of charge carriers and their mobility. When the amount of plasticizer is increased, the ions would transport mainly in the plasticizer-rich phase (Ibrahim et al., 2012). According to Bandara et al. (1998) by postulating the existence of separate ionic pathways for the migration of free ions through the plasticizer, it is possible to explain the enhancement of ionic conductivity when plasticizer is introduced. On the other hand, the decrement of conductivity is mainly due to the higher amount of glycerol which contributed to the overcrowd of ions thus reduces the number of charge carriers further gives limitation on the mobility of ions (Selvasekarapandian et al., 2005; Khiar et al., 2006). For further understanding on the ionic conductivity mechanism, the ionic conductivity of SPEs were tested at elevated temperature ranges from 313 to 393 K. From Figure 2, the relationship between conductivity and temperature of the SPEs are naturally Arrhenius behaviour. The regression values, R^2, obtained for the temperature dependence is $R^2 \sim 1$. Hence, proven the samples obey the Arrhenius law and it is thermally activated similar to the work done by Khiar et al. (2006); Nik Aziz et al. (2010) and Chai and Isa, 2012). The relation of Arrhenius law can be explained by:

$$\sigma = \sigma_o \exp(-E_a / kT) \tag{6}$$

Where σ_o is the pre-exponential factor, E_a is activation energy, k is Boltzmann constant and T is absolute temperature. The activation energy, E_a was calculated from the equation and shown in Figure 3.

Dielectric study

The dielectric constant is a measure of stored charge in a material (Khiar et al., 2006) where in polymer electrolytes, the charge carriers are ions. Meanwhile, dielectric loss explained the loss of energy which eventually produces a rise in temperature of a dielectric placed in an alternating electrical field. In the dielectric constant and dielectric loss plotted shown in Figure 4, no appreciable relaxation peaks observed in the frequency

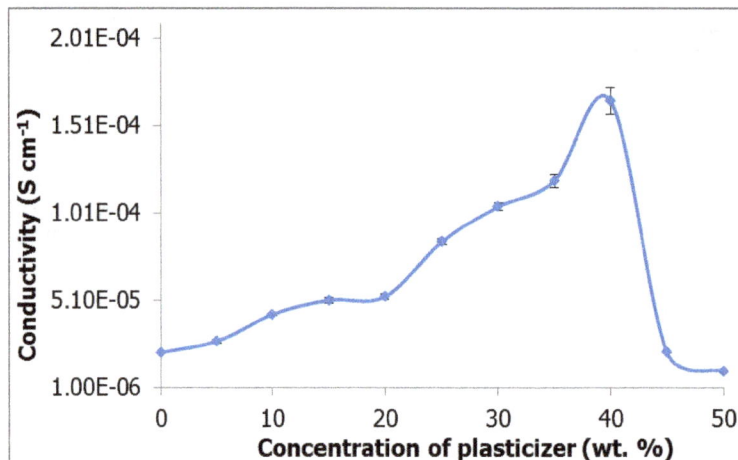

Figure 1. The conductivity of CMC-OA-glycerolSPEs at room temperature.

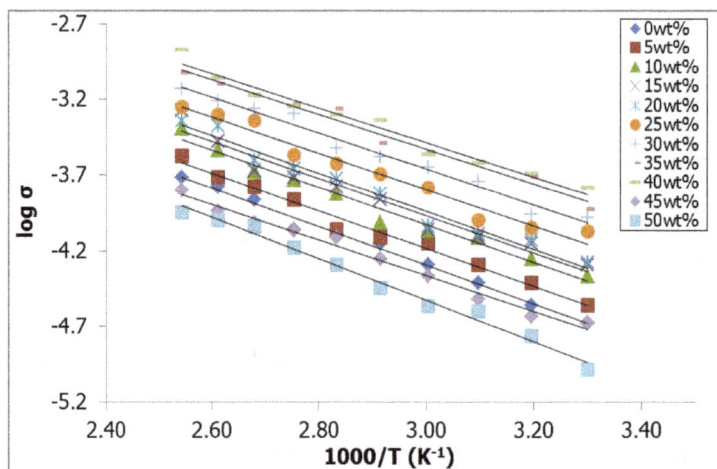

Figure 2. The Arrhenius plot of plasticized SPEs with different wt% of plasticizer.

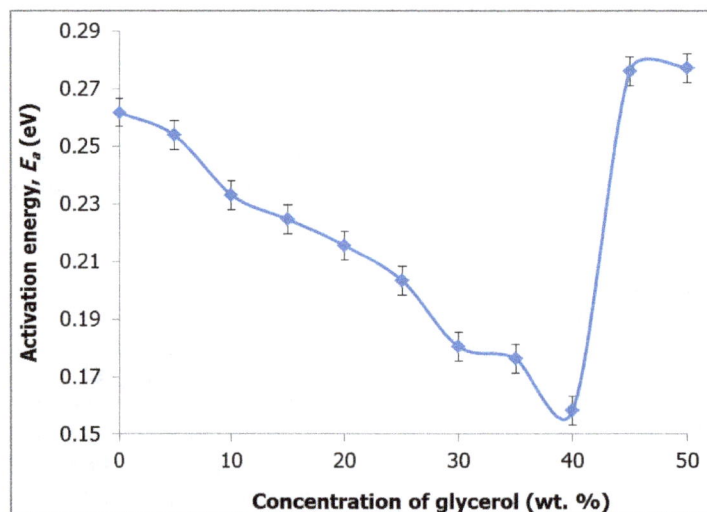

Figure 3. The activation energy of CMC-OA-glycerol SPEs.

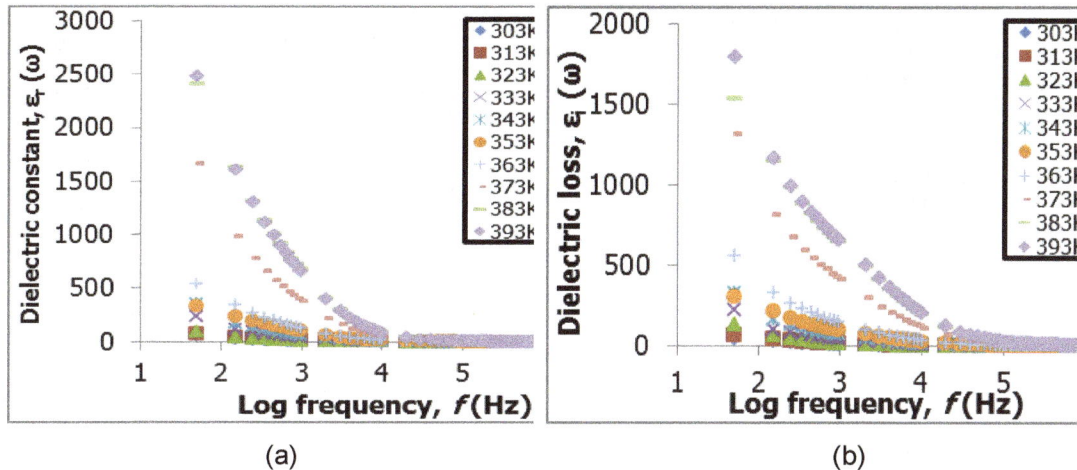

Figure 4. Frequency dependence of (a) dielectric constant, ε_r and (b) dielectric loss, ε_i at various temperatures for sample with 40 wt. % of glycerol.

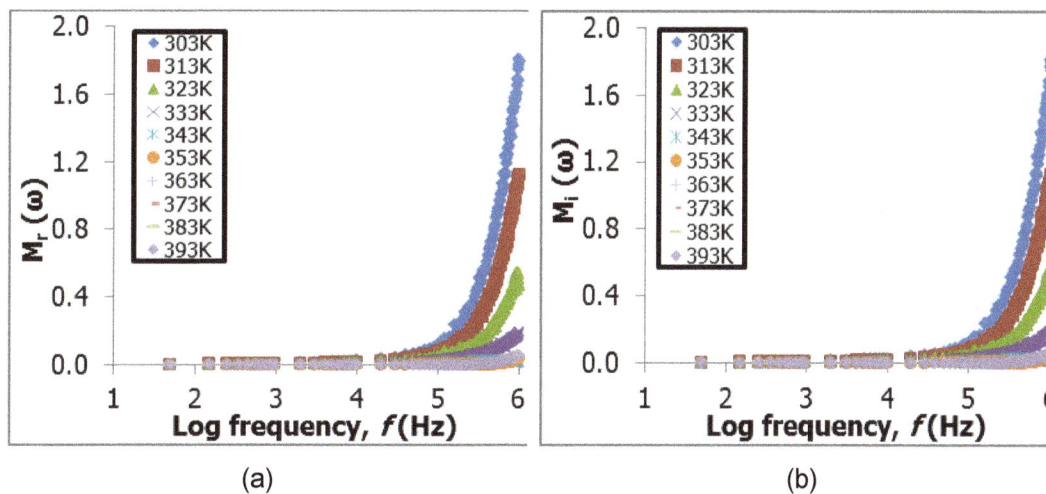

Figure 5. Frequency dependence of (a) real part, M_r and (b) imaginary part, M_i of modulus study at various temperatures for sample with 40 wt. % of glycerol.

range employed in this study. Both dielectric constant and loss rise sharply at low frequencies indicating that electrode polarization and space charge effects have occurred confirming non-Debye dependence. According to Khiar et al. (2006), this implied that the conductivity exhibits relaxation that is non-exponential in time. The dielectric constant in this present study indicates the increase in conductivity is mainly due to the increase in the number density of mobile ions which was also found in Majid and Arof (2007) and Khiar and Arof (2010). Based on their work and the similar result obtained in this current study, it can be deduced that as the frequency increases, the rate of reversal of the electric field also increases, as such; there is no charge build up at the interface which brings about a decrease in the values of

the dielectric loss due to the decrease of the polarization effect by the charge accumulated. Further analysis of the dielectric behaviour would be more successfully achieved using dielectric modulus, which suppresses the effect of electrode polarization.

Modulus study

The variations of real, M_r and imaginary, M_i parts of electrical modulus are shown in Figure 5 (a) and (b) respectively. It can be observed from Figure 5, both M_i and M_r value are approaching zero at low frequency and increases at the higher frequency but no relaxation peaks can be observed. According to Khiar et al. (2006) the presence of peaks in the modulus formalism at higher frequencies for all polymer systems and temperatures is an indicator that the polymer electrolyte films are ionic conductors. The value of both M_r and M_i at low frequencies indicates that electrode polarization is negligible. The appearance of a long tail at low frequencies indicates that there might be a large capacitance associated with the electrodes which further confirms non-Debye behavior in the samples (Nik Aziz et al., 2010).

Conclusion

Solid polymer electrolytes based on CMC-OA plasticized with glycerol were prepared. Sample with 40 wt. % of glycerol was found to have the highest ionic conductivity at room temperature (303 K) of 1.64×10^{-4} S cm^{-1}. The ionic conductivity results as a function of temperature exhibited Arrhenius rule and the value of activation energy is capsized of conductivity. Dielectric study and modulus study suggests that samples in this study show non-Debye behaviour.

Conflict of Interest

The authors have not declared any conflict of interest.

ACKNOWLEDGMENT

The authors would like to thank FRGS and ERGS grants for the financial supports given for this work.

REFERENCES

Bandara LRAK, Dissanayake MAKL, Mellander BE (1998). Ionic conductivity of plasticized (PEO)-LiCF3SO3 electrolytes. Electrochimica Acta. 43:1447-1451. http://dx.doi.org/10.1016/S0013-4686(97)10082-2

Chai MN, Isa MIN (2013). Characterization of Electrical And Ionic Transport Properties Of Carboxyl Methylcellulose- Oleic Acid Solid Polymer Electrolytes. Int. J. Polymer Anal. Character. 18:280-286. http://dx.doi.org/10.1080/1023666X.2013.767033

Guo L, Liu YH, Zhang C, Chen J (2011). Preparation of PVDF-based polymer inclusion membrane using ionic liquid plasticizer and Cyphos IL 104 carrier for Cr(VI) transport. J. Memb. Sci. 372:314–321. http://dx.doi.org/10.1016/j.memsci.2011.02.014

Ibrahim S, Mariah MYS, Meng NN, Roslina N, Rafie JM (2012) Conductivity, thermal and infrared studies on plasticized polymer electrolytes with carbon nanotubes as filler. J. Non-Crystalline Solids. 358:210–216. http://dx.doi.org/10.1016/j.jnoncrysol.2011.09.015

Khiar ASA, Puteh R, Arof AK (2006) Conductivity studies of a chitosan-based polymer electrolyte. Physica B. 373:23-27. http://dx.doi.org/10.1016/j.physb.2005.10.104

Nik Aziz NA, Idris NK, Isa MIN (2010) Solid Polymer Electrolytes Based on Methylcellulose: FT-IR and Ionic Conductivity Studies. Int. J. Polymer Anal. Character. 15:319-327. http://dx.doi.org/10.1080/1023666X.2010.493291

Chai MN, Isa MIN (2012) Investigation on the conduction mechanism of carboxyl methylcellulose- oleic acid natural solid polymer electrolyte. Int. J. Advanced Technol. Eng. Res. 2:36-39.

Khiar ASA, Arof AK (2010) Conductivity studies of starch- based polymer electrolytes. Ionics. 16:123-129. http://dx.doi.org/10.1007/s11581-009-0356-y

Majid SR, Arof AK (2007) Electrical behaviour of proton-conducting chitosan-phosphoric acid-based electrolytes. Physica B. 390:209-215. http://dx.doi.org/10.1016/j.physb.2006.08.038

Selvasekarapandian S, Hirankumar G, Kawamura J, Kuwata N, Hattori T (2005) H solid state NMR studies on the proton conducting polymer electrolytes. Mater. Lett. 59:2741-2745. http://dx.doi.org/10.1016/j.matlet.2005.04.018

Pai SJ, Bae YCand Sun YK (2005). Ionic conductivities of solid polymer electrolyte/salt systems for lithium secondary battery. Polymer. 46:3111-3118. http://dx.doi.org/10.1016/j.polymer.2005.02.041

19

Behavior architecture controller for an autonomous robot navigation in an unknown environment to perform a given task

Jasmine Xavier A.[1] **and Shantha Selvakumari R.**[2]

[1]Jayaraj Annapackiam CSI College of Engineering, Anna University, Chennai, Tamil Nadu India.
[2]Mepco Schlenk Engineering College, Anna University, Chennai, Tamil Nadu India.

The aim of this paper is to carry out navigation task in an unknown environment with high density obstacles using an autonomous mobile robot. Fuzzy logic approach is used for the robot planning because the output varies smoothly as the input changes. If the navigation environment contains one or more obstacles the robot must be able to avoid collisions. The robot uses the obstacle avoidance controller in order to reach the final destination safely without collision with these objects. The robot moves toward the goal and when an obstacle is detected in one of the three sides (front, left, right) the obstacle avoidance behavior is activated to generate the appropriate actions for avoiding these collisions.

Key words: Robot navigation, robot exploration, goal seeking.

INTRODUCTION

There is growing interest in applications of mobile robots. This is due to the fact that the robots are finding their way out of sealed working stations in factories to our homes and to populated places such as museum halls, office buildings, railway stations, department stores and hospitals (Shuzhi and Lewis, 2006). Mobile robots have been the object of many researchers over the last few years in order to improve their operational capabilities of navigation in an unknown environment which consist of the ability of the mobile robot to plan and execute a collision-free motion within its environment. However, this environment may be imprecise, complex and either partially or non-structured (Janglova, 2004). The path

planning problem of a mobile robot can be stated as: given the starting position of the robot, the target location and a description of its surrounding environment, plan a collision-free path between the specified points under satisfying an optimization criterion (Sugihara and Smith, 1997). The path planning in an unknown environment depends on the different sensory systems (cameras, sonar, etc.) which provide a global description of the surrounding environment of the mobile robot; therefore, this description might be associated with imprecision and uncertainty. Thus, to have a suitable path planning scheme, the controller must be robust to the imprecision of sensory measurements. Hence, the need for an approach such as fuzzy logic (Ehsan et al., 2011; Beom and Cho, 1995) which can deal with uncertainties is more suitable for this kind of situations. In real-world problem for autonomous mobile robot navigation, it should be capable of sensing its environment, understanding the sensed information to receive the knowledge of its location and surrounding environment, planning a real-time path from a starting position to goal position with hurdle avoidance, and controlling the robot steering angle and its speed to reach the target. Fuzzy Logic is used in the design of possible solutions to perform local navigation, global navigation, path planning, steering control and speed control of a mobile robot. Fuzzy Logic (FL) and Artificial Neural Network (ANN) are used to assist autonomous mobile robot move, learn the environment and reach the desired target (Velappa et al., 2009). Fuzzy logic was used in many works to design robust controllers for the navigation of mobile robots in a cluttered environments and it can solve such complex real world problems within a reasonable accuracy and a low computational complexity, due to their heuristic nature. In addition, genetic algorithms (Seng et al., 1999), neural networks (Kian et al., 2002) and their combinations were developed to construct the fuzzy logic controller automatically. However, the fusion of different behaviors remains to be difficult when they attempt to control the same actuator simultaneously. Many efforts have been devoted to solve the problem of fusion behavior methods. Because of the complexity of the surrounding environment to be characterized or modeled accurately, behavior architecture control applications become important for the mobile robots navigation. It decomposes the navigation system into specific behavior. Behavior architecture modules which are connected directly to sensors and actuators and operate in parallel. Simple behaviors are then combined in order to produce a complex strategy able to pursue the strategic goals while effectively reacting to any contingencies. Therefore, this architecture can act in real-time and has good robustness. Brooks (1986) proposed an architecture that has been applied successfully in mobile robot navigation, but its main drawback is the arbitration technique which allows only the activation of one behavior at one time. In many situations, the activation of two behaviors is required, for example, when the robot is moving toward the target and avoids obstacles at the same time, two behaviors should be combined to fulfill this task (Yung and Ye, 1999). The basic idea in behavior based navigation is to subdivide the navigation task into small easy to manage, program and debug behaviors (simpler well defined actions) that focus on execution of specific subtasks. For example, basic behaviors could be "avoiding obstacles", "goal seeking" or "wall following". This divide and conquer approach has turned out to be a successful approach, for it makes the system modular, which both simplifies the navigation solution as well as offers a possibility to add new behaviors to the system

without causing any major increase in complexity (Brooks, 1989; Saffiotti, 1997). The suggested outputs from each concurrently active behavior are then "blended" together according to some action coordination rule (Fatmi et al., 2006; Ye et al., 2003).

RELATED WORK

Yung and Ye (1999) presented a new method for behavior based control for mobile robots path planning in unknown environments using fuzzy logic. The main idea of this paper is to incorporate fuzzy logic control with behavior-based control. The basic behaviors are designed based on fuzzy control technique and are integrated and coordinated to form complex robotics system. More behaviors can be added into the system as needed. The output from the target steering behavior and the obstacle avoidance behavior are combined to produce a heading which takes a robot towards its target location while avoiding obstacles. Player/Stage simulation results show that the proposed method can be efficiently applied to robot path planning in complex and unknown environments by fusing multiple behaviors and the fuzzy behaviors made the robot move intelligently and adapt to changes in its environment. Seng et al. (1999) demonstrated a successful way of structuring the navigation task in order to deal with the problem of mobile robot navigation. Issues of individual behavior design and action coordination of the behaviors were addressed using fuzzy logic. The coordination technique employed in this work consists of two layers. A layer of primitive basic behaviors, and the supervision layer which based on the context makes a decision as to which behavior(s) to Fuzzy Logic Based Navigation of Mobile Robots process (activate) rather than processing all behavior(s) and then blending the appropriate ones, as a result time and computational resources are saved. Simulation and experimental studies were done to validate the applicability of the proposed strategy. Yang et al. (2005) proposed an approach which utilizes a hybrid neuro-fuzzy method where the neural network effectively chooses the optimum number of activation rules time for real-time applications. Initially, a classical fuzzy logic controller has been constructed for the path planning problem. The inference engine required 625 if-then rules for its implementation. Then the neural network is implemented to choose the optimum number of the activation rules based on the input crisp values. Simulation experiments were conducted to test the performance of the developed controller and the results proved that the approach to be practical for real time applications. The proposed neuro-fuzzy optimization controller is evaluated subjectively and objectively with other fuzzy approaches and also the processing time is taken into consideration. Samsudin et al. (2011) dealt with the reactive control of an autonomous mobile robot which should move safely in a crowded unknown environment to reach a desired goal. A successful way of structuring the navigation task in order to deal with the problem is within behavior based navigation approaches. In this study, issues of individual behavior design will be addressed using fuzzy logic approach. Simulation results show that the designed fuzzy controllers achieve effectively any movement control of the robot from its current position to its end motion without any collision. Wang and Liu (2008) proposed a new behavior-based fuzzy control method for mobile robot navigation. This method takes angular velocities of driving wheels as outputs of different behaviors. Fuzzy logic is used to implement the specific behaviors. In order to reduce the number of input variables, we introduced a limited number of intermediate variables to guarantee the consistency and completeness of the fuzzy rule bases. To verify the correctness and effectiveness of the proposed approach, simulation and experiments were performed. Seraji and Howard (2002) presented a simple fuzzy logic controller which involves searching target and path planning with obstacle avoidance. In this contest, fuzzy logic controllers are constructed for target searching behavior and obstacle avoidance behavior based on the distance and angle between the robot and the target as inputs for the first behavior and the distance between the robot and the nearest obstacle for the second behavior; then a third fusion behavior is developed to combine the outputs of the two behaviors to compute the speed of the mobile robot in order to fulfill its task properly. Simulation results show that the proposed approach is efficient and can be applied to the mobile robots moving in unknown environments. Selekwa et al. (2005) proposed navigation and obstacle avoidance in an unknown environment using hybrid neural network with fuzzy logic controller. The overall system is termed as Adaptive Neuro-Fuzzy Inference System (ANFIS). ANFIS combines the benefits of fuzzy logic and neural networks for the purpose of achieving robotic navigation task. Abdessemed et al. (2004) presented a Mamdani type minimum rule base fuzzy logic system which has been used successfully in a control system for robot hurdles avoidance in cluttered environment. The fuzzy logic will collect the sonar measurement data as inputs, and select an action for the robot so that it can navigate in the environment successfully.

Mobile robots have expected a substantial concentration from early research community, up to this instant. Today a fully automated robot is expected to travel, detect objects and explore any unknown environments. One such interest is focused on the predicament of generating a map for a functioning environment depiction for navigational tasks. A robot explores while travelling on a trajectory, with its sensors it senses the obstacles and generates a map. The most common sensors used are sonar and laser scanners, which detect the distance of an obstacle within the range

of the sensors by transmitting out signal and compute the time till the resonance of the signal income. The ultrasound rangefinders have a very wide possibility of exploitation due to their ease of functioning, low cost and modest realization. In the majority of circumstance the signal will bounce against the nearby obstacle in the course of the sensor and as a result the calculated distance will be the distance to the nearby obstacle. But however, there are some occasions where there occur measurement failures, and as a result the calculated distance becomes flawed. These measurement failures may occur due to the uncertainties provided by the range measuring sensors. These uncertainties are origin by the characteristics of air such as its temperature, humidity, turbulence and pressure.

One such ambiguity results from the promulgation of the ultrasonic signal to the space in the form of a cone with an axis in the scanning course. So the exact angular position of the object reflecting the echo might not be determined, because it may occur somewhere along the arc with the radius of the measured distance. A further cause of ambiguity is a experience of numerous reflections, that occurs in the case that the incidence angle of signal to the obstacle is larger than a so called critical angle, which is strongly reliant on the exterior distinctiveness. In this occurrence the reflection of the signal is mainly specular and the sensor may perhaps receive the ultrasonic beam after numerous reflections, what is called a elongated reading, or it may even get lost. Consequently, to return a momentous range reading, the angle of incidence on the object exterior has to be lesser than the critical angle.

Steering and obstacle evading are very significant issues for the doing well use of a sovereign mobile robot. Computing the configuration succession, allow the robot to move from one location to a further location. When the surroundings of the robot are obstacle free, the predicament becomes not as much of complex to handle. But as the surroundings becomes a complex, movement planning need much more effective to allow the robot to move between its in progress and closing configurations without any collision with the surroundings. A flourishing approach of configuring the steering assignment to deal with the dilemma is within behavior based navigation approaches.

The fundamental scheme in behavior based steering is to subdivide the navigation task into diminutive simple to supervise, course and sort out behaviors that focus on implementation of explicit sub schemes. For illustration, fundamental behaviors could be obstacle avoiding, target seeking, or wall following. This split and triumph over approach has turned out to be a flourishing approach, for it makes the scheme modular, which both make simpler the steering way out as well as tender a prospect to insert new behaviors to the system without causing any major increase in complexity.

The intention of this dissertation is to be evidence for

how to conduct an autonomous mobile robot in unfamiliar surroundings by means of fuzzy logic approach and to build map based on the range readings obtained from the sensors. Fuzzy logic control (FLC) is an appealing contrivance to be useful to the dilemma of conduit arrangement given that the output varies efficiently as the input adjusts. In this exertion, we will discuss a fuzzy conduit arrangement controller design based on connoisseur understanding and information that was applied to a mobile robot. The fuzzy inference system is based on a person driver reminiscent of interpretation in a indoor surroundings that is vacant or surround obstacles.

The paper is structured as follows: At first, the sculpt of the mobile robot is presented and the essential background of fuzzy logic method and a epigrammatic of fuzzy behavior based steering is presented. The proposed organizers, the map building algorithms are introduced and elucidated. Simulation results for illustration of movement of the robot in unknown surroundings are demonstrated.

DESIGN OF FUZZY BEHAVIOR BASED NAVIGATION METHOD

Mobile robot kinematics

In this exploration, a differentially ambitious mobile robot is used; its kinematic illustration is show in the Figure 1. The kinematic model of the mobile robot has two rear driving wheels and a passive front wheel. The inputs of the scheme are the steering angle α of the front wheel and the linear velocity V_R. The outputs are the coordinates of the robot (X_R, Y_R and θ_R). In ideal sticking together circumstances, this kinematic model can be described by the following equations:

$$X_R = V_R * \cos(\theta_R) \tag{1}$$

$$Y_R = V_R * \sin(\theta_R) \tag{2}$$

$$\theta_R = V_R * tg(\alpha)/l \tag{3}$$

Where X_R, Y_R are the position coordinates, θ_S angle error between the robot axis and goal vector and θ_R is the orientation angle of the robot. l is the robot length. In our work, we suppose that the simulated mobile robot is able to detect the coordinates of the final goal and it is equipped by sensors for perceiving its environment.

Fuzzy logic approach

The premise of fuzzy logic scheme is motivated by the significant human ability to rationale with perception based information. Rule based fuzzy logic afford a proper methodology for linguistic rules ensuing from interpretation and decision making with ambiguous and indefinite information. The building block illustration of a fuzzy control scheme is shown in Figure 2. The fuzzy controller is composed of fuzzification interface, a rule base, an interface mechanism. The fuzzification interface renovates the actual controller inputs into information so as to the inference mechanism can straightforwardly exercise to make active and relate rules. A rule-base has a set of If-Then rules which enclose a fuzzy logic quantification of the connoisseur's linguistic depiction of how to

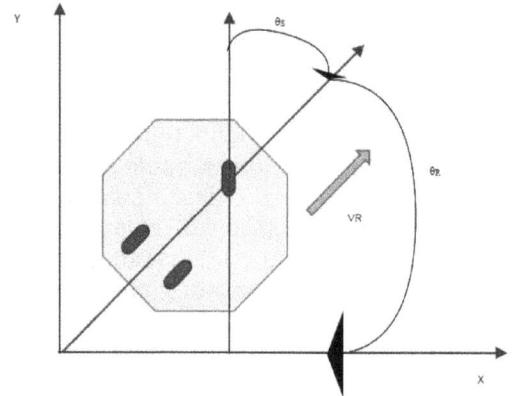

Figure 1. Mobile robot kinematics.

accomplish superior control. An inference mechanism emulates the connoisseur's decision making in interpreting and concern understanding about how preeminent to direct the plant. defuzzification interface renovate the conclusions of the inference mechanism into actual inputs for steering the course. Mamdani and Takagi-Sugeno model are two popular models used mostly in fuzzy logic control. In this paper we use zero order Takagi-Sugeno model owing to its simplicity and effectiveness to the course control.

Design of robot steering and velocity control

The motion control variables of the mobile robot are the angular velocity of the front wheel and the velocity of the rear wheels. The angular velocity is represented by θ_R. The vehicle velocity is determined by the rear wheels speed which is denoted by V_R. The position of the vehicle is denoted by (X_R, Y_R).

The left steering angle is represented by a three-variable linguistic fuzzy input membership set {S, LO1, LO2} which define the distance of the obstacles in three different levels from farthest to the closet respectively, the obstacle distances are estimated from the ultrasound sensor range readings, and the output membership set {ST, L1, L2} which define the steering actions to the left of the vehicle in the different levels from straight steering to intense left turning actions respectively.

Similarly the right steering angle is represented by a three-variable linguistic fuzzy input membership set {S, RO1, RO2} which define the distance of the obstacles in three different levels from farthest to the closet respectively, and the output membership set {ST, R1, R2} which define the steering actions to the right of the vehicle in the different levels from straight steering to intense left turning actions respectively. The rule base of the steering behavior is summarized in Table 1.

Similarly the motor speed of the rear wheel is represented by a five-variable linguistic fuzzy input membership set {D1, D2, D3, D4, D5}, defines the five different levels of obstacle distance from the front end of the mobile robot from very near to the farthest position respectively and the output membership {V1, V2, V3, V4, V5} which define the velocity actions such as too slow, slow, medium, fast, fastest most respectively.

Design of goal seeking behavior

The mission of the robot is to arrive at a preferred position in the surroundings called a goal. This goal seeking behavior is anticipated to line up the robot's cranium with the course of the goal

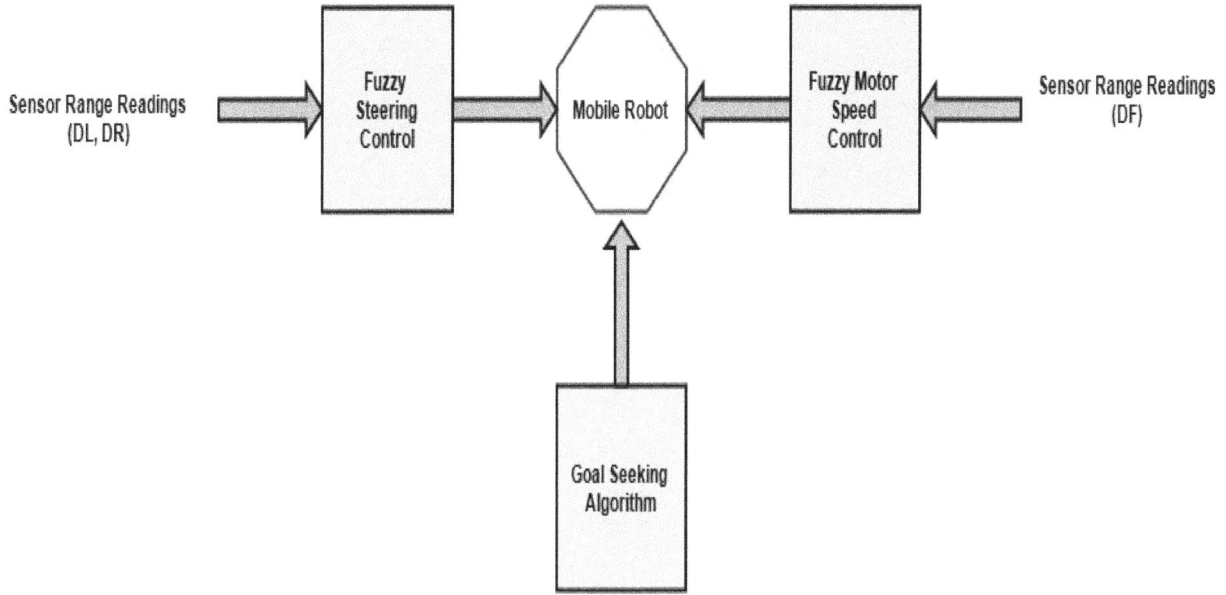

Figure 2. Block Diagram of the proposed scheme.

Table 1. Fuzzy rule set.

Fuzzy rule base for steering control	Fuzzy rule base for motor control
1. If (input1 is LO2) then (output1 is L2)	1. If (input1 is D1) then (output1 is V1)
2. If (input1 is LO1) then (output1 is L1)	2. If (input1 is D2) then (output1 is V2)
3. If (input1 is S) then (output1 is ST)	3. If (input1 is D3) then (output1 is V3)
	4. If (input1 is D4) then (output1 is V4)
1. If (input1 is RO2) then (output1 is R2)	5. If (input1 is D5) then (output1 is V5)
2. If (input1 is RO2) then (output1 is R1)	
3. If (input1 is S) then (output1 is ST)	

coordinates. The calculation module compares the actual robot coordinates with the coordinates of the target using mathematical equations. The outputs are the angle noted θ_D and the distance between the robot and the goal (position error) noted D_{RG}. The angle value is compared with the orientation of the robot delivered by the odometry module in order to compute the angle error θ_{ER} between the robot axis and the goal vector. Prearranged a mobile robot, it must be capable to engender a course between two specific position, the start nodule and the target nodule. The course ought to be free of collision and be required to persuade convinced optimization criterion i.e. least time consuming course. The only information offered to the robot is it's in progress location and the location of the goal in the grid map. The robot has to constantly be in motion from the in progress position until it reaches the goal by avoiding the obstacles detected on course. Occupancy grids are used for the representation of the environment.

At this juncture each cell in the grid encloses information concerning its circumstances, which is premeditated depending on the probability of occupancy of that particular cell. Consequently a cell which is engaged by an obstacle will cover a very high probability of occupancy cost returned by the sensor, which makes it engaged for the robot to pass through. Frontier based heuristic exploration algorithm is the chief province of the dilemma. It is

helpful in situations wherever no preceding preparation is viable and all decisions are taken at instantaneous.

To begin with the robot executes a bursting surrounding look into of its surroundings and updates the occupancy cost G_O of its four neighboring cells, one in each course: top, right, bottom and left. These four cells sensed on each scan are termed as the cells in the in progress sensing area. A frontier cell is a cell explored by the robot which is having at least one unexplored cell as its neighboring cell. After each scanning operation, the newly detected frontier cells are assigned heuristic cost value known as Goal Seeking Index $G_{SI}(X_R, Y_R)$. The cost of moving to a cell (X_R, Y_R) is found as the product of its occupancy cost G_O and the distance of the cell (X_R, Y_R) from the in progress position of the robot. Calculating the cost based on occupancy value is explained well in the GSI for each frontier cell is found out with the help of Equation (4):

$$G_{SI}(X_R, Y_R) = ((D_{MAXIMUM} - D_{((X,Y),TARGET)}) - C_{((X,Y),CURRENT)}) \quad (4)$$

Where $D_{MAXIMUM}$ is the principal distance probable amid any cell in the grid and the goal location, $D_{((X,Y),TARGET)}$ is the distance amid the given frontier cell (X_R, Y_R) and the goal cell, and $C_{((X,Y),CURRENT)}$ is product of the occupancy cost (G_O) of the frontier cell (X_R, Y_R) and its distance from the current location of the robot.

Table 2. Simulation parameters.

Specification	Description
Operating System	Windows Seven
Simulation Tool	MATLAB 2009
Number of Robots	1
Simulation Area	50m x 50m
No of wheels	3 (2 Rear + 1 front steering)
Minimum – Maximum Motor Speed	0-30 RPM

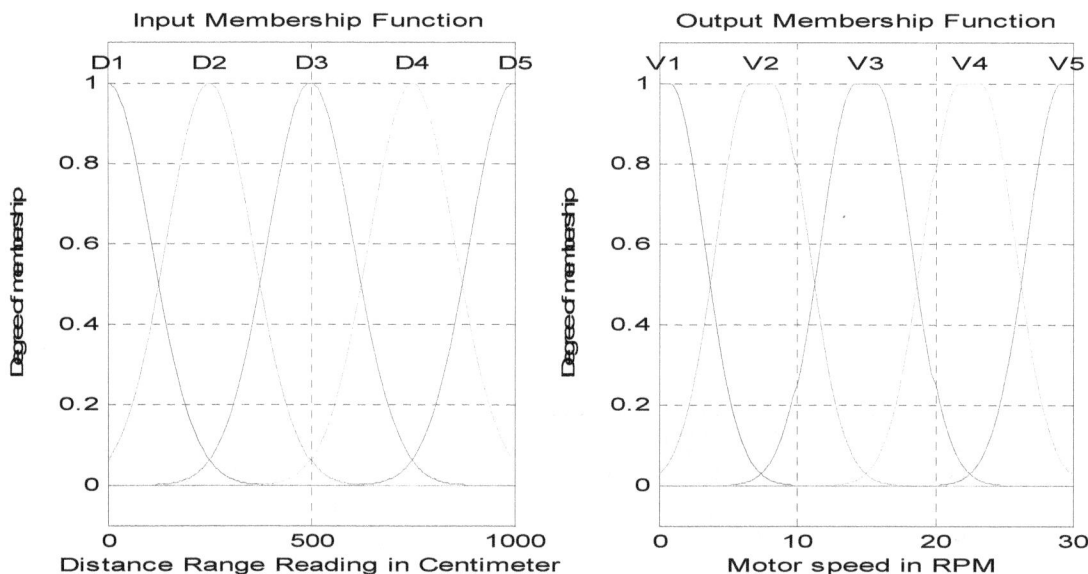

Figure 3. Input and output membership function for Motor speed control.

RESULTS AND DISCUSSION

For the proposed scheme, simulation has been done with MATLAB. The simulation parameters are described in Table 2. Figure 3 shows a five-variable linguistic fuzzy input membership set {D1, D2, D3, D4, D5}, defines the five different levels of obstacle distance from the front end of the mobile robot from very near to the farthest position respectively and the output membership {V1, V2, V3, V4, V5} which define the velocity actions such as too slow, slow, medium, fast, fastest most respectively. Figure 4 shows the left steering angle which is represented by a three-variable linguistic fuzzy input membership set {S, LO1, LO2} which define the distance of the obstacles in three different levels from farthest to the closet respectively, the obstacle distances are estimated from the ultrasound sensor range readings, and the output membership set {ST, L1, L2} which define the steering actions to the left of the vehicle in the different levels from straight steering to intense left turning actions respectively. Similarly the right steering

angle is represented by a three-variable linguistic fuzzy input membership set {S, RO1, RO2} which define the distance of the obstacles in three different levels from farthest to the closet respectively, and the output membership set {ST, R1, R2} which define the steering actions to the right of the vehicle in the different levels from straight steering to intense left turning actions respectively. Figure 5 shows the robot steering control achieved in turn angle with respect to the distance of the obstacle sensed in the left and right of the robot. Figure 6 shows the robot speed control achieved in rotations per minute with respect to the obstacle distance sensed by the front sensor range readings. Figures 7, 8 and 9 shows the robot goal seeking in a MATLAB simulated environment with a high obstacle density similar to a maze like environment. Here, the robot is supposed to move from the start point $(X_I, Y_I) = (45, 45)$ to the goal $(X_T, Y_T) = (5, 5)$ in two different environments. The initial orientation of the robot is $\theta m = \pi/2$ and there are many obstacles in the environment. The obtained results show the efficiency of the proposed control method. In all the

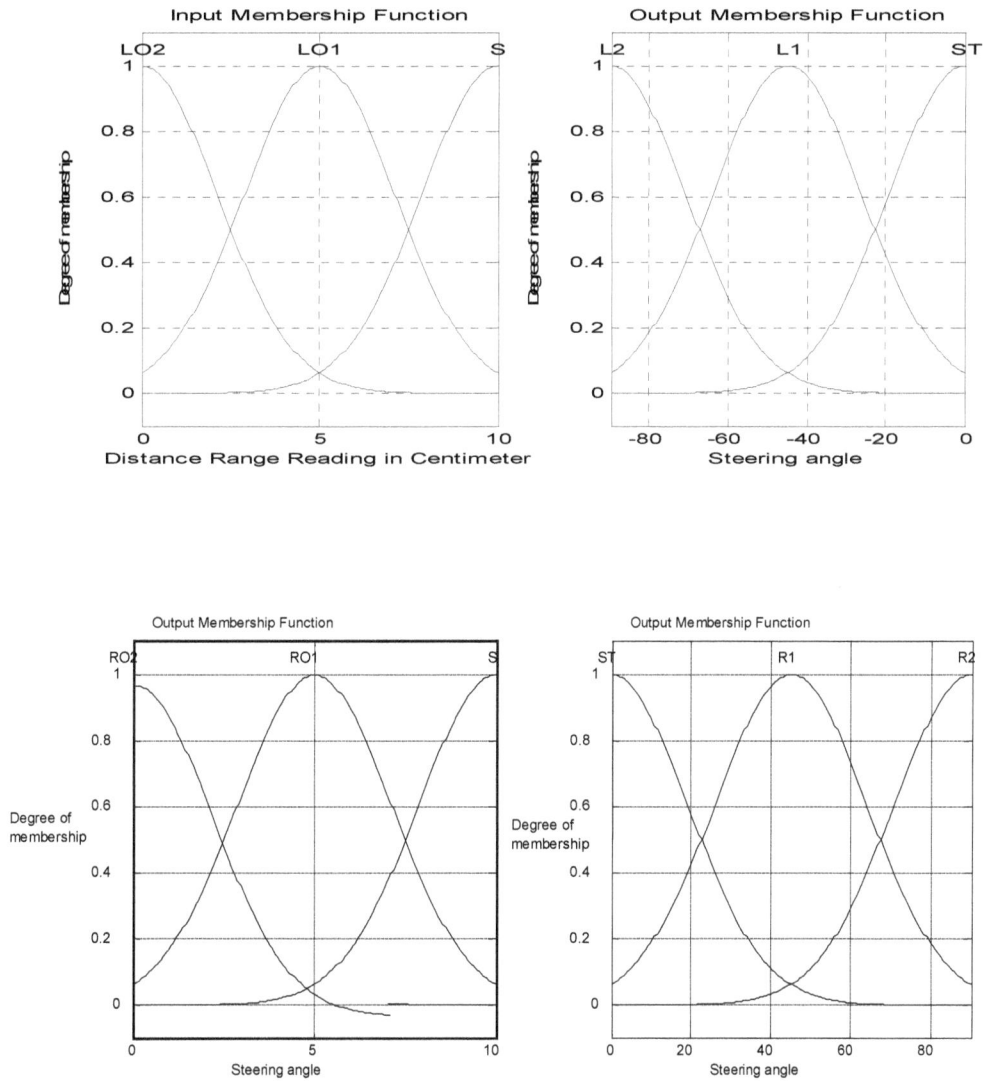

Figure 4. Input and output membership function for motor steering control.

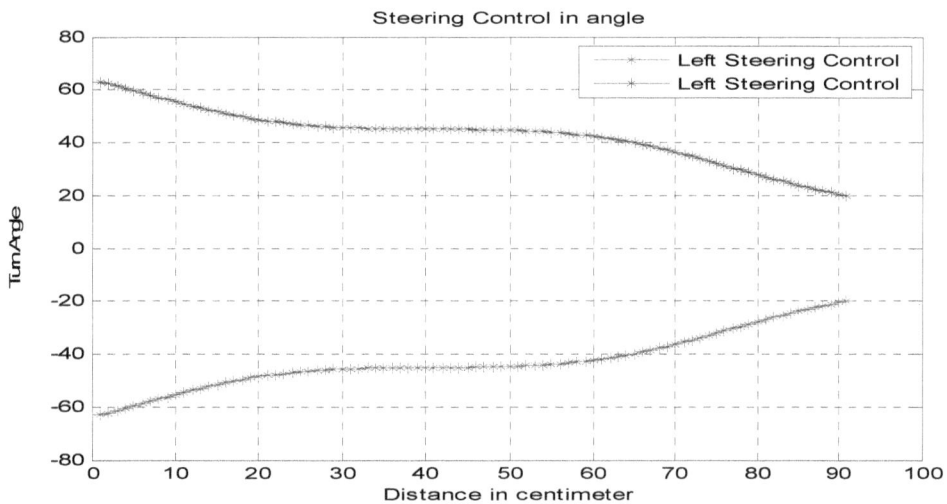

Figure 5. Robot steering control distance vs. turn angle.

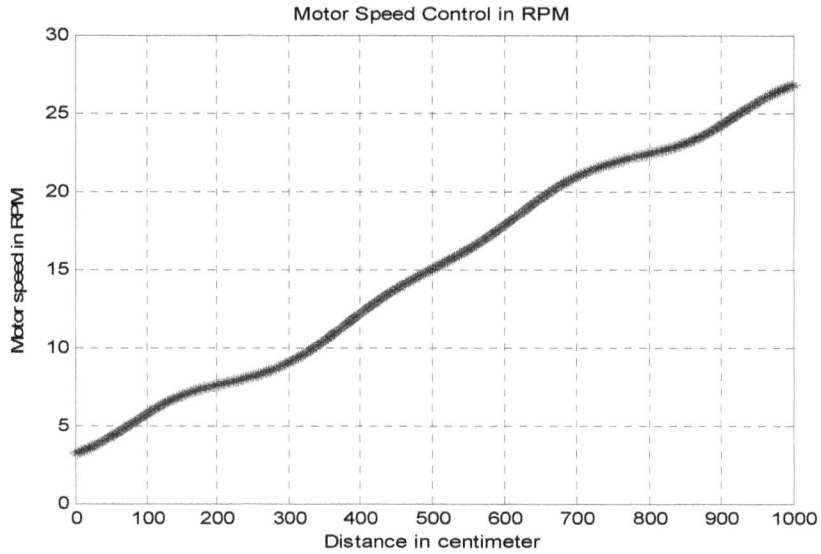

Figure 6. Robot speed control distance vs. motor speed in RPM.

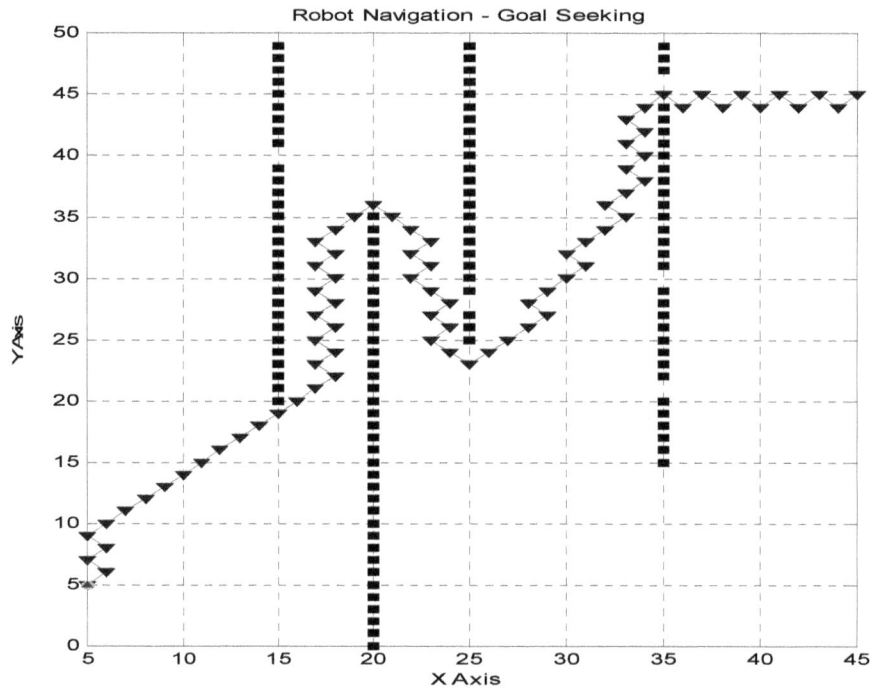

Figure 7. Robot goal seeking in Environment 1.

different configuration of the environment, the robot is able to reach the goal point.

CONCLUSION

This work presented a method that can be efficiently used to design behaviors based steering scheme. An uncomplicated harmonization scheme is used to toggle between steering procedures according to outputs of apiece manners. The outcome attained illustrates the effectiveness of the proposed control scheme. In every scenario, the robot is capable to arrive at the goal in diverse configurations of the surroundings by avoiding

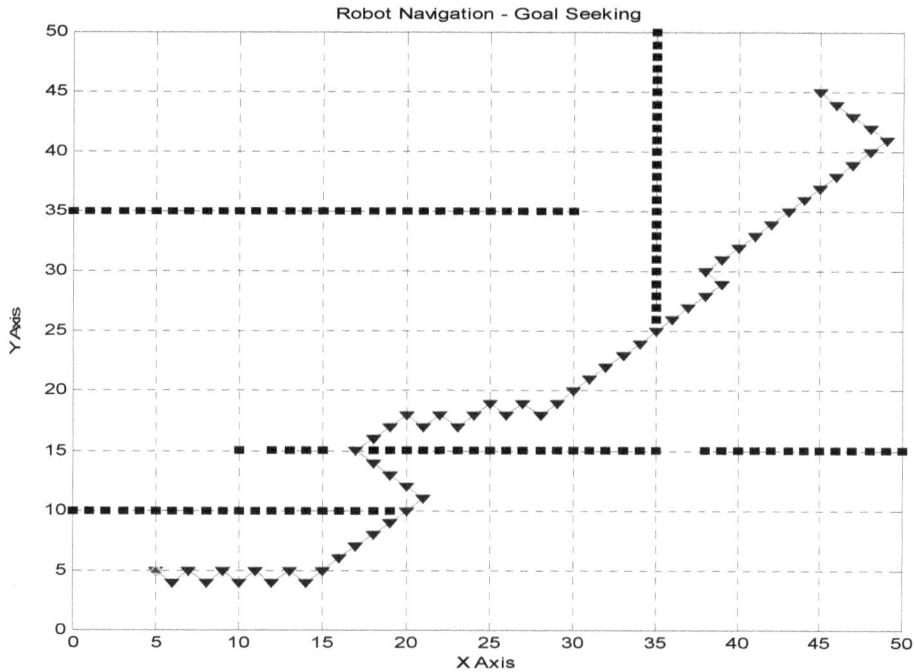

Figure 8. Robot goal seeking in Environment 2.

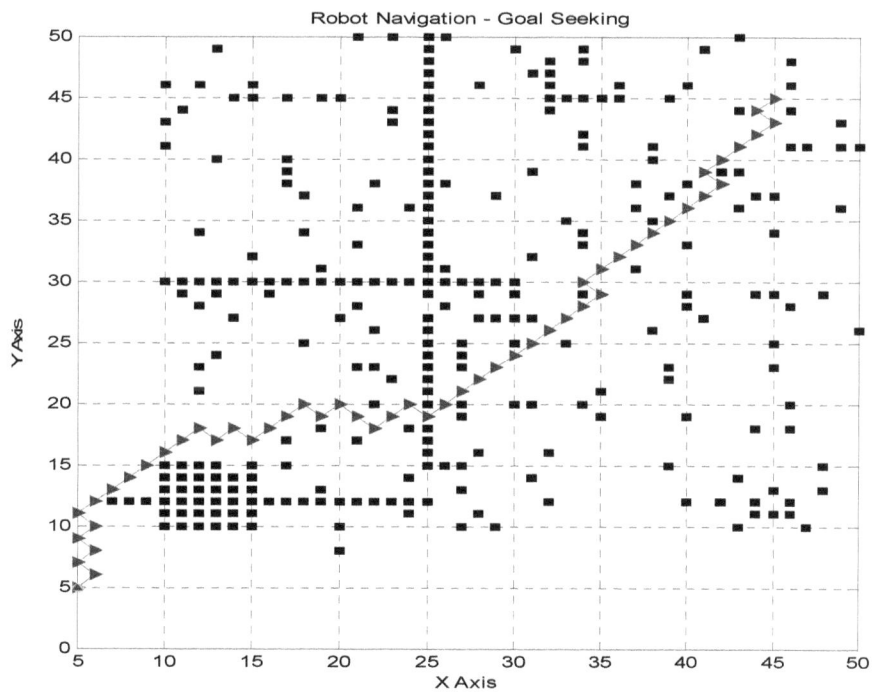

Figure 9. Robot goal seeking in Environment 3.

obstacles. It is of an immense significance to highlight on the attained efficiency of the robot activities. In prospect the attention will be certain to the development of an inclusive steering scheme counting other behaviors like wall following and avoiding moving obstacles. In future works, the interest will be given to moving obstacles.

Conflict of Interest

The authors have not declared any conflict of interest.

REFERENCES

Abdessemed F, Benmahammed K, Monacelli E (2004). A fuzzy based reactive controller for a non-holonomic mobile robot. Robotics and Autonomous Systems. 47(1):31–46.

Beom HR, Cho HS (1995). A sensor-based navigation for a mobile robot using fuzzy Logic and Reinforcement Learning. IEEE Tran. Syst. Man. Cyber. 25(3):464-477.

Brooks RA (1989). A Robot that Walks; Emergent Behavior from a Carefully Evolved Network. IEEE International Conference on Robotics and Automation. Scottsdale; AZ. pp. 292–296.

Brooks RA (1986). A Robust Layered Control System for a Mobile Robot. IEEE J. Robotics Autom. RA-2(1):14–23.

Ehsan H, Maani GJ, Navid GJ (2011). Model based PI-fuzzy control of four-wheeled omni-directional mobile robots. Robot. Autom. Syst. 59(11):930-942.

Fatmi AS, Yahmedi AI, Khriji L, Masmoudi N (2006). A Fuzzy Logic based Navigation of a Mobile robot. World academy Sci. Eng. Technol. 22:169-174.

Janglova D (2004). Neural networks in mobile robot motion. Int. J. Adv. Robot. Syst. 1(1):15-22.

Kian HL, Wee KL, Jr. Ang MH (2002). Integrated planning and control of mobile robot with self-organizing neural network. Proceeding of18th International Conference onRobotics and Automation (ICRA '02). May 11-15, 4:3870-3875.

Saffiotti A (1997). The uses of fuzzy logic for autonomous robot navigation. Soft Comput. 1(4):180-197.

Samsudin KF, Ahmad A, Mashohor S (2011). A highly interpretable fuzzy rule base using ordinal structure for obstacle avoidance of mobile robot. Appl. Soft Computing J. 11(2):1631–1637.

Selekwa MF, Damion D, Collins Jr. EG (2005). Implementation of multi-valued fuzzy behavior control for robot navigation in cluttered environments. Proceedings of the 2005 IEEE International Conference on Robotics and Automation, Barcelona, Spain. pp. 3699-3706.

Seng TL, Khalid MB, Yusof R (1999). Tuning a Neuro-Fuzzy Controller by Genetic Algorithm. IEEE Trans. Syst. Man. Cybernetics 29(2):226-236.

Seraji H, Howard A (2002). Behavior - based robot navigation on challenging Terrain: A Fuzzy Logic Approach. IEEE Trans. Rob. Autom. 18(3):308-321.

Shuzhi SG, Lewis FL (2006). Autonomous Mobile Robots, Sensing, Control, Decision, Making and Applications, CRC, Taylor and Francis Group.

Sugihara K, Smith J (1997). Genetic algorithms for adaptive motion planning of an autonomous mobile robot. Proceedings of the IEEE International Symposium on Computational Intelligence in Robotics and Automation. pp.138-146.

Velappa G, Soh CY, Jefry Ng (2009). Fuzzy and neural controllers for acute obstacle avoidance in mobile robot navigation. IEEE/ASME International Conference on Advanced Intelligent Mechatronics Suntec Convention and Exhibition Center. pp.1236-1241.

Wang M, Liu JNK (2008). Fuzzy logic-based real-time robot navigation in unknown environment with dead ends. Robotics Autonomous Syst. 56(7):625–643.

Yang SX, Moallem M, Patel RV (2005). A layered goal-oriented fuzzy motion planning strategy for mobile robot navigation. IEEE transactions on systems, man, and cybernetics—part b: cybernetics. 35(6):1214-1224.

Ye CN, Yung HC, Wang D (2003). A fuzzy controller with supervised learning assisted reinforcement learning algorithm for obstacle avoidance. IEEE Trans. Syst. Man. Cybern. B. 33(1):17-27.

Yung NHC, Ye C (1999). An intelligent mobile vehicle navigator based on fuzzy logic and reinforcement learning. IEEE Trans. Syst. Man. Cybern. 29(2):314-321.

20

Comparative performance of Raman-SOA and Raman-EDFA hybrid optical amplifiers in DWDM transmission systems

V. Bobrovs, S. Olonkins, A. Alsevska, L. Gegere and G. Ivanovs

Institute of Telecommunications, Riga Technical University, Azenes Str. 12, Rīga, LV-1048, Latvia.

To combine the benefits and compensate for the drawbacks of different optical amplifier types, a hybrid amplifier can be composed. The authors consider two most frequently used hybrid amplifiers that can provide better performance: a semiconductor optical amplifier (SOA), and an erbium doped fiber amplifier (EDFA), both in combination with a distributed Raman amplifier (DRA). To compare performance of the hybrid amplifiers, the eye diagrams of detected signals were analyzed and the maximum transmission distances were found. The results obtained show that even under the conditions advantageous for a SOA-DRA hybrid, the EDFA-DRA combination will produce less distortions of the amplified signal.

Key words: Dense wavelength division multiplexing, hybrid amplifier, semiconductor optical amplifier.

INTRODUCTION

During the last decade the evolution of available multimedia services and the rapid growth in the number of worldwide internet users has given rise to the demand for high capacity networks; this, in turn, causes a major shift in the evolution of optical transmission systems. Nowadays, one on the most typical solutions for raising transmission capacity is the use of wavelength division multiplexing (WDM), where different optical signal frequencies are used in order to achieve simultaneous transmission of a definite number of optical channels over a single fiber. It is also important to maintain the required level of system performance over a longer transmission distance. Such multichannel systems – in addition to linear effects such as optical attenuation and chromatic dispersion – are highly sensitive to the fiber non-linearity, the presence of which may result in serious signal distortion thus causing a dramatic degradation of a system's performance. Still, the effect that puts the greatest limitations on the transmission distance is the optical signal attenuation (Bobrovs et al., 2011a, b; Olonkins et al., 2012).

To compensate for optical signal attenuation, two ways are known: the use of signal repeaters and optical signal amplification. The former solution is not the best for WDM systems, because it requires demultiplexing, conversion, processing and regenerating of signals of all 16 channels; therefore, it is too complex and expensive (Agrawal, 2002). At the same time, by amplifying the optical signal we raise its power during transmission without conversion into any other form; the method is therefore simpler and much cheaper than those using repeaters. In some of the types of optical amplifiers optical signal gain is provided through stimulated emission, and in the others fiber nonlinearity is used.

In the amplifier medium also a spontaneous emission also occurs, which is amplified together with the transmitted signal. This results in the amplified spontaneous emission (ASE) noise, which in some cases can seriously limit the total transmission distance.

In WDM transmission systems the following types of optical amplifiers are used: semiconductor optical amplifiers (SOAs), doped fiber amplifiers (DFAs), discrete (lumped), and distributed Raman amplifiers (LRAs and DRAs). In the nearest future, parametrical amplifiers will also become available for multichannel systems. Each of these amplifier types has its own benefits and drawbacks. The main problem with SOAs is that they produce the largest amount of ASE and their gain dynamics can cause serious signal distortions. The DFAs can provide signal amplification with considerably less signal impairments than in the SOA case; however, their gain spectrum is highly frequency-dependent due to the characteristics of the doped material. Raman amplifiers can provide the most noiseless amplification; in this case the gain spectrum can easily be changed by varying the number of pumps and their frequencies, while to achieve a high enough gain a very powerful pump is needed, the use of which is not economically reasonable (Agrawal, 2002). To compensate for the drawbacks and combine the benefits of different amplifiers, these can be used together, forming a hybrid amplifier. In modern transmission systems a great variety of such combinations can be used; we, therefore, decided to carry out research on the hybrid amplifiers.

INVESTIGATION ON THE AMPLIFIER TYPES

As already mentioned, for amplification of optical signals the stimulated emission is used. In SOAs, the electrical energy is applied as a pump to achieve the population inversion, and amplification is achieved via the stimulated recombination luminescence. The spontaneous carrier lifetime in the active region of material is times smaller than in other amplifier types, so it is highly important for the SOA to work close to the saturated mode in order to keep the ASE level low. The amplifier gain dynamics, which is determined by the quick carrier recombination lifetime, for SOAs is faster than in other types of amplifiers. Consequently, the amplifier will respond relatively quickly to the changes in the input optical signal power. This may cause severe signal distortions, especially in multichannel systems (Connely, 2004). Because pulses from different channels are amplified simultaneously, the pulse belonging to one channel may drain the total higher energy level population, thus resulting in smaller optical gain for a pulse that corresponds to another channel; this process is called cross-gain modulation (Agrawal, 2002). The main

advantages of using SOAs are their broad amplification bandwidth (that is, -3 dB up to 70 nm) and relatively low price (Agrawal, 2002).

DFAs make use of rare-earth elements for doping some silica fibers during the manufacturing process. For this purpose, many different rare-earth elements can be used (erbium, thulium, neodymium, ytterbium, chromium etc.) (Cheng and Huang, 2013). The most usable element is erbium, because it allows optical amplifiers to operate in the C-band, (that is from 1530 to 1565 nm). In order to achieve efficient pumping in erbium- doped fiber amplifiers (EDFAs) the 980 and 1480 nm semiconductor lasers are applied, while population inversion is achievable using co-propagating, counter-propagating and bi-directional pumps. The gain spectrum of EDFAs is determined by the molecular structure of the doped fiber, and is strictly wavelength-dependent. The main disadvantage of EDFAs is that their wavelength-dependent gain spectrum bandwidth is only about 40 nm; besides, it is not flat. On the other hand, it determines amplification of individual channels when a WDM signal is amplified, so no cross-gain saturation occurs. Due to a relatively long spontaneous carrier lifetime in silica fibers, this allows achieving high gain for a weak signal with low noise figure, which represents the difference in signal-noise ratio at the input and output of the device under consideration (Agrawal, 2002). This is the main reason why the EDFAs are most frequently used for optical amplification.

Nowadays, Raman amplifiers are being deployed in most of the new long-haul and ultra-long haul fiber optic transmission systems, placing them among the first widely commercialized nonlinear optical devices in telecommunications (Islam, 2004a). In Raman amplifiers a small signal gain arises from stimulated Raman scattering (SRS) – the energy transfer from a powerful pumping optical beam to the amplified signal. In silica fibers, the peak amplifications correspond to the signal frequency that is ~ 13.2 THz lower than the pumping one; this frequency difference is called the Stokes shift. Such downshift is defined by the energy of optical phonons which represent the vibration mode of medium (Islam, 2004a). Despite the fact that the spontaneous Raman scattering spectrum is broad, the coherent nature of the process implies that the small signal radiation becomes coherently amplified by the SRS. The main advantage of Raman amplifiers is that the gain spectrum is very broad, and its shape can be changed by varying the number of pumps and their wavelengths (Mustafa et al., 2013). The relatively low noise figure of Raman amplifiers also is a significant benefit. It is these two aspects that make Raman amplifiers the main component of hybrid amplifiers, as they can be used to enhance the gain of a particular amplifier, and to broaden and equalize the gain spectrum, adding very little noise to the amplified signal. The main disadvantages of Raman amplifiers are the poor pumping efficiency at lower signal power (Tragarajan and Ghatak, 2007), and the use of expensive powerful lasers capable of delivering great powers into single-mode fibers.

In the systems with optical amplification the intensity of amplified signal can reach the level high enough to cause fiber nonlinearity, which may result in serious inter-channel crosstalk, thus also in a dramatic decrease in the transmission quality. For the systems that are highly sensitive to fiber nonlinearity (such as dense WDM (DWDM) systems with equal channel spacing) it is very important to keep track of the inter-channel crosstalk produced by the four-wave mixing (FWM). Indeed, such FWM induces spectral components with frequencies that may coincide with those of transmitted signal channels, thus limiting the amplifier gain for which the required quality of service is maintained. In such cases, the ASE and other amplifier-produced signal distortions may have a great impact on the maximum achievable transmission distance. This means that SOAs are not the preferable type of optical amplifier for such a system.

With the mentioned gain limitations the Raman amplifiers may cause too much inter-channel crosstalk, while the DFAs also can raise the signal intensity level significantly enough to cause inter-channel crosstalk, and, due to their gain- frequency dependence, they may not provide equivalent amplification for all of the system's channels so this needs to be equalized. The use of a SOA-DRA or, alternatively, of an EDFA-DRA hybrid may help to overcome these problems (Islam, 2004b).

In general, two types of hybrid amplifiers are known: the wideband hybrid amplifier (WB-HA) and the narrowband hybrid amplifier (NB-HA) (Islam, 2004b). In the former a wider band for the gain is obtained using combinations of different amplifier types, while in the NB-HA such combinations are meant for obtaining a compound with lower ASE-produced noise and higher gain of the amplified signal.

Raman amplifiers are an essential component of hybrid amplifiers. Obviously, hybrid SOA-EDFAs can be used in the cases where it is necessary to widen the gain spectrum of an EDFA, which could be done applying the most cost-effective solution (Zimmerman and Spiekman, 2004); however, such a combination generates a greater amount of ASE than in the cases of EDFA-DRAs or SOA-DRAs. This significantly affects the total system performance in the case of a nonlinearity-sensitive transmission system, where, due to the limitations on signal amplification caused by nonlinearity, the received optical power penalty plays a great role as it affects the receiver's sensitivity needed for achieving a definite bit error rate (BER). Therefore, normally it is not applied in long haul or DWDM systems. Our previous studies show that the noise figure of Raman amplifiers is much lower than that of EDFAs, and definitely lower than in the cases where SOAs are used. So the best way to achieve a higher gain with lower noise figure or a wider amplification band is to use a SOA or an EDFA in combination with a distributed Raman amplifier (DRA). For this purpose, we can also use another type Raman amplifiers – the discrete ones (Islam, 2004b); however, due to a small effective area and a high nonlinearity coefficient of the HNLFs and dispersion compensating fibers employed as the amplifier medium in LRAs, the discrete RAs generate a multitude of nonlinearity-related distortions in the cases when the intensity level of a weak signal to be amplified is relatively high.

Therefore, it is unclear whether SOA-DRA or EDFA-SOA combinations would provide better system performance in the case where the impact of fiber nonlinearity is strong. Our main goal was therefore to find out which of these combinations can ensure good enough signal amplification with less distortions (that is, with longer transmission distance) and without loss in the operational quality of a nonlinearity-sensitive transmission system.

SIMULATION SCHEME AND MEASUREMENT TECHNIQUE

Here, we will describe the simulation and measuring schemes used for performance estimation of hybrid optical amplifiers. To compare the performance of SOA-DRA and EDFA-SOA combinations a quality-characterizing parameter is to be evaluated. In our case, the most efficient way to assess the quality of transmission is to analyze the eye diagrams, which show patterns of the electrical signal after detection, and to evaluate BER values of the transmitted signal as a parameter featuring best the signal distortions arising during transmission. To estimate the transmitted signal distortions caused by fiber nonlinearity, we will observe its optical spectrum, while the level of ASE-generated noise will be assessed by noise figures.

To obtain the experimental results we needed a strong mathematical tool. For this purpose, the OptSim 5.2 simulation software was chosen, so that this all-optical network simulator can handle complex simulations and introduce high-accuracy results without imposing high requirements on the relevant hardware. This simulation tool uses the split-step method to perform integration of the fiber propagation equation (OptSim 5.2 User Guide, 2010):

$$\frac{\partial A(t,z)}{\partial z} = \{L + N\}A(t,z) \qquad (1)$$

where $A(t,z)$ is the optical field, L is the linear operator (for calculation of such linear effects as attenuation and dispersion), and N is the nonlinear operator (accounting for fiber nonlinearity).

The calculation is done dividing the whole optical link (fiber) into Δz-long spans, and deriving the L and N operators separately (Zimmerman and Spiekman, 2004). Two variants of the split-step method are applied: time domain split step (TDSS) and frequency domain split step (FDSS). These two differ only in the way the L operator is calculated: in the TDSS method – in the time domain, while in the FDSS – in the frequency domain. The nonlinear operator in both cases is obtained in the time domain. The former method gives highly precise results, however it is difficult to implement. The FDSS is easier to implement, but intrinsic errors (that decrease dramatically the precision of the results (OptSim 5.2 User Guide, 2010) can arise during the calculation process. Therefore, for our simulation the TDSS method was chosen.

For studying the signal distortions caused by hybrid amplifier a 10 Gbps 16- channel DWDM transmission system was designed,

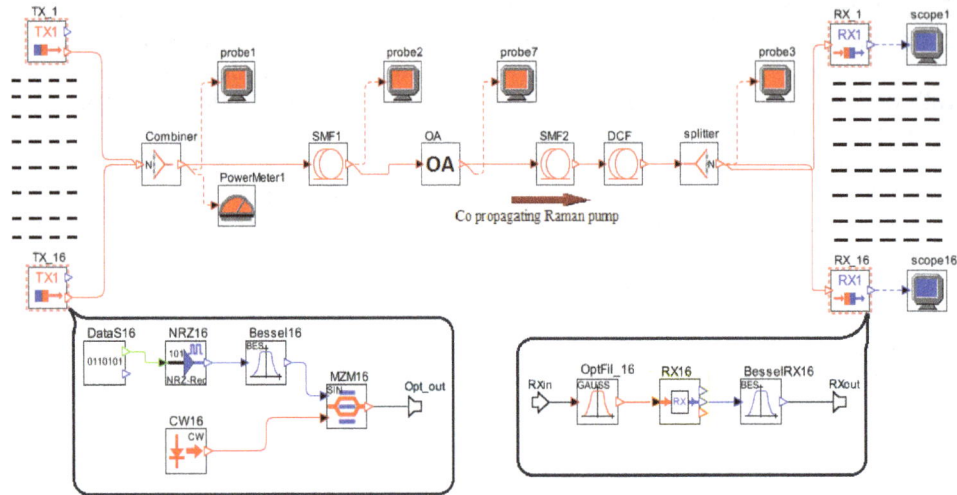

Figure 1. Simulation scheme of a 10 Gbit/s 16 channel DWDM system.

with the non-return-to-zero (NRZ) encoding, the on-off keying (OOK) intensity modulation format (less tolerant to the influence of fiber nonlinearity than advanced modulation formats (Bobrovs et al., 2011c), and a 50 GHz channel spacing. Such system configuration was chosen to purposefully to cause the Kerr effect in order to impact the amplified signal. This nonlinear effect (arising in systems with equal and relatively small channel spacing) produces strong inter-channel crosstalk, thus limiting the maximum intensity level of the transmitted signal and, therefore, the total amplification. In such a system the amplifier-produced signal impairments will directly influence the achievable transmission distance. Therefore it is easier to assess the performance of the amplifiers by comparing the achieved transmission distances. The simulation scheme comprising three main blocks: the transmitter block, the optical link and the receiver block, are shown in Figure 1.

The transmitter block consists of 16 NRZ-OOK externally modulated channel transmitters, each of them operating at its own frequency in the range from 193.05 to 193.8 THz. Each transmitter contains a pulse pattern generator (PPG), an NRZ driver, an electrical filter, a continuous wave (CW) laser, and a Mach-Zender's modulator. The continuous optical signal is externally modulated by NRZ-coded electrical pulses via an electro-optical MZM. Then all of the 16 generated optical signals are combined and transmitted through the optical link.

The signal first overcomes 72 km of a single mode fiber (SMF) with 0.2 dB/km attenuation and 16 ps/nm/km chromatic dispersion. The SMF length is dictated by the required optical signal power at the input of the optical amplifier, which is very important due to its saturation effect – especially when SOA is used. For an EDFA this parameter is also relevant, but it has been optimized for the semiconductor amplifier (due to the high level of noise produced by SOA and its gain dynamics). The weak signal power level for each channel at the amplifier input is around -22.4 dBm. Then the signal is amplified by an in-line SOA or an EDFA.

The two hybrid amplifiers (SOA-DRA and EDFA-DRA) will be compared as in-line amplifiers, because such amplifiers not only cause signal impairments and raise the intensity level significantly enough to cause fiber nonlinearities, but also amplify the nonlinearity-caused signal distortions accumulated during transmission. This makes the requirements for the total acceptable amount of amplifier noise stricter.

The SOA pumping current is optimized in order to minimize the amplifier-produced signal impairments. The EDFA parameters are chosen in such a manner that its gain spectrum irregularities would be easy to compensate with a single Raman amplifier pump, keeping in mind the total gain limitation caused by FWM. Then the amplified signal enters another SMF where it is amplified by a low-power DRA, the power of which allows for achievement of the maximum signal gain without causing too much nonlinearity-produced distortions. The length of this second SMF is variable in order to obtain the maximum transmission distance. At the end of optical link the signal enters a dispersion compensation fiber (DCF), the length of which will also be varied so as to find a balance between the dispersion compensation and the DCF insertion loss. After propagating through the DCF, the optical signal enters the receiver block. It is divided among 16 receivers, where the optical signal is filtered, detected and converted into electrical current. The DCF attenuation at 1550 nm is 0.55 dB/km, and the dispersion at this wavelength is -80 ps/nm/km.

RESULTS AND DISCUSSION

We will focus our attention on the results obtained with the simulation scheme described above. Besides, the amplifier optimization results will be presented, which may provide a good basis for estimating the cause of amplifier performance limitations. As already mentioned, in order to estimate the system performance the eye diagrams should be analyzed. The eye diagram is a powerful time domain tool for assessing the quality of the received signal and for analyzing the signal distortions. It can give much information on the timing jitter, the system rise time, and the signal amplitude distortions (Bobrovs and Ivanovs, 2008). First, we will discuss the configurations of SOA-DRA and EDFA-DRA allowing the maximum transmission distance to be achieved. The active layer parameters of the SOA and its other geometrical and material parameters were found in Singh and Kaler (2007), where a semiconductor optical amplifier was optimized for a similar system. They are specified in Table 1.

Table 1. The SOA parameters.

Parameter	Value	Unit
Amplifier length	750	μm
The active layer width	2	μm
The active layer thickness	0.2	μm
Confinement factor	0.41	μm
Transparency carrier density	$1.5 \cdot 10^{18}$	cm^{-3}
Differential gain constant	$2.1 \cdot 10^{-16}$	cm^2
Carrier recombination time	0.3	Ns
Input and output coupling losses	3	dB

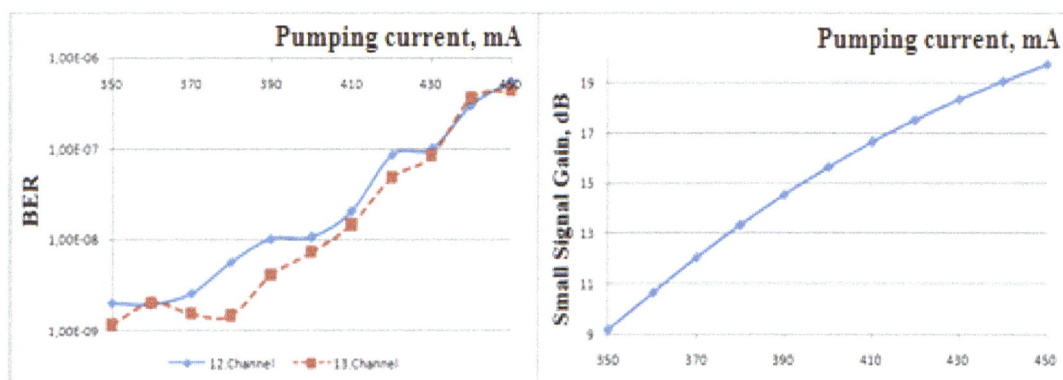

Figure 2. BER values at the output of SOA for channels 12 and 13 of the transmitted signal with gain compensation vs. pumping current (left); the optical gain vs. the pumping current (right).

The current to be used for pumping in the semiconductor amplifier should be chosen from the considerations of achieving the maximum amplification with the minimum noise. With this purpose in mind, we found the BER level in two channels for the signal before and after amplification at different current values (from 350 to 450 mA). The channels were purposely chosen with the highest and the lowest optical power level at the SOA output – the 12th and the 13th channel, respectively. It is important to note that in order to avoid the impact of the amplifier gain on the BER values of the two channels, the optical signal was intentionally attenuated, so that a weak signal's gain would be completely compensated. We also obtained the dependence of the amplified signal gain on the pumping current and its increase with every additional 10 mA. The results are shown in Figure 2.

The BER value of the 12th channel at the amplifier input was 2.04•10-9, and of the 13th channel – 9.96•10-10, that is, lower, despite the fact that the optical power of the 12th channel is slightly higher. This can be explained by the closeness of the 12th channel to the center of the transmitted signal spectrum, which increases (also slightly) its inter-channel crosstalk. For the pump current starting from 380 mA, the BER values of the two

channels under consideration experience a significant increase. This evidences that for the pump current values over 370 mA the amount of amplifier-produced signal distortions starts to grow. Therefore, we took the 370 mA pumping current as optimal for this system.

In Figure 2 it could be seen that with increase in the pumping current the amplifier gain increment is becoming smaller. This evidences that the amplifier slowly reaches the maximum level of population inversion; thus, due to the short spontaneous carrier lifetime, the generated ASE also experiences an increment with the increase in pumping current. For its value of 370 mA the SOA provides a small signal gain of 12.1 dB. The fact that SOAs provide a very broad gain bandwidth is confirmed by another fact – that the difference in the optical gain values for all 16 channels is only 0.02 dB. The rest of the amplifier gain will be provided by a noiseless distributed Raman amplifier.

For the hybrid EDFA-DRA we have chosen a bi-directionally pumped EDFA, with 980 nm co-propagating and 1480 nm counter-propagating pumps. In our earlier research such combination of pump wavelengths showed the best result from those for many single and multiple pump combinations. The pump powers were chosen with the purpose to make the gain spectrum irregularity of the

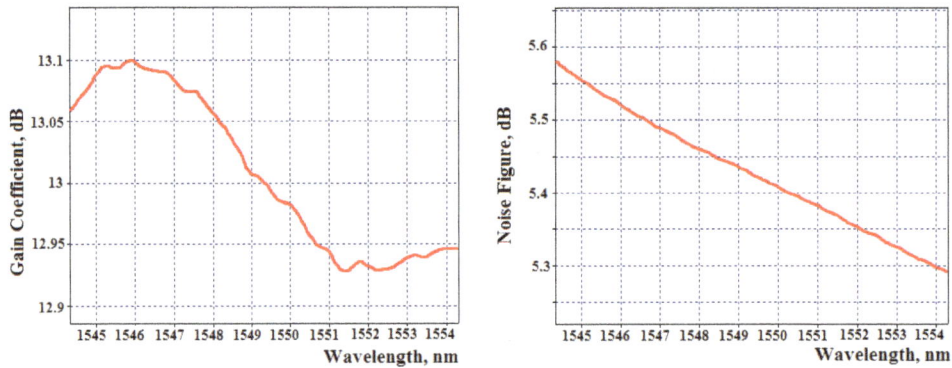

Figure 3. Gain spectrum (left) and noise figure (right) of the EDFA with 5 m long doped fiber and 10 dBm 980nm co-propagating and 16 dBm 1480 nm counter-propagating pumps.

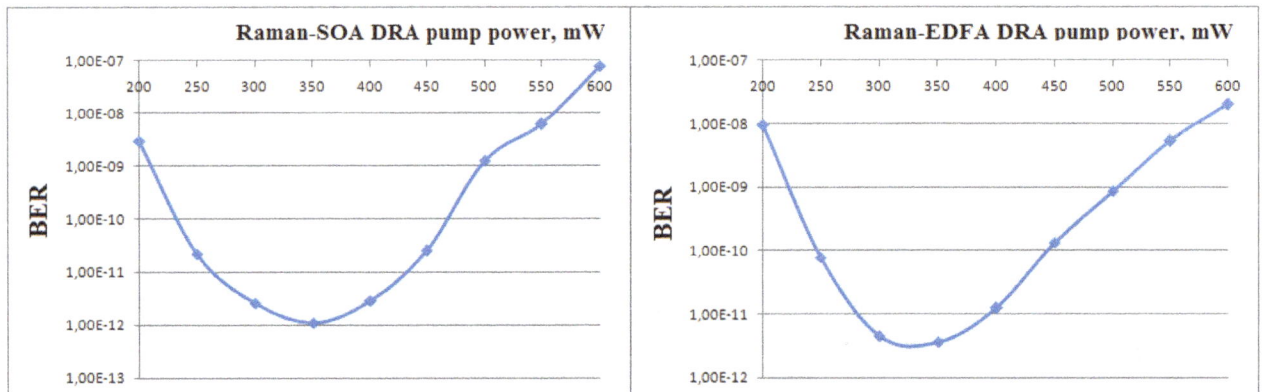

Figure 4. System's BER dependence on the power of a 1451.8 nm SOA-DRA (left) and EDFA- DRA (right) co-propagating pump.

EDFA easier to compensate with a single-pump Raman amplifier. In our case, the 16 channels occupy a ~ 6 nm bandwidth, in the limits of which a low-power single-pump DRA gain difference is < 0.5 dB. Taking this fact into account, we decided that a 5 m long doped fiber should be used for our EDFA, with the population inversion achieved using a combination of 10 dBm 980 nm co-propagating and 16 dBm 1480 nm counter-propagating pumps. The obtained gain spectrum and noise figure are shown in Figure 3, where the gain spectrum obtained for wavelengths from 1547 to 1553 nm varies from 12.93 dB to 13.1 dB. This but minor gain unevenness and the specific shape of the gain spectrum allowed us to conveniently equalize the obtained gain-wavelength dependence by applying a single-pump Raman amplifier. The noise figure obtained for the wavelengths under attention varies from 5.33 to 5.49 dB, which is rather a large increment for the EDFAs operating at high levels of population inversion. For the optimal amplifier configuration the noise figure close to 3 dB mark is achievable (Agrawal, 2002). So the EDFA configuration was not optimal, still the obtained noise figure is lower than the theoretical for a SOA.

After configuration of EDFA and SOA this procedure was carried out for the co-propagating Raman pump wavelengths and power in both cases. For the SOA-DRA hybrid the main requirement to the Raman pump was to ensure the minimum difference in the signal gain for all 16 channels. To achieve this, the center of the amplifier gain should coincide with the central wavelength of the transmitted signal. We found out that a 1451.8 nm pump is most suitable for this purpose. In the case of hybrid EDFA-DRA a pump is needed to ensure that the Raman amplification maximum coincides with the EDFA amplification minimum, which, in turn, corresponds to the wavelength of 1551.5 nm (or 193.23 THz frequency). It was found that to satisfy this criterion a 1453.1 nm pump is to be used.

To find the optimal pump power, we considered the BER values of all 16 channels and obtained the maximum one (further in the text the system's BER) and its dependence on the co-propagating pump power in both cases (Figure 4). From the results obtained it can be seen that in both cases for the pump power over 350 mW the amount of FWM-generated crosstalk exceeds the permissible value and seriously deteriorates the total

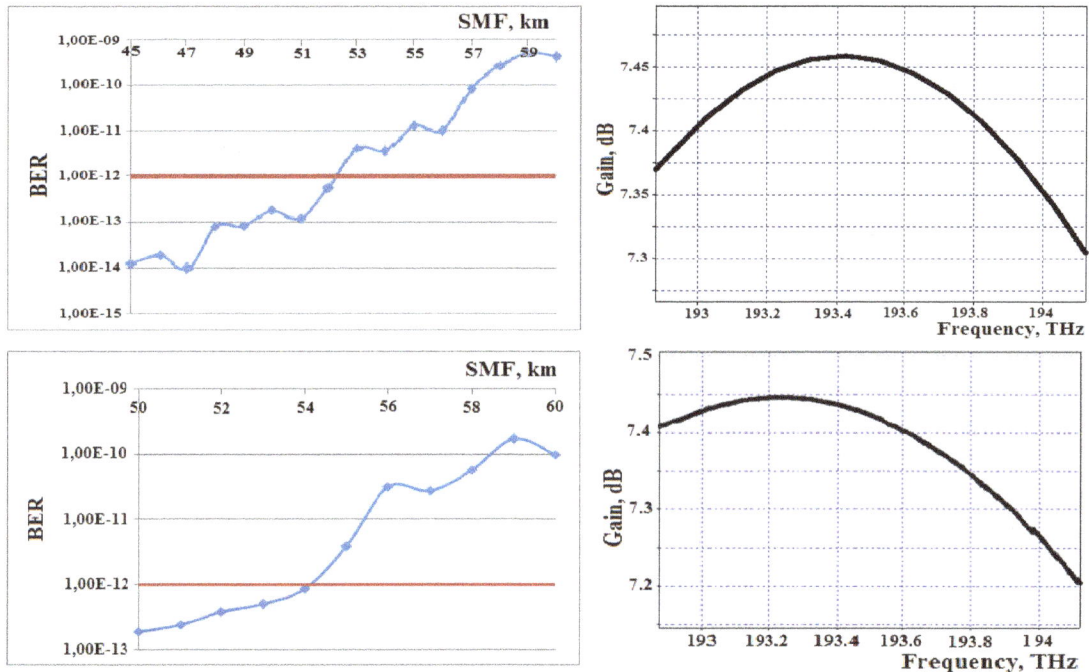

Figure 5. Dependences of the system's BER on the SMF length between the amplifier and the receiver block (left), and the DRA produced gain spectra (right) in the cases of SOA-DRA (above) and EDFA-DRA (below).

system performance. Therefore, 350 mW was found to be the most suitable value in the existing conditions. Since the SMF for DRA plays the role of amplifier medium, to obtain the total DRA gain we should first define the SMF length between the amplifier and the receiver block, thus also deriving the maximum transmission distance. For this, it is required to obtain the optimal DCF length, which, on the one hand, is determined by the total accumulated chromatic dispersion, while, on the other, is limited by the signal attenuation caused by DCF insertion. In both cases, the optimal DCF length was found to be 17 km. To find the maximum transmission distance we obtained the dependence of the system's BER on the sought-for SMF section length shown in Figure 5 along with the DRA gain spectra. The dependences shown in Figure 5 evidence that in the case of SOA-DRA the maximum SMF length between the amplifier and the receiver block providing the system's BER < 10-12 is 52 km, thus the overall transmission distance will be 124 km. For the EDFA-DRA based system this fiber length is 54 km, and the transmission distance – 126 km. It is important to add that we have obtained also the maximum transmission distance for the system in which no amplification was applied, and it was equal to 69 km. This means that the SOA-DRA combination was able to extend this distance by 55 km and the EDFA-DRA – by 57 km. The optical signal values at the receiver block input for the system with no amplification varied from -23.32 to -23.57 dBm, for the SOA-DRA based system from -21.78 to

-21.44dBm, and for that with EDFA-DRA – from -21.31 to -21.05 dBm. To identify the factors that limit transmission in each of the three cases the eye diagrams of the channels with the worst BER were analyzed. These, together with the relevant inter-channel crosstalk, are shown in Figure 6.

As was expected, in the case with no amplification the main limitation factor is the optical signal attenuation. It also can be seen that even without amplification the optical signal intensity is high enough to initiate FWM, and the produced minor inter-channel crosstalk also affects the BER value. From the eye diagram of the 7th channel of the EDFA-DRA based system it can be seen that the FWM produced inter-channel crosstalk is the main limiting factor for transmission, since the FWM harmonics are clearly seen on the level of logical "1", and the critical BER value was reached with a higher level of the detected signal power than in the other two cases. For the SOA-DRA based system this inter-channel crosstalk is also quite high, though lower than in the case with EDFA-DRA, due to ~ 0.8 dB difference in the amplification in both cases. Still, the BER limit was reached at a shorter transmission distance, and not due to the mentioned difference, because the level of the detected signal was high enough to ensure the required quality of transmission. If we compare the optical signal power levels at the input of the receiver block, it can be seen that the average difference is ~ 0.4 dB, while the difference in the amplification is ~ 0.8 dB. Therefore it is clear that the SOA produces more ASE noise than the

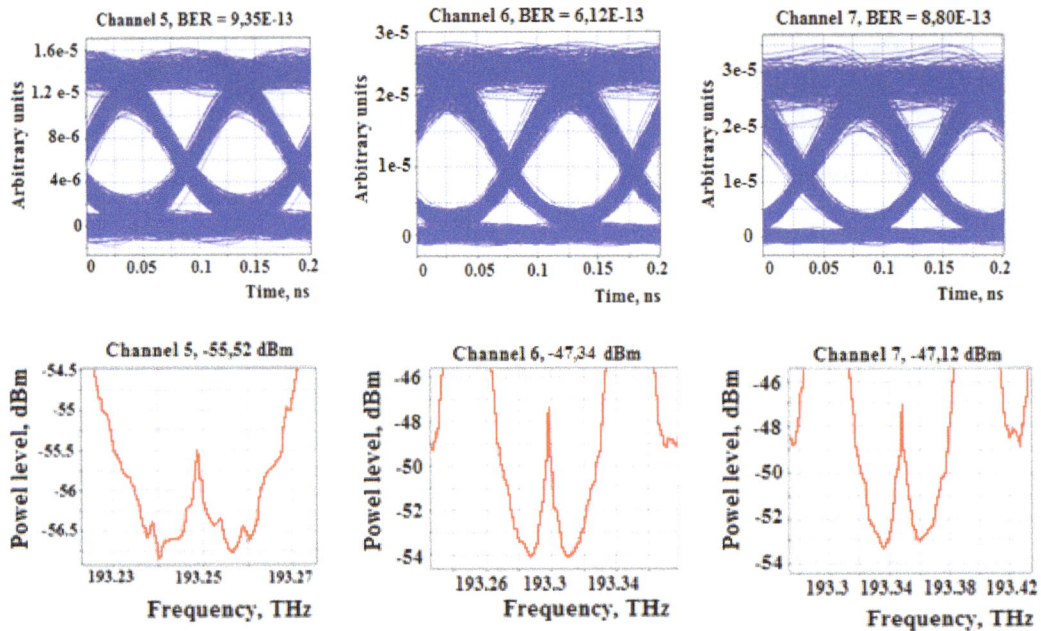

Figure 6. Eye diagrams for the channels with the worst BER (above) in the system with no amplification (left), with SOA-DRA (center), and with EDFA-DRA (right); the inter-channel crosstalk in the corresponding channels (below).

EDFA, which, in addition to the inter-channel crosstalk, increases the detected signal power penalty at the receiver.

Conclusions

Based on the results obtained in this work, the following conclusions can be drawn:

1. The fiber nonlinearity in the implemented 16-channel DWDM transmission system has been found to exert a strong influence on the quality of transmission, which allowed the performance of narrowband SOA-DRA and EDFA-DRA hybrid amplifiers to be compared.

2. Testing the amplifiers under severe conditions has given a clear view on the amplified signal distortions. The parameters of SOA were adjusted so that it would produce higher amplification with less signal distortions. It was observed that increasing the SOA pumping current (from 370 mA on) leads to a signal's BER growing after amplification, which points to greater signal distortions generated by SOA.

3. Implementation of hybrid amplification can provide more equal gain for all channels of the system under attention. In the case of the EDFA-DRA solution the parameters of the EDFA were adjusted to obtain the gain spectrum which could easily be equalized by a single-pump Raman amplifier. The introduced EDFA configuration ensured 0.17 dB gain difference among all

16 channels, but after supplementing the EDFA with a DRA we obtained the gain spectrum with only 0.05 dB maximal difference in amplification.

4. Even a non-optimally configured EDFA produces less signal distortions than the SOA. The input signal power was adjusted specially for the SOA, thus the EDFA was not optimally configured; still the EDFA-DRA hybrid amplifier showed better results and provided transmission over a longer optical link than the SOA-DRA (126 and 124 km, respectively). The SOA-DRA provided an average gain of 19.6 dB, and the EDFA-DRA – of 20.4 dB.

5. Since the EDFA generated less signal distortions, the EDFA-DRA solution ensured better quality of amplification than the SOA-DRA. In both cases, the main factor that limited transmission was the FWM-produced inter-channel crosstalk, with the DRA pumping power being the same. So the difference in the total transmission quality can be explained only by the performance of the SOA and the EDFA. In the case of SOA-DRA the total amplification was slightly lower, thus also lower inter-channel crosstalk could be expected. Still, the non-optimally configured EDFA-DRA showed better results. This can be explained by heavier signal distortions that those produced by SOA, even though it was configured in a manner to obtain more gain with less noise.

So it is clear that even though the gain spectrum of the EDFA can be quite uneven, the EDFA-DRA hybrid can ensure better quality of transmission than the SOA-DRA one, even in the conditions favorable for the latter. Of course, if the signal power level at the input of amplifier was lower, also its fiber nonlinearity produced distortions would have been smaller, with much greater gain and longer transmission distance provided by both hybrid amplifiers. Still the results obtained give quite clear view on their performance. The main conclusion therefore is: despite greater signal distortions produced by SOA-DRA, due to the spectral limitations of EDFAs it still is the preference solution for broad coarse WDM (CWDM) transmission systems.

ACKNOWLEDGMENT

This work has been supported by the European Regional Development Fund in Latvia within the project Nr. 2010/0270/2DP/2.1.1.1.0/10/APIA/VIAA/002.

REFERENCES

Agrawal GP (2002). Fiber-optic communication systems. New York : John Wiley & Sons Inc. 546:226-260.

Bobrovs V, Ivanovs G (2008). Comparison of Different Modulation Formats and Their Compatibility With WDM Transmission System. Latv. J. Phys. Tech. Sci. 45(2):3-16.

Bobrovs V, Ivanovs G, Spolitis S (2011a). Realization of Combined Chromatic Dispersion Compensation Methods in High Speed WDM Optical Transmission Systems. Electronics Electrical Eng. 7:101-106.

Bobrovs V, Ivanovs G, Udalcovs A, Spolitis S, Ozolins O (2011b). Mixed Chromatic Dispersion Compensation Methods for Combined HDWDM Systems. IEEE CPS Proceedings, 6th Int. Conf. Broadband Wireless Computing, pp. 313-319.

Bobrovs V, Spolitis S, Udalcovs A, Ivanovs G (2011c). Schemes for Compensation of Chromatic Dispersion in Combined HDWDM Systems. Latv. J. Phys. Tech. Sci. 48(5):30-44.

Cheng WH, Huang YC (2013). 300-nm Broadband Chromium-Doped Fiber Amplifiers. OFC/NFOEC Technical Digest, OTh4C.4.

Connely MJ (2004). Semiconductor Optical Amplifiers. Dordrecht: Kluwer Academic Publishers, 169:127-132.

Islam MN (2004a). Raman Amplifiers for Telecommunications. New York: Springer, 298:1-39.

Islam MN (2004b). Raman Amplifiers for Telecommunications 2, New York: Springer, 730:413-424.

Mustafa FM, Khalaf AAM, Elgeldawy FA (2013). Multi-pumped Raman Amplifier for Long-Haul UW-WDM Optical Communication Systems: Gain Flatness and Bandwidth Enhancements. IEEE CPS Proc. 15th Int. Conf. Adv. Communication Technol. pp.122-127.

Olonkins S, Bobrovs V, Ivanovs G (2012). Comparison of Semiconductor Optical Amplifier and Discrete Raman Amplifier Performance in DWDM Systems. Electronics Electrical Eng. 7:133-136.

OptSim 5.2 User Guide (2010). RSoft design group Inc.

Singh S, Kaler RS (2007). Simulation and Optimization of Optical Amplifiers in Optical Communication Networks. Thapar University, India. Doctoral Thesis. 68-71.

Tragarajan K, Ghatak A (2007). Fiber Optics Essentials. New Jersey: John Wiley & Sons Inc. 242:151-160

Zimmerman DR, Spiekman H (2004). Amplifiers for the Masses: EDFA, EDWA, and SOA Amplets for Metro and Access Applications. J. Lightwave Technol. 22(1):63-70.

Permissions

All chapters in this book were first published in IJPS, by Academic Journals; hereby published with permission under the Creative Commons Attribution License or equivalent. Every chapter published in this book has been scrutinized by our experts. Their significance has been extensively debated. The topics covered herein carry significant findings which will fuel the growth of the discipline. They may even be implemented as practical applications or may be referred to as a beginning point for another development.

The contributors of this book come from diverse backgrounds, making this book a truly international effort. This book will bring forth new frontiers with its revolutionizing research information and detailed analysis of the nascent developments around the world.

We would like to thank all the contributing authors for lending their expertise to make the book truly unique. They have played a crucial role in the development of this book. Without their invaluable contributions this book wouldn't have been possible. They have made vital efforts to compile up to date information on the varied aspects of this subject to make this book a valuable addition to the collection of many professionals and students.

This book was conceptualized with the vision of imparting up-to-date information and advanced data in this field. To ensure the same, a matchless editorial board was set up. Every individual on the board went through rigorous rounds of assessment to prove their worth. After which they invested a large part of their time researching and compiling the most relevant data for our readers.

The editorial board has been involved in producing this book since its inception. They have spent rigorous hours researching and exploring the diverse topics which have resulted in the successful publishing of this book. They have passed on their knowledge of decades through this book. To expedite this challenging task, the publisher supported the team at every step. A small team of assistant editors was also appointed to further simplify the editing procedure and attain best results for the readers.

Apart from the editorial board, the designing team has also invested a significant amount of their time in understanding the subject and creating the most relevant covers. They scrutinized every image to scout for the most suitable representation of the subject and create an appropriate cover for the book.

The publishing team has been an ardent support to the editorial, designing and production team. Their endless efforts to recruit the best for this project, has resulted in the accomplishment of this book. They are a veteran in the field of academics and their pool of knowledge is as vast as their experience in printing. Their expertise and guidance has proved useful at every step. Their uncompromising quality standards have made this book an exceptional effort. Their encouragement from time to time has been an inspiration for everyone.

The publisher and the editorial board hope that this book will prove to be a valuable piece of knowledge for researchers, students, practitioners and scholars across the globe.

List of Contributors

Yousef Farid
Department of Electrical Engineering (EE), Imam Khomeini International University, Qazvin, Iran

Nooshin Bigdeli
Department of Electrical Engineering (EE), Imam Khomeini International University, Qazvin, Iran

Karim Afshar
Department of Electrical Engineering (EE), Imam Khomeini International University, Qazvin, Iran

Bobrovs Vjaceslavs
Institute of Telecommunications, Riga Technical University, Azenes Str. 12, Rīga, LV-1048, Latvia, Europe

Spolitis Sandis
Institute of Telecommunications, Riga Technical University, Azenes Str. 12, Rīga, LV-1048, Latvia, Europe

Ivanovs Girts
Institute of Telecommunications, Riga Technical University, Azenes Str. 12, Rīga, LV-1048, Latvia, Europe

Ahmed Moumena
Electronic Institute, LATSI Laboratory, Blida University, Road of Soumaa, PB 270, Blida, Algeria

Abderezzak Guessoum
Electronic Institute, LATSI Laboratory, Blida University, Road of Soumaa, PB 270, Blida, Algeria

Abdoul K. MBODJI
Centre International de Formation et de Recherche en Energie Solaire (C.I.F.R.E.S), ESP BP 5085 Dakar-Fann, Sénégal

Mamadou L. NDIAYE
Centre International de Formation et de Recherche en Energie Solaire (C.I.F.R.E.S), ESP BP 5085 Dakar-Fann, Sénégal

Papa A. NDIAYE
Centre International de Formation et de Recherche en Energie Solaire (C.I.F.R.E.S), ESP BP 5085 Dakar-Fann, Sénégal

F. E. Ismael
CoE Telecommunication Technology, University of Technology, Malaysia (UTM), Johor, 81310 UTM Skudai, Malaysia

S. K. Syed-Yusof
CoE Telecommunication Technology, University of Technology, Malaysia (UTM), Johor, 81310 UTM Skudai, Malaysia

M. Abbas
Head of Wireless Communication Cluster, MIMOS Berhadi, Malaysia

N. Fisal
CoE Telecommunication Technology, University of Technology, Malaysia (UTM), Johor, 81310 UTM Skudai, Malaysia

N. Muazzah
CoE Telecommunication Technology, University of Technology, Malaysia (UTM), Johor, 81310 UTM Skudai, Malaysia

Yusuf Arif Kutlu
Department of Geophysics, Faculty of Engineering and Architecture, Nevsehir University, 50300, Nevsehir-Turkey

Nilgün Say
Department of Geophysics, Faculty of Engineering, Karadeniz Technical University, 61080, Trabzon-Turkey

Johann Christiaan Pretorius
Department of Electrical Electronic and Computer Engineering, University of Pretoria, Pretoria, South Africa

B. T. Maharaj
Department of Electrical Electronic and Computer Engineering, University of Pretoria, Pretoria, South Africa

Mohamed E. Wahed
Faculty of Computers and Informatics, Suez Canal University, Ismailia, Egypt

Mohamed K. Hussein
Faculty of Computers and Informatics, Suez Canal University, Ismailia, Egypt

Mohamed H. Mousa
Faculty of Computers and Informatics, Suez Canal University, Ismailia, Egypt

Mohamed Abdel Hameed
Faculty of Computers and Informatics, Suez Canal University, Ismailia, Egypt

R. Pushpa Lakshmi
PSNA College of Engineering and Technology, Dindigul, Tamilnadu, India

A. Vincent Antony Kumar
PSNA College of Engineering and Technology, Dindigul, Tamilnadu, India

Abdulsalam Ya'u Gital
Department of Computer Systems and Communications, Faculty of Computer Science and Information Systems, Universiti Teknologi, 81310, UTM Johor, Bahru, Malaysia

Abdul Samad bn Ismail
Department of Computer Systems and Communications, Faculty of Computer Science and Information Systems, Universiti Teknologi, 81310, UTM Johor, Bahru, Malaysia

Shamala Subramaniam
Department of Computer Technology and Network, Faculty of Computer Science and Information Technology, Universiti Putra Malaysia, 43400 Serdang, Malaysia

J. C. Garcia Infante
Mechanical and Electrical Engineering School IPN, Col. San Francisco Culhuacan Del. Coyoacán, D. F. ext.73092 Mexico

J. J. Medel Juarez
Computing Research Centre, Av. 100 m, esq., Venus, Col. Nueva Industrial Vallejo, C. P. 07738 D. F. ext. 56570 Mexico

J. C. Sanchez Garcia
Mechanical and Electrical Engineering School IPN, Col. San Francisco Culhuacan Del. Coyoacán, D. F. ext.73092 Mexico

Narayanan Arumugam
Anna University, Chennai, India

Chakrapani Venkatesh
Faculty of Engineering, EBET Group of Institutions,Kankayam, Tamilnadu, India, Member IEEE, India

Junita Mohd Nordin
School of Computer and Communication Engineering, University Malaysia Perlis, 01000 Perlis, Malaysia

Syed Alwee Aljunid
School of Computer and Communication Engineering, University Malaysia Perlis, 01000 Perlis, Malaysia

Anuar Mat Saf
School of Computer and Communication Engineering, University Malaysia Perlis, 01000 Perlis, Malaysia

Amir Razif Arief
School of Computer and Communication Engineering, University Malaysia Perlis, 01000 Perlis, Malaysia

Rosemizi Abd Rahim
School of Computer and Communication Engineering, University Malaysia Perlis, 01000 Perlis, Malaysia

R. Badlishah Ahmad
School of Computer and Communication Engineering, University Malaysia Perlis, 01000 Perlis, Malaysia

Naufal Saad
School of Electric and Electronic Engineering, Universiti Teknologi Petronas, Tronoh, Perak, Malaysia

Chamran Asgari
Department of Computer Engineering, Payame Noor University, Iran

Rohollah Rahmati Torkashvand
Department of Computer Engineering, Islamic Azad University, Borujerd Branch, Lorestan, Iran

Masoud Barati
Department of Computer,Islamic Azad University, Kangavar branch, Kangavar, Iran

Alka Kalra
Department of Electronics and Communication Engineering Haryana College of Technology and Management, Kaithal Haryana, India

Rajesh Khanna
Department of Electronics and Communication Engineering, Thapar University, Patiala, Punjab, India

Jihan K. Raoof
Institute of Visual Informatics, UKM-Malaysia

Halimah Badioze Zaman
Institute of Visual Informatics, UKM-Malaysia

Azlina Ahmad
Institute of Visual Informatics, UKM-Malaysia

Ammar Al-Qaraghuli
Institute of Visual Informatics, UKM-Malaysia

Chun-Chia Hsu
Department of Cultural Creativity and Digital Media Design, Lunghwa University of Science and Technology, Gueishan, Taoyuan County, Taiwan, Republic of China

Chih-Yung Linand
Department of Multi-media and Game Science, Lunghwa University of Science and Technology, Gueishan, Taoyuan County, Taiwan, Republic of China

Chin-Ping Fung
Department of Mechanical Engineering, Oriental Institute of Technology, Panchiao, New Taipei City, Taiwan, Republic of China

M. N. Chai
School of Fundamental Sciences, University Malaysia Terengganu, 21030 Kuala Terengganu, Terengganu, Malaysia

M. I. N. Isa
School of Fundamental Sciences, University Malaysia Terengganu, 21030 Kuala Terengganu, Terengganu, Malaysia
Center of Corporate Communication and Image Development, Chancellery, Universiti Malaysia Terengganu, 21030 Kuala Terengganu, Terengganu, Malaysia

A. Jasmine Xavier
Jayaraj Annapackiam CSI College of Engineering, Anna University, Chennai, Tamil Nadu India

R. Shantha Selvakumari
Mepco Schlenk Engineering College, Anna University, Chennai, Tamil Nadu India

V. Bobrovs
Institute of Telecommunications, Riga Technical University, Azenes Str. 12, Rīga, LV-1048, Latvia

S. Olonkins
Institute of Telecommunications, Riga Technical University, Azenes Str. 12, Rīga, LV-1048, Latvia

A. Alsevska
Institute of Telecommunications, Riga Technical University, Azenes Str. 12, Rīga, LV-1048, Latvia

L. Gegere
Institute of Telecommunications, Riga Technical University, Azenes Str. 12, Rīga, LV-1048, Latvia

G. Ivanovs
Institute of Telecommunications, Riga Technical University, Azenes Str. 12, Rīga, LV-1048, Latvia

www.ingramcontent.com/pod-product-compliance
Lightning Source LLC
Chambersburg PA
CBHW050458200326

41458CB00014B/5225